Lecture Notes in Mechanical Engineering

Lecture Notes in Mechanical Engineering (LNME) publishes the latest developments in Mechanical Engineering—quickly, informally and with high quality. Original research or contributions reported in proceedings and post-proceedings represents the core of LNME. Volumes published in LNME embrace all aspects, subfields and new challenges of mechanical engineering.

To submit a proposal or request further information, please contact the Springer Editor of your location:

Europe, USA, Africa: Leontina Di Cecco at Leontina.dicecco@springer.com
China: Ella Zhang at ella.zhang@springer.com
India: Priya Vyas at priya.vyas@springer.com
Rest of Asia, Australia, New Zealand: Swati Meherishi at swati.meherishi@springer.com

Topics in the series include:

- Engineering Design
- Machinery and Machine Elements
- Mechanical Structures and Stress Analysis
- Automotive Engineering
- Engine Technology
- Aerospace Technology and Astronautics
- Nanotechnology and Microengineering
- Control, Robotics, Mechatronics
- MEMS
- Theoretical and Applied Mechanics
- Dynamical Systems, Control
- Fluid Mechanics
- Engineering Thermodynamics, Heat and Mass Transfer
- Manufacturing Engineering and Smart Manufacturing
- Precision Engineering, Instrumentation, Measurement
- Materials Engineering
- Tribology and Surface Technology

Indexed by SCOPUS, EI Compendex, and INSPEC.

All books published in the series are evaluated by Web of Science for the Conference

Proceedings Citation Index (CPCI).

To submit a proposal for a monograph, please check our Springer Tracts in Mechanical Engineering at https://link.springer.com/bookseries/11693.

Rajendra K. Ray · Swaroop Nandan Bora ·
Dipak Kumar Maiti
Editors

Advances in Theoretical and Applied Mechanics

Proceedings of ISTAM 2022

Springer

Editors
Rajendra K. Ray
School of Mathematical and Statistical Sciences
Indian Institute of Technology Mandi
Mandi, Himachal Pradesh, India

Swaroop Nandan Bora
Department of Mathematics
Indian Institute of Technology Guwahati
Guwahati, Assam, India

Dipak Kumar Maiti
Department of Aerospace Engineering
Indian Institute of Technology Kharagpur
Kharagpur, West Bengal, India

ISSN 2195-4356 ISSN 2195-4364 (electronic)
Lecture Notes in Mechanical Engineering
ISBN 978-981-97-0417-0 ISBN 978-981-97-0418-7 (eBook)
https://doi.org/10.1007/978-981-97-0418-7

This Springer imprint is published by the registered company Springer Nature Singapore Pte Ltd.
The registered company address is: 152 Beach Road, #21-01/04 Gateway East, Singapore 189721, Singapore

Preface

The Indian Society of Theoretical and Applied Mechanics (ISTAM) was established in the year 1955 under the dynamic leadership of Prof. B. R. Seth as its Secretary and Prof. K. S. Krishnan as its President, with its headquarter at Department of Mathematics, Indian Institute of Technology Kharagpur, India. The 67th Congress of the Indian Society of Theoretical and Applied Mechanics (ISTAM) an International Meet, was held at Indian Institute of Technology Mandi, India during 14–16 December 2022. After two years of online version of this meet, due to COVID-19 pandemic, ISTAM Congress was organized physically. I take this opportunity to thank the administration of IIT Mandi, for coming forward to host this congress. There was an overwhelming response for this congress. Over 300 researchers from all over India and abroad participated in this congress. On behalf of the Society, I take this opportunity to express my sincere gratitude and thanks to all these eminent researchers for readily agreeing to take part in this congress.

The conference was held offline mode with more than 300 participants and 20 invited speakers from all over the globe. The conference was inaugurated by the Chief Patron of the conference and the Honorable Director of IIT Mandi, Prof. Laxmidhar Behera. In his inaugural lecture, Prof. Behera quoted "Our great ancestors have truly believed in the idea of *Vasudhaiva Kutumbakam* which means the whole world is one family. With similar belief, irrespective of nationality, ethnicity or colour–we researchers are also one single family. As academicians, the same sense of togetherness is what we need to feel to do something great, to achieve our respective goals and overall, to contribute to the very society, that nurtures us". Professor Swaroop Nandan Bora, President, ISTAM, in his Presidential Address, emphasized on the role of mechanics in real-life problems starting from blood flow in human to satellites. In this conference, two **Plenary Lectures** were delivered by Prof. Vijay P. Singh (Texas A&M University, USA) and Dr. Makrand Joshi (Scientist, DRDO, Pune, India) and five **Memorial Lectures** were delivered in honor of **Prof. G. I. Taylor, Prof. B. R. Seth, Prof. P. L. Bhatnagar, Prof. B. Karunesh**, and **Prof. A. S. Gupta** by Prof. J. N. Reddy (Texas A&M University, USA), Prof. D. Pham Van Bang (University of Québec, Canada), Prof. C. Kong (Chosun University, South Korea), Prof. Sarit K. Das (IIT Madras, India), and Prof. Sanjay Mittal (IIT Kanpur, India), respectively. Numerous **invited lectures** were also organized to discuss the recent trends in Mechanics. Those lectures were delivered by renowned experts from India and abroad. Overall, 25 **Technical Sessions** were conducted during three days of conference, which included more than 180 paper presentations. A special Technical Session was conducted for the **Young Scientist Award**. Ms. Deepali Goyal, a Ph.D. student from IIT Ropar, was awarded the Young Scientist Award at the Congress for her outstanding research contribution. Out of all presented papers, 36 papers are selected for inclusion in this conference proceedings after two layers of review process.

We sincerely thank the ISTAM members, presenters, committee members, reviewers, technical session chairs, faculty members, and the students and staffs of our institute, who contributed so magnificently to the success of the conference. We sincerely thank the Science *and Engineering Research Board (SERB, Government of India),* the *Board of Research in Nuclear Sciences* (BRNS, *Government of India*), and the *Council of Scientific and Industrial Research* (*CSIR, Government of India*) for providing financial supports to organize this mega event. We also acknowledge the financial support of different corporate houses and local vendors. We tried hard to ensure that the conference would be well organized. We hope that it met the expectations. We are grateful to the presenters and participants for their thought-provoking contributions. We extend our very best wishes to them wherever they may be around the globe.

On behalf of the Organizing Committee
Organizing Chair, ISTAM-2022
and Editorial Board members

Mandi, India

Prof. Rajendra K. Ray, Ph.D.

Contents

About the Editors

Dr. Rajendra K. Ray is currently a Professor in the School of Mathematical and Statistical Sciences, Indian Institute of Technology Mandi, Himachal Pradesh, India. He obtained his B.Sc. (Math.) and M.Sc. (Math.) degrees from Visva-Bharati University, Shantiniketan, West Bengal, India in the years 2001 and 2003, respectively. He obtained his Ph.D. from Indian Institute of Technology Guwahati, Assam, India in 2009. He did his post-doctoral research from Indian Institute of Science Bangalore, India, and then from École des Ponts Paris Tech, France. His major research interests include Development of higher order accurate numerical methods for solving PDEs and their applications to study complex fluid flow problems, heat and mass transfer, moving interface problems and mathematical models for image processing. He has developed a number of efficient higher order accurate numerical schemes for solving complex fluid flow and heat transfer problems and a number of PDE-based mathematical models for image denoising. Professor Ray has published more than 50 papers in internationally reputed journals and has delivered invited talks in a number of national and international conferences. He is currently the Vice-President *Indian Society of Theoretical and Applied Mechanics* (ISTAM). He is serving as a reviewer in more than 20 internationally reputed journals.

Dr. Swaroop Nandan Bora is currently a professor at the Department of Mathematics, Indian Institute of Technology (IIT) Guwahati. He pursued his B.Sc. (Math.) from Cotton College, Gauhati University, India, M.Sc. (Math.) from University of Delhi, and Ph.D. from Technical University of Nova Scotia, Canada. His major areas of research interests include wave-structure interactions, scattering and trapping of water waves, flows in porous media, sloshing, river mechanics, theoretical aspects of fractional differential equations, mathematical biology. Prof. Bora has published more than 110 papers in internationally reputed journals and has given many invited/plenary talks in national and international conferences. He was the President of *Indian Society of Theoretical and Applied Mechanics* (ISTAM) for the year 2022.

Dr. Dipak Kumar Maiti is currently a professor in the Department of Aerospace Engineering, Indian Institute of Technology Kharagpur, West Bengal, India. He did his B.E. (Civil) from University of Calcutta, India in the year 1991, M.Tech. and Ph.D. (Aero-Structure) from IIT Kharagpur, India in the year 1993 and 1997, respectively. His major areas of research interests include aeroelastic, analysis of Smart Lifting Surface, analysis of Smart Landing Gear, Structural Health Monitoring, FE analysis of Smart Multidirectional, Composites Nonlinear Analysis of Smart Composite, Plates and Shells Reliability Analysis of Composite, Wing Subjected to various Loads. Professor Maiti has published more than 100 papers in internationally reputed journals and has given many invited/plenary talks in national and international conferences. Currently, he is the Secretary of Indian Society of Theoretical and Applied Mechanics (ISTAM).

Fluid Mechanics

Modal Analysis of Undamped MDOF Vibratory System

Dip D. Shah and Sachin S. Naik

Abstract This paper deals with the numerical estimation of natural frequencies and mode shapes using basic vibration theory for two and three Degree of freedom (DOF) vibratory system. The procedure can be extended to an '*n*' DOF vibratory system easily. The equations of motion are dependent on each coordinate. The principle of orthogonality of mode shapes is used to convert the physical coordinate system into a generalised coordinate system which gives rise to an n-uncoupled differential equations. Response for free vibrations for each DOF can be obtained for different initial conditions to analyse the system. It can be realised that the response for an arbitrary initial displacement to the masses is the superposition of responses for each mode of vibration.

Keywords Modal analysis · Natural frequency · Mode shapes · MDOF systems

1 Introduction

Mechanical vibration is a repetitive motion of a body or a dynamic system about the static position. This oscillatory motion can be either periodic, non-periodic or transient. Presence of vibratory motion can cause unwanted noise, displacement and even excessive stresses leading to loss of performance and reliability of mechanical components. Accidents caused thereof cause loss of infrastructure and/or life. Vibration control and mitigation therefore becomes an essential part of mechanical design of many systems. On the other hand, mechanical vibration is very useful in diverse applications such as musical instruments, shakers, and cardiac pacemakers, etc. Therefore, it is imperative to model and analyse the behaviours of vibrating systems [1].

D. D. Shah
M. Tech Machine Design, VJTI, Mumbai, India

S. S. Naik (✉)
Department of Mechanical Engineering, VJTI, Mumbai, India
e-mail: ssnaik@me.vjti.ac.in

R. K. Ray et al. (eds.), *Advances in Theoretical and Applied Mechanics*, Lecture Notes in Mechanical Engineering, https://doi.org/10.1007/978-981-97-0418-7_1

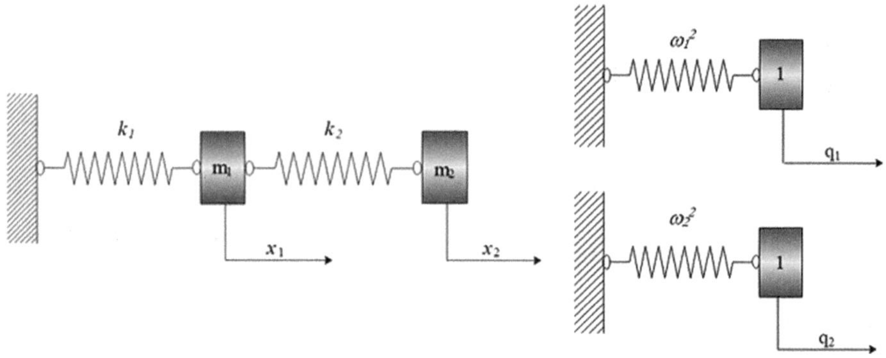

Fig. 1 Illustration of decoupling equation of motion using modal analysis [3]

In the past, various failures in mechanical system have occurred due to resonance. Therefore, it is necessary to know the natural frequencies of a particular system. Most of the real-life mechanical systems are continuous systems and hence they can be treated at discrete MDOF systems and their natural frequencies are obtained by modal analysis.

Modal analysis is a tool for the determination of natural frequencies, mode shapes and damping factor which are inherent characteristics of a system in the prevailing conditions in terms of boundary condition, loading, soundness of health, etc. These are important factors in Engineering design and control because they are indicators of resonance conditions represented by the system [2].

In modal analysis, N-DOF vibratory system can be simplified into N numbers of equivalent single DOF system using orthogonality property of mode shape. So, it becomes very convenient to do vibration analysis [3]. The conversion of a coupled system to an uncoupled system is illustrated in Fig. 1.

2 Methodology

For performing modal analysis, initially, natural frequency and corresponding normal mode shapes have to be found out, then using orthogonality property of modes shapes, physical coordinates are converted to generalised coordinates, so the vibration analysis can be carried out using the uncoupled systems of equation readily.

Let us consider two DOF vibration system as shown in Fig. 2,

Selecting $x_1(t)$ and $x_2(t)$ as coordinates to describe the displacement of vibration of system, by using Newtons second law of motion, the equation of motion can be derived as,

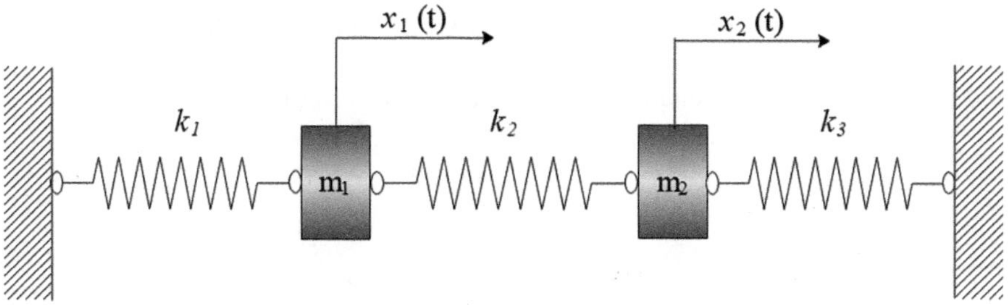

Fig. 2 Two DOF vibratory system [1]

$$m_1\ddot{x}_1(t) + (k_1 + k_2)x_1(t) - k_2x_2(t) = 0 \tag{1}$$

$$m_2\ddot{x}_2(t) - k_2x_1(t) + (k_2 + k_3)x_2(t) = 0 \tag{2}$$

In matrix form, both the equation can be written as,

$$\begin{bmatrix} m_1 & 0 \\ 0 & m_2 \end{bmatrix}\begin{bmatrix} \ddot{x}_1(t) \\ \ddot{x}_2(t) \end{bmatrix} + \begin{bmatrix} k_1 + k_2 & -k_2 \\ -k_2 & k_2 + k_3 \end{bmatrix}\begin{bmatrix} x_1(t) \\ x_2(t) \end{bmatrix} = \begin{bmatrix} 0 \\ 0 \end{bmatrix} \tag{3}$$

which is in form:

$$[m][\ddot{x}] + [k][x] = [0] \tag{4}$$

By assuming that the complete system is oscillating harmonically with the same frequency (ω) and same phase angle ($\emptyset$),

$$\begin{aligned} x(t) &= X\cos(\omega t + \emptyset) \\ \ddot{x}(t) &= -X\omega^2\cos(\omega t + \emptyset) = -\omega^2 x(t) \end{aligned} \tag{5}$$

Therefore, equation can be written as,

$$\begin{bmatrix} k_1 + k_2 - m_1\omega^2 & -k_2 \\ -k_2 & k_2 + k_3 - m_2\omega^2 \end{bmatrix}\begin{bmatrix} X_1 \\ X_2 \end{bmatrix} = \begin{bmatrix} 0 \\ 0 \end{bmatrix} \tag{6}$$

To have non-trivial solution of $\{X\}$, matrix $\left([K]-\omega^2[M]\right)$ has to be singular so that: $\left|[K]-\omega^2[M]\right| = 0$. This yields one algebraic equation in ω^2, which is known as the characteristic equation for the system. Substitution of the obtained values of (ω^2), corresponding normal mode shapes can be obtained using Eq. (6).

Equations (1) and (2) are coupled. To simplify this problem, coupled equations can be decoupled by using modal analysis as follows. Here obtained normal mode shapes are not unique as their multiples are equally valid mode shapes. Therefore, orthogonalizing mode shapes with respect to mass matrix are followed, such that

$\vec{X}^{(i)^{\mathrm{T}}}[m]\vec{X}^{(i)} = 1$, where '$i$' is a particular mode of vibration. Later the mode shape matrix is obtained as $[\boldsymbol{X}]^{\mathrm{T}} = \left[\vec{X}^{(1)}\vec{X}^{(2)}\right]$.

Let $\vec{x}(t)$ be given by generalised coordinate $\vec{q}(t)$ as follows.

$$\vec{x}(t) = [\boldsymbol{X}]\vec{q}(t) \tag{7}$$

$$\text{and } \ddot{\vec{x}}(t) = [\boldsymbol{X}]\ddot{\vec{q}}(t) \tag{8}$$

Substituting above equations in Eq. (4) gives,

$$[m][\boldsymbol{X}]\ddot{\vec{q}} + [k][\boldsymbol{X}]\vec{q} = 0 \tag{9}$$

By pre-multiplying Eq. (9) by $[\boldsymbol{X}]^{\mathrm{T}}$, we get,

$$[\boldsymbol{X}]^{\mathrm{T}}[m][\boldsymbol{X}]\ddot{\vec{q}} + [\boldsymbol{X}]^{\mathrm{T}}[k][\boldsymbol{X}]\vec{q} = 0 \tag{10}$$

As normal modes are normalised with respect to mass matrix,

$$[\boldsymbol{X}]^{\mathrm{T}}[m][\boldsymbol{X}] = [I]$$

$$[\boldsymbol{X}]^{\mathrm{T}}[k][\boldsymbol{X}] = \left[\nwarrow \omega^2 \searrow\right].$$

$$\therefore \ddot{\vec{q}}(t) + \left[\nwarrow \omega^2 \searrow\right]\vec{q}(t) = 0 \tag{11}$$

Equation (11) gives a set of uncoupled differential equations. In scaler form, it becomes

$$\ddot{q}_i(t) + \omega_i^2 q_i(t) = 0 \tag{12}$$

where, $i = 1$ and 2. The solution of (12) is,

$$q_i(t) = q_i(0)\cos\omega_i t + \frac{\dot{q}_i(0)}{\omega_i}\sin\omega_i t \tag{13}$$

where, $q_i(0)$, $\dot{q}_i(0)$ are the initial values of $q_i(t)$ and $\dot{q}_i(t)$ respectively that are unknown, and ω_i is corresponding natural frequency. The initial values of $q_i(0)$, $\dot{q}_i(0)$ are obtained as:

$$\vec{q}(0) = \begin{Bmatrix} q_1(0) \\ q_2(0) \end{Bmatrix} = [\boldsymbol{X}]^{\mathrm{T}}[m]\vec{x}(0)$$

$$\dot{\vec{q}}(0) = \begin{Bmatrix} \dot{q}_1(0) \\ \dot{q}_2(0) \end{Bmatrix} = [\boldsymbol{X}]^{\mathrm{T}}[m]\dot{\vec{x}}(0) \tag{14}$$

where, $\vec{x}(0)$, $\dot{\vec{x}}(0)$ are the column matrix of physical initial condition in terms of displacement and velocity respectively.

Substituting the initial conditions in Eq. (13), we get the response in terms of generalised coordinates. Such response does not have any physical significance unless generalised coordinates are reconverted into physical coordinates using Eq. (7).

3 Results and Discussion

Modal analysis is carried out for two and three DOF vibration systems.

3.1 Two DOF Vibration System

The system parameters as shown in Fig. 3 are selected for two DOF vibration system as: $m_1 = 0.52$ kg, $m_2 = 0.75$ kg, $k_1 = 315$ N/m, $k_2 = 242$ N/m and $k_3 = 518.6$ N/m. The natural frequencies and corresponding mode shapes are obtained as 25.567, 37.826 rad/s and $[1 \quad 0.896]^T$, $[1 \quad -0.774]^T$ respectively. Response of the two DOF system for 1st mode shape as the initial condition in generalised coordinates is:

$$q_1(t) = 1.059\cos(25.567t) \tag{15}$$

$$q_2(t) = 0 \tag{16}$$

And in physical coordinates is

$$x_1(t) = 1\cos(25.567t) \tag{17}$$

$$x_2(t) = 0.896\cos(25.567t) \tag{18}$$

It can be realised from Eqs. (17) and (18) that, when initial condition to the masses is the particular nth mode shape itself then the whole system oscillates with only the nth natural frequency. Response graph for initial conditions as 1st mode of vibration is shown in Fig. 3.

Similarly, for an arbitrary excitation, say $x_1(0) = 1$, $x_2(0) = 0$, $\dot{x}_1(0) = 0$, $\dot{x}_2(0) = 0$, the solution in generalised coordinate is obtained as

$$q_1(t) = 0.491\cos(25.567t) \tag{19}$$

$$q_2(t) = 0.528\cos(37.826t) \tag{20}$$

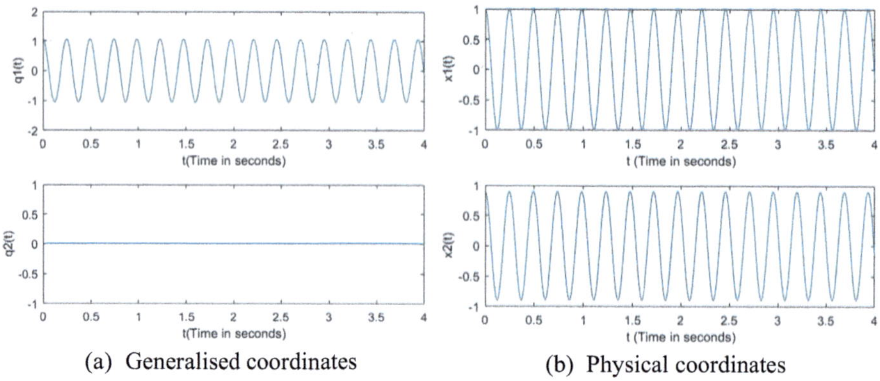

Fig. 3 Response for 1st mode of vibration as the initial displacements

And, solution obtained in physical coordinates as,

$$x_1(t) = 0.463\cos(25.567t) + 0.537\cos(37.826t) \tag{21}$$

$$x_2(t) = 0.415\cos(25.567t) - 0.415\cos(37.826t) \tag{22}$$

It can be observed from Eqs. (21) and (22) that for an arbitrary initial condition, the system response is essentially the superposition of each mode (refer Fig. 4).

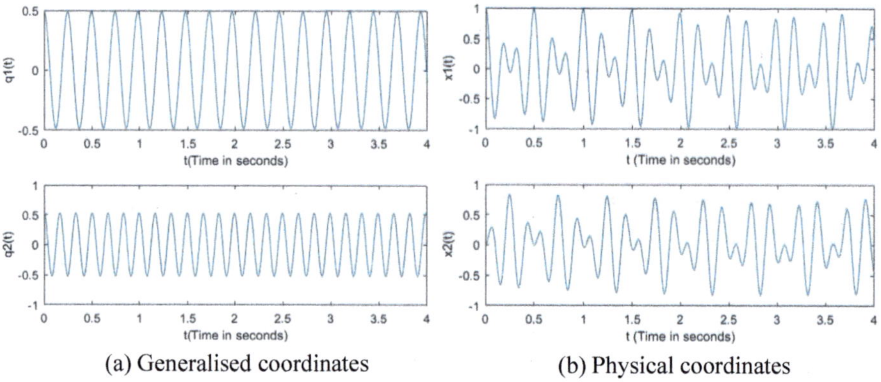

Fig. 4 Response graph for initial condition as $x_1(0) = 1$, $x_2(0) = 0$, $\dot{x}_1(0) = 0$, $\dot{x}_2(0) = 0$

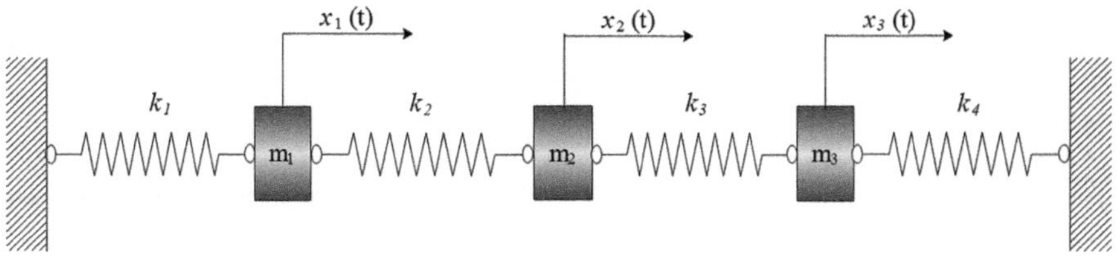

Fig. 5 Three DOF vibration system

3.2 Three DOF Vibration System

The system parameters as shown in Fig. 5 are selected for three DOF vibration system as $m_1 = 0.055$ kg, $m_2 = 0.083$ kg, $m_3 = 0.107$ kg, and $k_1 = 83.66$ N/m, $k_2 = 67.64$ N/m, $k_3 = 67.64$ N/m and $k_4 = 105.97$ N/m.

The natural frequencies for the selected system parameters are obtained as 25.194, 44.161, 58.467 rad/s and the corresponding mode shapes are $[1 \quad 1.721 \quad 1.101]^T$, $[1 \quad 0.651 \quad -1.256]^T$, $[1 \quad -0.543 \quad 0.191]^T$ respectively.

When 1st mode shape is given as an initial condition, i.e. $x_1(0) = 1, x_2(0) = 1.721$, $x_3(0) = 1.101$, $\dot{x}_1(0) = 0$, $\dot{x}_2(0) = 0$, $\dot{x}_3(0) = 0$, solution obtained in generalised coordinates is:

$$q_1(t) = 0.656 \cos(25.194t) \tag{23}$$

$$q_2(t) = 0 \tag{24}$$

$$q_3(t) = 0 \tag{25}$$

In physical coordinates the solutions are,

$$x_1(t) = 1 \cos(25.194t) \tag{26}$$

$$x_2(t) = 1.721 \cos(25.194t) \tag{27}$$

$$x_3(t) = 1.101 \cos(25.194t) \tag{28}$$

In general, it can be realised that when the nth mode shape itself is an initial condition, all the masses in the system oscillate with only nth natural frequency. Response graph for initial conditions as 1st mode vibration is shown in Fig. 6.

Similarly, for an arbitrary initial condition, $x_1(0) = 1$, $x_2(0) = 0$, $x_3(0) = 0$, $\dot{x}_1(0) = 0$, $\dot{x}_2(0) = 0$, $\dot{x}_3(0) = 0$, the solution in generalised coordinate is:

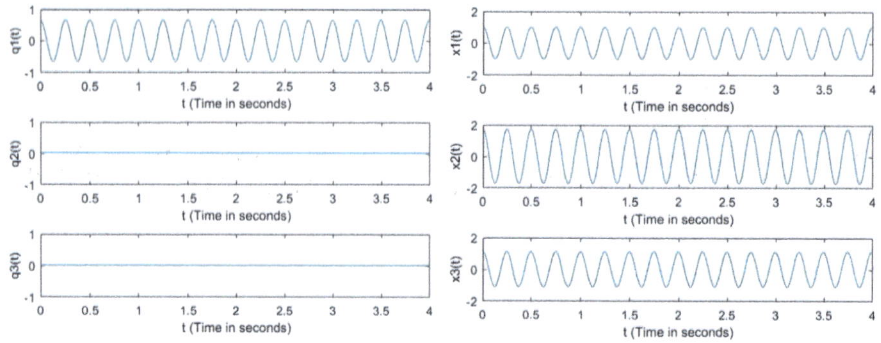

(a) Generalised coordinates (b) Physical coordinates

Fig. 6 Response graph for 1st mode of vibration as the initial condition

$$q_1(t) = 0.084\cos(25.194t) \tag{29}$$

$$q_2(t) = 0.108\cos(44.161t) \tag{30}$$

$$q_3(t) = 0.191\cos(58.467t) \tag{31}$$

And, solution obtained in physical coordinates is obtained as.

$$x_1(t) = 0.128\cos(25.194t) + 0.212\cos(44.161t) + 0.660\cos(58.467t) \tag{32}$$

$$x_2(t) = 0.220\cos(25.194t) + 0.138\cos(44.161t) - 0.358\cos(58.467t) \tag{33}$$

$$x_3(t) = 0.141\cos(25.194t) - 0.267\cos(44.161t) + 0.126\cos(58.467t) \tag{34}$$

Response graph is shown in Fig. 7,

Equations (32)–(34) show the system response obtained by superposition of the three modes of vibration.

4 Conclusions

From the results of modal analysis of two DOF and three DOF vibratory system are summarised as follows:

(1) When the initial displacements to the masses is the particular nth mode shape itself then the whole system oscillates with the nth natural frequency only.

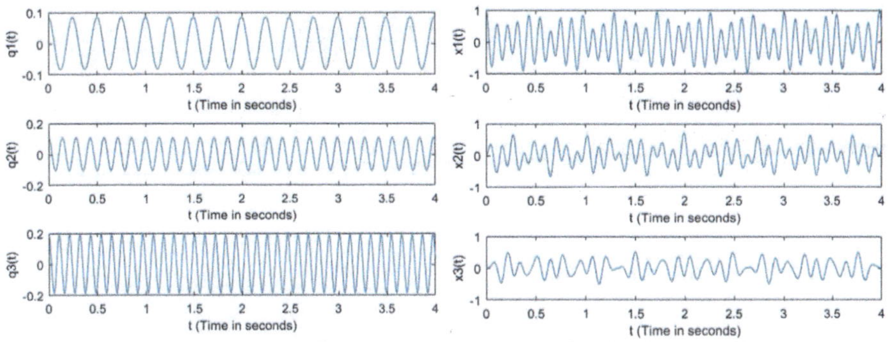

(a) Generalised coordinates (b) Physical coordinates

Fig. 7 Response for initial condition as $x_1(0) = 1$, $x_2(0) = 0$, $x_3(0) = 0$, $\dot{x}_1(0) = 0$, $\dot{x}_2(0) = 0$, $\dot{x}_3(0) = 0$

(2) For an arbitrary initial condition, the response of the system consists of superposition of each mode of vibration.
(3) Generalised coordinates are only time dependent whereas physical coordinates depend on time as well as the mode of vibration.

References

1. Rao SS (2001) Mechanical vibrations, 5th ed. p. cm
2. Fu ZF, He J (2001) Modal analysis. Elsevier, Amsterdam
3. Inman DJ, Singh RC (1994) Engineering vibration, vol 3. Prentice Hall, Englewood Cliffs, NJ

Comparative Study of CDP and JH2 Damage Model of Koyna Dam Under Blast Loading by CEL Technique

Tanuja Khati and Smita Kaloni

Abstract Present study aims to analyze the dynamic response of Koyna dam with reservoir interaction under the influence of underwater explosion in ABAQUS/CAE. For dam water interaction, water in the reservoir is modeled by volume of fluid method in coupled Eulerian Lagrangian (CEL) technique in which each element of material is meshed in Eulerian Volume fraction (EVF). Concrete is modeled as lagrangian mesh while the TNT, air and water are modeled as eulerian mesh. The response of a dam subjected to underwater explosion is studied for different targets with respect to stagnation and detonation points. Non-Linearity of concrete is considered using Johnson-Holmquist 2 (JH-2) model and Concrete Damage Plasticity model. Results are evaluated and compared based on pressure history curves subjected to different TNT weights. Effect of foundation is neglected in the present study.

Keywords ABAQUS · CEL technique · Underwater explosion · Concrete Damage Plasticity model and Johnson-Holmquist 2 (JH-2) model

1 Introduction

For many years, dams due to its complicated configuration have always been a complex problem in case of blast loads. Blasts in case of terrorist activities can severely damage dam, reservoir and its foundation. There is a great need for safety of these important lifeline structures against such loading conditions as its failure can result in many life and property loss. Population in the downstream area is mostly affected by the failure of dams so there is a great need for research in this area. Due to the vulnerability of such large hydraulic structures against dynamic

T. Khati (✉) · S. Kaloni
Department of Civil Engineering, National Institute of Technology, Uttarakhand, Srinagar, Garhwal, India
e-mail: tanukhati@gmail.com

S. Kaloni
e-mail: smitakaloni@nituk.ac.in

R. K. Ray et al. (eds.), *Advances in Theoretical and Applied Mechanics*, Lecture Notes in Mechanical Engineering, https://doi.org/10.1007/978-981-97-0418-7_2

loading like seismic loads and blast loads which develop cracks, there is a great need to study different damage model parameters like Concrete damage plasticity model, RHT model, JH2 model etc. These models can anticipate crack initiation and help in understanding the crack pattern in case of any impact event. In case of large/ extreme deformations, while attaining simulation there may be some problems while meshing, as lagrangian meshing is not convenient for such deformations. In such a case eulerian meshing is more convenient. In the eulerian method, if EVF is 1, that means the element is completely filled with the material. If EVF is 0, that means no material is present in the element. Coupled Eulerian Lagrangian (CEL) technique uses the advantages of both lagrangian and eulerian methods. Non-linearity of materials plays an important role while depicting the crack propagation under impact loading. Johnson and Holmquist introduced Johnson-Holmquist 2 model to describe quassi brittle material. Strength, damage and pressure are the three main damage parameters of JH2 model. In case of blast loading, the structure may undergo either air blast or underwater explosion. From the studies, it was suggested that the damage effect on the dam will be severe in case of underwater explosion as compared to air blast [1]. To study the response of structure due to blast loads, most of the researchers have studied full or scale down models. In 1940, a scale model of Mohne dam was undergone through a series of trial blast loads to check the stability of the dam [2]. Estimating the response of a full-scale gravity dam in case of blast loading is a complex task. Experimental test of blast load on gravity dam may not be expedient as it has higher risk due to the involvement of higher explosives. Due to advancements in computer technology, computational simulation of damage prediction of gravity dams is much feasible now. Yu [3] investigated the response of concrete gravity dams by using the ALE algorithm. Wang [4] proposed a model in LS-DYNA to study the interaction of shock waves with dam and foundation. Zhang et al. [5] investigated concrete gravity dam underwater explosion (UNDEX) for different water height, stagnation distance and detonation point. Zhang et al. [6] depicted the role of negative pressure subjected to underwater explosion at dam water interaction. Wang [7] investigated Guandi gravity dam under seismic loads, evaluating the effect of opening of contraction joints in the CDP model. Martin Weiland [8] illustrated the safety of dams in terms of slippage of faults, ground movement etc. A Coupled method can operate on the benefits of both the Lagrangian and Eulerian methods [9] that simplifies the meshing problems that arise due to large deformation.

The present study aims to anticipate the performance of Koyna Dam under the influence of underwater explosion due to blast loading with different TNT weights and stagnation distance for JH-2 and CDP damage model. CEL technique is used for the simulation in ABAQUS. Pressure history curves are plotted against detonation time for underwater explosion.

2 Model Description

A three-dimensional Koyna Dam of height 103 m and base width 70.2 m is taken into consideration with a reservoir of 91.75 m height of water. Width of non-overflow of dam is taken as 15 m. Computational simulation of Koyna Dam under the effect of blast loading is done by CEL technique. Lagrangian mesh is used for dam, while eulerian mesh is used for explosion, air and water. Explosion is modeled as the cubical charge of weight 500 and 1000 kg located at a standoff and detonation distance of 10 m towards upstream. Johnson-Holmquist-2(JH2) model is used to define non-linearity in the concrete of dam (Fig. 1).

2.1 *Properties of Concrete*

The concrete damaged plasticity model is employed in the ABAQUS software to assess tensile cracks in the concrete dam caused by UNDEX. The damage variable d, which is used in the CDP model, specifies the concrete damages that are responsible for the loss of rigidity. The value of the damage variable d is taken as 0 and 1 for undamaged and damaged states respectively.

Stress–strain relationship is given by:

$$\sigma = (1 - d)\overline{\sigma} = (1 - d)D_0 : \left(\varepsilon - \varepsilon^{\mathrm{pl}}\right) \tag{1}$$

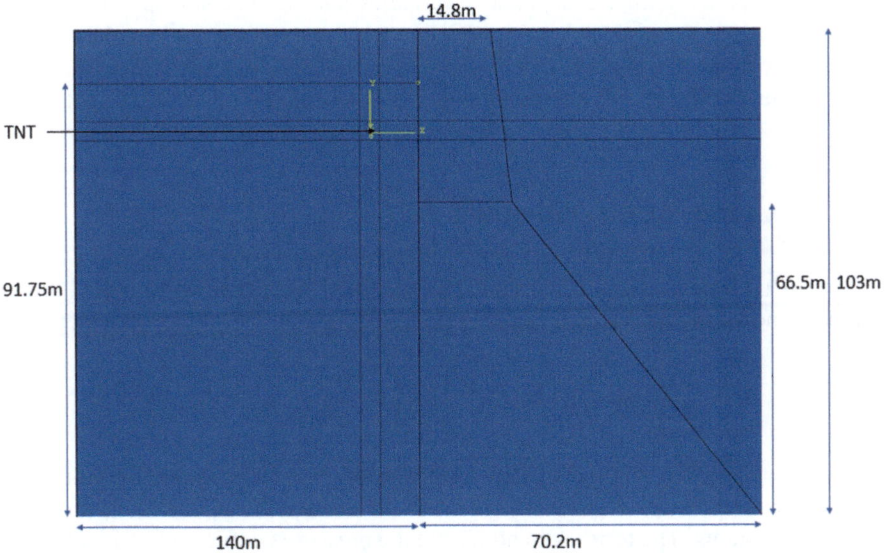

Fig. 1 FEM model of Koyna dam with reservoir

Table 1 Material property of CDP model

Density	Dilation angle	Elastic modulus (Pa)	Poisson's ratio	Dilation angle	Eccentricity	fb0/fc0	k	Viscosity parameter
2640	36.31^0	$31.027*10^6$	0.2	36.31	0	0	0	0

where, fb0/fc0 = ratio of the strengths in the biaxial state to the uniaxial state. k = ratio of the hydrostatic axis distance from the compression meridian to its distance from the tension meridian in the deviatoric cross section (Fig. 2).

where D_0 is the initial elastic stiffness matrix of the material; σ is the stress tensor, and $\overline{\sigma}$ is the effective stress tensor; d is the scalar damage variable; $\varepsilon^{\rm pl}$ is the plastic strain.

In CDP model, two failure modes of concrete are assumed i.e., tensile cracking and compressive crushing:

$$\sigma_t = (1 - d_t) D_0 \times \left(\varepsilon_t - \varepsilon_t^{\rm pl}\right) \quad (2)$$

$$\sigma_c = (1 - d_c) D_0 \times \left(\varepsilon_c - \varepsilon_c^{\rm pl}\right) \quad (3)$$

where, $\varepsilon_t^{\rm pl}$ = equivalent plastic strain in tension, $\varepsilon_c^{\rm pl}$ = equivalent plastic strain in compression, d_t and d_c are tensile and compressive scalar damage variables respectively (Table 1).

Concrete is also modeled asJH-2 model developed by Johmnson-Holmquist 2 model [9]. For modeling, 8 dependent variables and 32 mechanical constants are taken. Eight dependent variables are equivalent plastic strain, equivalent plastic strain rate, ductile damage initiation criterion, damage variable, pressure increment due to bulking, yield strength, volumetric strength and material point status[10, 11]

2.2 *Properties of Air Water and TNT*

The most common equation of state for modeling explosive is Standard Jones, Wilkins, and Lee (JWL) which can be written as:

$$P = C_1\left(1 - \frac{\omega}{R_1\vartheta}\right)e^{-R_1\vartheta} + C_2\left(1 - \frac{\omega}{R_2\vartheta}\right)e^{-R_2\vartheta} + \frac{\omega E}{\vartheta} \quad (4)$$

where, ϑ = specific volume of detonation products over the specific volume of undetonated explosive, E is the specific internal energy, $C1$, $C2$, $R1$, $R2$ and x are material constants. The terms $C1$ and $C2$ are the pressure coefficients; $R1$ and $R2$ are the principal and secondary eigenvalues, respectively; and ω is the fractional part of the normal Tait equation adiabatic exponent. TNT properties are specified in Table 2.

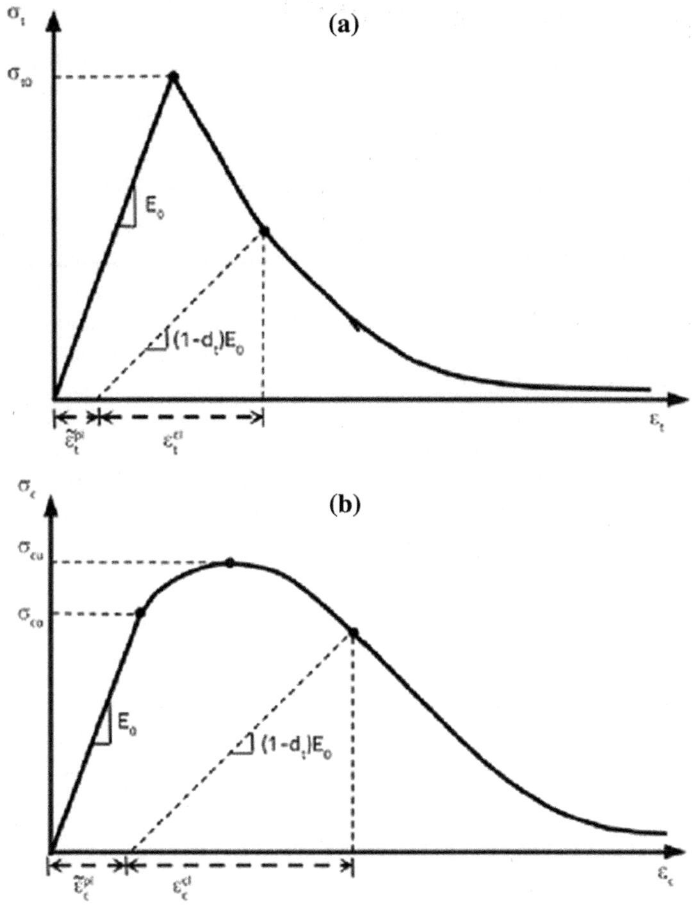

Fig. 2 Concrete response to loading in **a** tensile, **b** compression

Table 2 Properties of TNT

Parameters	Weight	Eos	C_1	C_2	R_1	R_2	ω
Value	500 and 1000 kg	JWL	$3.73e^{11}$ Pa	$3.747e^{9}$ Pa	4.15	0.9	0.35

Water is modeled as a linear us-up equation of state. Material properties are as follows: density = 998 kg/m^3, Newtonian fluid and dynamic viscosity is 0.001. Air is modeled as ideal gas equation of state and the properties of air are shown in Table 3.

Table 3 Properties of air

Parameters	Density	Eos	Gas constant	Ambient pressure
Value	1.2 kg/m^3	Ideal gas	287	101.33 Pa

2.3 Meshing and Boundary Condition of Materials

Dynamic explicit step time 0.04 s to be taken for the analysis. Bottom of the dam is taken as fixed in all directions. For eulerian boundary flow type is outflow and non-reflecting so that waves cannot be reflected back at eulerian boundary. Boundary condition for air is free. For meshing, explicit and element deletion is used. An 8-node linear brick, reduced integration, hourglass control is used for meshing of lagrangian and eulerian parts. Johnson-Holmquist damage model (JH-2) is used to analyze the behavior of concrete under blast load. Luccioni et al. [11] considered the effect of meshing in fluid–structure interaction and stated that the dispersion of explosives can be measured accurately if element grid size is 100 mm.

3 Results and Discussion

In the present study, a concrete gravity Koyna Dam subjected to underwater explosion is taken into consideration. Two charge weight i.e., 500 and 1000 kg is adopted for the analysis. Different target points are placed inside the water in order to investigate the characteristics of shock wave propagation, and the typically recorded pressure history curves at selected targets along the direct path from the charge center to the dam structure are shown in Figs. 3, 4 and 5. From Fig. 4 it can be seen that as the distance from the charge point increases, the shock wave pressure in the water rapidly decreases. Figure 5 depicts the pressure history at some distance from charge of center to the frees surface. As can be seen, as the distance from the charge center increases, the attenuation of the shock wave peak pressure is also quick. Figure 6 shows the comparison of JH-2 and CDP models for different TNT weights.

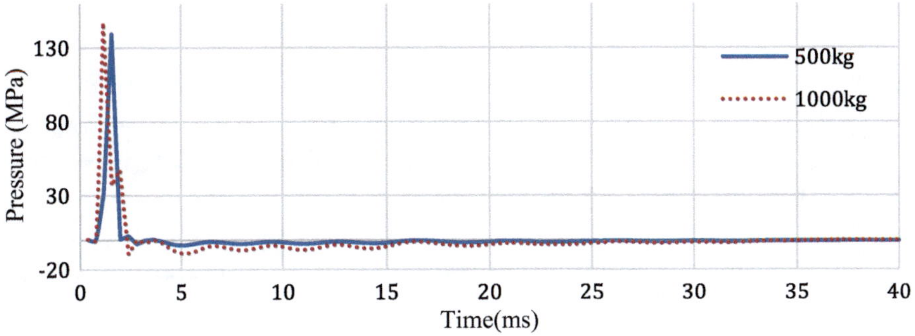

Fig. 3 Pressure diagram at the center of charge

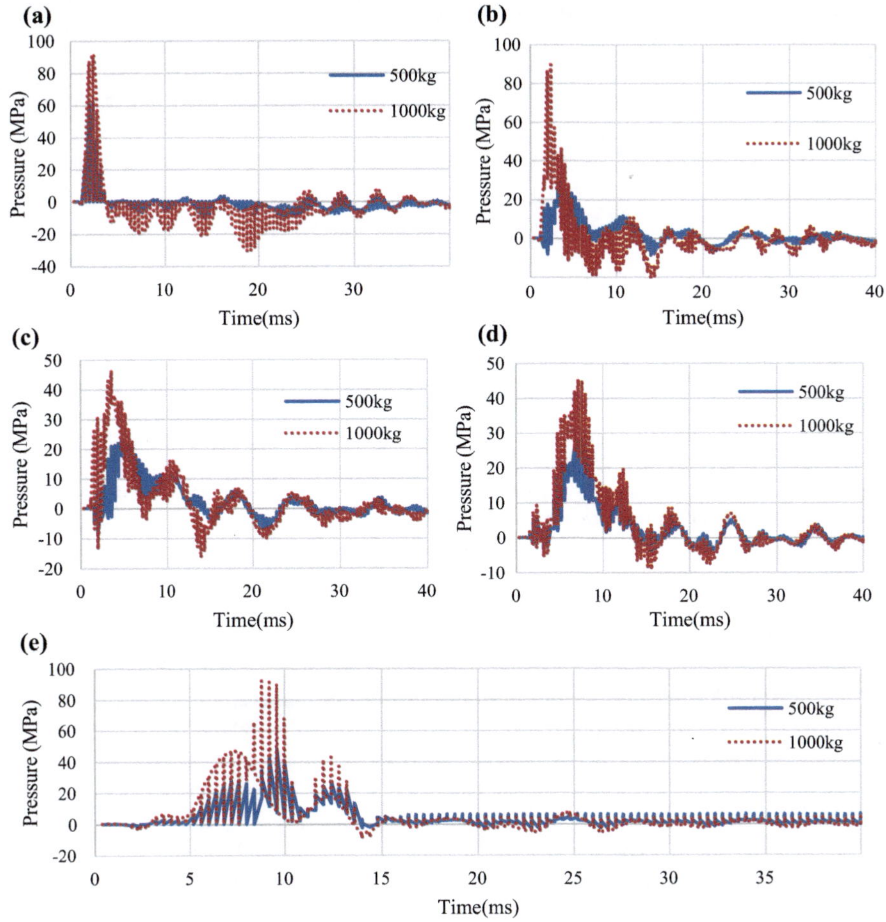

Fig. 4 Pressure history diagram from center of charge to face of dam for different TNT weight **a** 2 m, **b** 4 m, **c** = 6 m, **d** 8 m, **e** 10 m

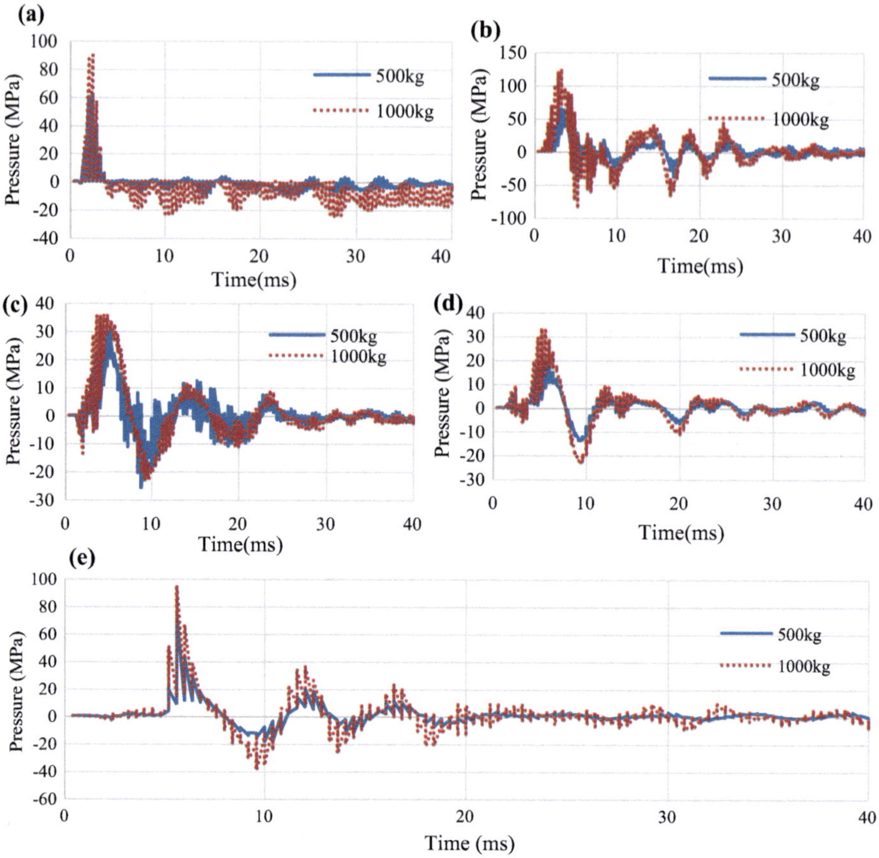

Fig. 5 Pressure history diagram from center of charge to free surface of reservoir for different TNT weight **a** 2 m, **b** 4 m, **c** 6 m, **d** 8 m, **e** 10 m

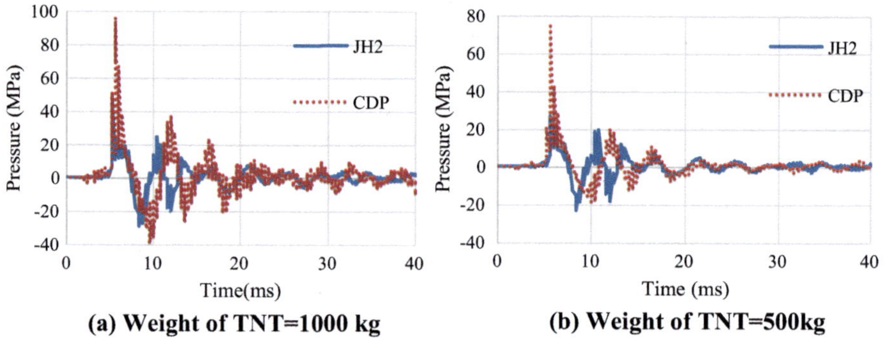

Fig. 6 Comparison of Pressure of JH-2 model and CDP model at the free surface of reservoir

4 Conclusion

A numerical model depicting the damage of Koyna Dam subjected to underwater explosion is presented in this paper. In the simulation, a series of standoff distance i.e., 2–10 m from the face of the dam and frees surface are considered to observe the response of dam under impact loading. From the results, it is clearly shown that the response of dam is greatly influenced by increasing the weight of TNT. Upstream face of dam is seemed to get more damage when charge center is near the face of the dam. In Fig. 6 it is seen that CDP model gives greater pressure value at the surface of reservoir irrespective of TNT weight, however, it is obvious that 1000 kg TNT gives more pressure value than 500 kg TNT. The presented conclusions and observations are based on a particular type of gravity dam based on analysis of dam under impact load and may differ for different types of concrete gravity dams with different interactions. However, conclusions may remain same to some extent in the broad sense.

References

1. Rajendran R, Lee JM (2009) Blast loaded plates. Mar Struct 22(2):99–127
2. Falconer J (2020) The Dam Buster story. The History Press, Cheltenham
3. Yu T (2009) Dynamical response simulation of concrete dam subjected to underwater contact explosion load. In: 2009 WRI world congress on computer science and information engineering, vol 1. IEEE
4. Wang X et al (2020) Blast-induced damage and evaluation method of concrete gravity dam subjected to near-field underwater explosion. Eng Struct 209:109996
5. Zhang S et al (2014) Numerical simulation of failure modes of concrete gravity dams subjected to underwater explosion. Eng Fail Anal 36:49–64
6. Zhang Q-L et al (2021) Role of negative pressure in structural responses of gravity dams to underwater explosion loadings: the need to consider local cavitation. Eng Fail Anal 122:105270
7. Wang G et al (2017) Deterministic 3D seismic damage analysis of Guandi concrete gravity dam: a case study. Eng Struct 148:263–276
8. Wieland M (2016) Safety aspects of sustainable storage dams and earthquake safety of existing dams. Engineering 2(3):325–331
9. Oucif C, Mauludin LM (2019) Numerical modeling of high velocity impact applied to reinforced concrete panel. Underground Space 4(1):1–9
10. Benson DJ (1992) Computational methods in Lagrangian and Eulerian hydrocodes. Comput Methods Appl Mech Eng 99(2–3):235–394
11. Luccioni B, Ambrosini D, Danesi R (2006) Blast load assessment using hydrocodes. Eng Struct 28(12):1736–1744

Prediction of Bifurcation Phenomenon Inside a Sudden Expansion Pipe Using Neural Network

Swapnil Maiti and Somnath Roy

Abstract The current study investigates internal flow through a sudden symmetrical expansion pipe of different expansion ratios. The flow pattern is studied using Finite Volume based computational tool which trained a neural network to predict bifurcation phenomena at different expansion ratios and Reynolds Number. This study focuses on the training of neural network using the dataset generated by the Finite Volume based computational tool and the predicted dataset is validated by comparing it with the results of experimental investigation obtained from relevant literature. Latin Hypercube Sampling is used to generate random values of Reynolds Number, which is given as an input to the solver, from where we get outputs for vortex lengths. This process is repeated in a loop for the sample space of Reynolds Number and varying expansion ratios. A neural network is trained by the dataset generated by the solver using Reynolds Number as input and vortex lengths as output.

Keywords Internal flow · Sudden expansion pipe · Reynolds number · Bifurcation phenomena · Expansion ratio · Neural network

1 Introduction

Studying the flow through a sudden expansion pipe is of practical importance as the geometry is required in engineering applications, such as in pipe networks and mixing devices. It can also be found in biological systems blood flow past stenoses [1]. The Coandă Effect is the phenomenon in which a fluid jet attaches itself to a surface in its vicinity and remains attached to it even when the surface curves away from the initial jet direction. Inside past a symmetrical sudden expansion region, the

S. Maiti (✉)
Department of Mechanical Engineering, Indian Institute of Engineering Science and Technology, Shibpur, West Bengal 711103, India
e-mail: 511019006.swapnil@students.iiests.ac.in

S. Roy
Department of Mechanical Engineering, Indian Institute of Technology, Kharagpur, West Bengal 721302, India

R. K. Ray et al. (eds.), *Advances in Theoretical and Applied Mechanics*, Lecture Notes in Mechanical Engineering, https://doi.org/10.1007/978-981-97-0418-7_3

fluid has a tendency to flow along the edge of the pipe, creating separation regions on both sides of the centerline within a certain range of length of the expanded region.

The Bifurcation phenomenon is owing to the Coandă Effect, separation regions will be created on both sides of the centerline of the pipe. At low Reynolds Number of flow, the separation regions created are of same lengths. But, it has been observed that after a certain Critical Reynolds Number within the laminar range, separation regions of different lengths are created [2]. This asymmetricity may arise from differences in friction factors between regions of the surface of the pipe. We are investigating bifurcation phenomena inside a sudden expansion pipe of specific dimensions under laminar conditions. The fluid used in this study is water.

Considerable amount of work has already been reported on two-dimensional sudden expansion flows. The flow through a sudden symmetric expansion in a pipe is of both fundamental and practical interest. This geometry arises in many engineering applications ranging from heat exchangers to combustion chambers and is also of relevance in physiological problems such as the flow past stenoses [1].

Double sided expansion is characterised by expansion ratio, which is the ratio of the diameters of the thinner section to the thicker section, and is an important parameter used in this study.

It has been observed that even though the geometry is symmetric, asymmetric flow develops at a certain Reynolds Number for the particular geometric conditions [3]. We are going to study the critical Reynolds Number for the series of geometric conditions. At low Reynolds Number, the lengths of separation regions are observed to increase while increasing Reynolds Number. At higher Reynolds Number, separation regions of unequal lengths develop and the asymmetry remains even in the turbulent region.

Mullin et al. [1] made a physical attempt to visualise flow patterns achieved by Magnetic Resonance Imaging and Simple Flow Visualisation by using reflective platelets to study both steady and temporal bifurcation phenomena. It was observed that there is steady symmetry breaking bifurcation in the flow through a 1:2 sudden axisymmetric expansions as Reynold Number was increased. The initially symmetric downstream flow evolves smoothly and rapidly into an asymmetric steady state. But recent attempts to calculate the bifurcation point numerically have failed with calculations indicating that the symmetric flow becomes linearly unstable at a Reynolds Number significantly greater than what was reported in the study.

Battaglia et al. [2] demonstrated instability of low Reynolds Number expanding planar flows leading to several numerical studies that investigated the symmetry breaking bifurcation using linear stability approach to locate the bifurcation points of the transition between symmetric and asymmetric flows. Linear stability analysis showed that the vortex disturbances were due to the instability of a vortex core along the centreline. Numerical simulations of symmetric flow for expansion ratio 1:2 shows that at low Reynolds Number = 171, the flow downstream of the expansion is observed to develop symmetrically along the centre plane. Velocity profiles show that the flow develops symmetrically downstream for both experiments and simulations. Numerical simulations for asymmetric flow showed that for larger expansion ratio, the flow becomes more unstable at lower Reynolds Number. Consequently, there is a significant error in the predictions of bifurcation. For expansion ratio 1:2, a critical Reynolds Number of 217 was evaluated according to the experiments. The error between numerical simulation and Particle Image Velocimetry is expected to be large near critical Reynolds Number, which is difficult to measure. However, it was found that for well-developed symmetric and asymmetric flows away from critical Reynolds Number, the error was significantly less.

Drikakis et al. [3] have discussed that an accurate method of convergence criterion is comparing skin friction distribution and size of separation bubbles during iterations. When these variables do not change any further, a steady solution is achieved. By using 4th order of accuracy, it was found that for expansion ratio 1:2, critical Reynolds Number was 216, for expansion ratio 1:4, critical Reynolds Number was 53, and for expansion ratio 1:6, critical Reynolds Number was 33.

After comparing various studies, an open question still exists regarding the reliability of CFD methods to predict bifurcation phenomena. Differential equations which are represented by difference equations introduce numerical instabilities due to numerical scheme, time step, initial and boundary conditions, and time interval till which the calculation proceeds.

Neural networks are a useful tool helping to reduce experimental burden and computational time. Neural network structure replicates that of the biological neural structure. It establishes the relation between input set and output set, which are fed as training. They are able to modify their internal neural structure in relation to the function objective, which is derived by itself based on the input and output values [4]. The current study focuses on Deep Learning, which represents multiple stacked layers of the network, that connect each node with a distinct neuron, which establishes mathematical relations between two nodes in adjacent layers. It can solve complex non-linear problems and establish high level relations between input and output by the help of numerous hidden layers stacked between input layer and output layer [5]. The applications of neural network in the area of fluid flow include simulation of fluid flow and transport, computing friction factor in pipeline flow and prediction of flow field using hybrid scheme of neural networks [6].

The objective of this work is to create a neural network model which predicts the bifurcation phenomena in sudden expansion flows. The neural network will be trained by feeding Reynolds Number as input and corresponding separation region lengths of both upper and lower regions of the pipe as outputs. Using this training, we will predict corresponding separation region lengths for the desired values of Reynolds Number.

2 Methodology

The process is divided into three sections which are mentioned below. The process is explained by taking an example of the model for expansion ratio 1:5.

2.1 Geometry, Meshing and Boundary Conditions

The CFD analysis is conducted on a two-dimensional model of the sudden expansion pipe. The geometry and meshing are generated in ANSYS module.

This geometry is the specific case for expansion ratio 1:5 and is created in ANSYS SpaceClaim as shown in Fig. 1.

The inlet thinner section has a fixed height of 5 mm and a fixed length of 10 mm. Rest of the dimensions are set according to the expansion ratio.

Geometry is created for all the expansion ratios from the set {1:2; 1:2.5; 1:3; 1:3.5; 1:4; 1:4.5; 1:5; 1:6}.

Mesh files are generated for the geometries with varying expansion ratio. Mesh created from geometry for expansion ratio 1:5 is given below in Fig. 2.

The fluid medium was selected as liquid water from ANSYS Fluent database and viscous-laminar model was selected. The inlet was selected as the left-most end of the pipe, where an inlet velocity parameter was assigned. Parabolic velocity profile was assigned to the inlet velocity profile. The outlet was selected as the right-most end of the pipe, where the gauge pressure was assigned as zero. The accuracy of the solver for x-velocity, y-velocity and continuity was set to 10^{-6}. A sample calculation was run for Reynolds Number = 50 for expansion ratio 1:5.

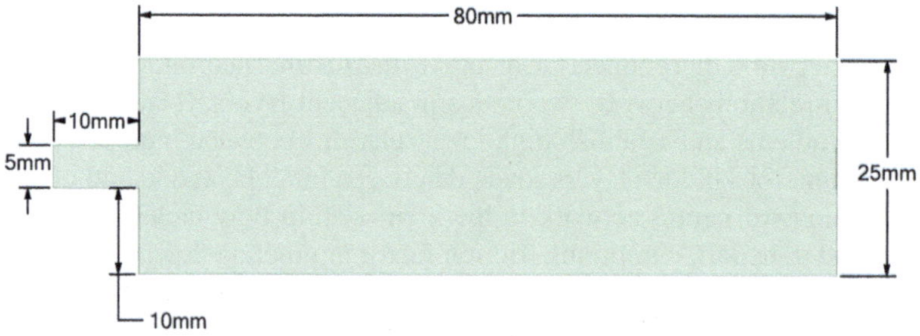

Fig. 1 Geometry used for analysis of internal flow through a sudden expansion pipe of expansion ratio 1:5

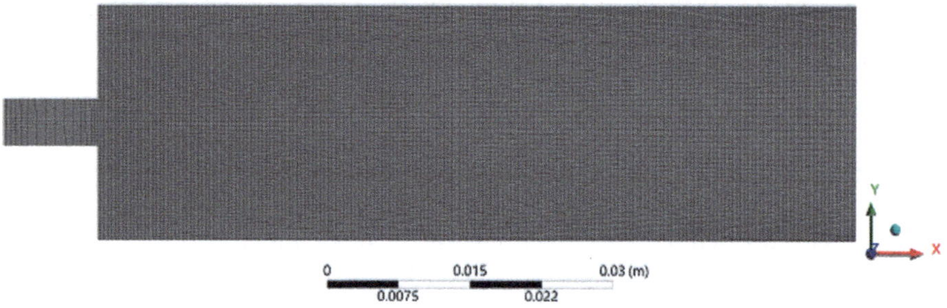

Fig. 2 Mesh used for analysis of internal flow through a sudden expansion pipe of expansion ratio 1:5. Number of nodes in the mesh is 15,165 and the number of elements in the mesh is 14,877

The incomplessible, steady state, laminar flow is described by continuity equation and Navier–Stokes equations.

Mass conservation or continuity equation:

$$\nabla \cdot (\rho \overrightarrow{v}) = 0 \quad (1)$$

Momentum conservation equation:

$$\rho \vec{v} \nabla \cdot \vec{v} = \mu \nabla^2 \vec{v} - \nabla P \quad (2)$$

2.2 Creating Journal Files and Running ANSYS Fluent in Batch Mode Using MATLAB and Training a Neural Network

ANSYS Fluent can be run in batch mode by calling ANSYS Fluent in -g mode. The required steps for the flow analysis are written in a journal file, where the input parameter for velocity is changed according to the Reynolds Number. This journal file is run in MATLAB to generate the desired output. Using Latin Hypercube Sampling, a random dataset for Reynolds Number is generated, which will be given as input to ANSYS Fluent as the parameter for inlet velocity. ANSYS Fluent will generate solution data, which is read and sorted by MATLAB to get the values for vortex lengths corresponding to the input Reynolds Number. This process is run in a loop to get corresponding output for the sample space of Reynolds Number.

2.3 Training a Neural Network

It is required to build a regression model which predicts the values of vortex lengths for a given Reynolds Number. To achieve this, a neural network was built in Keras Functional API with two dense layers between an input and output layer. Both the dense layers contain 128 units and use activation function 'relu'. The input dataset is split into training and testing datasets, with training as 80% of the input dataset and 20% of the input dataset is used for testing. A schematic sample architecture of the model is given in Fig. 3.

The training dataset consists of randomised Reynolds Number and corresponding vortex lengths. The vortex lengths are represented as **X1**, which is the length of the vortex in the upper section and **X2**, which is the length of the vortex in the lower section. **H** is the height of the thinner section which is of fixed value 5 mm. Output columns are denoted as **XR1/H** and **XR2/H** which are the ratio of vortex lengths to

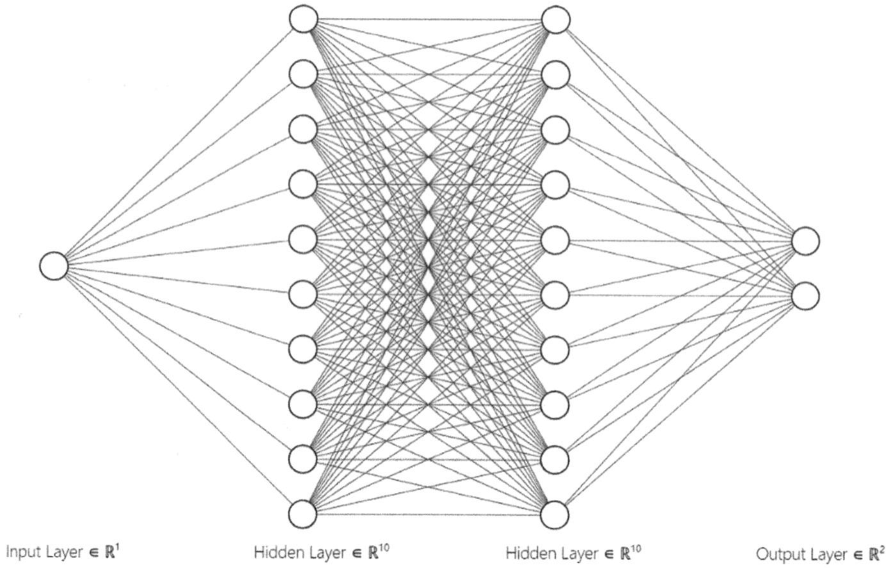

Fig. 3 A sample schematic diagram of the neural network architecture that will be used in the study

the height of the thinner section. The results from training the neural network will be discussed in the following section.

3 Results

Geometric models of different expansion ratios were used to simulate the interior flow of fluid using a Computational Fluid Dynamics Solver. Using Latin Hypercube Sampling, random values of Reynold Number were generated. For the case of expansion ratio 1:5, the range of Reynolds Number was required between 25 and 130. This randomised Reynolds Number was given as input to ANSYS Fluent, from where we received the values of vortex lengths as output. Considering Reynolds Number as input parameter and the vortex lengths as two output parameters, we trained a neural network. The results of training the neural network for varying expansion ratios will be discussed in the coming sections.

A sample result of flow analysis is shown to explain the underlying process of giving Reynolds Number as input to the solver and getting the lengths of the vortices as output. The results of the flow analysis for the particular case for Reynolds Number = 50 and expansion ratio of 1:5 are shown in Figs. 4 and 5.

To calculate the lengths of vortices, we need to locate the reattachment points. The method used here is to locate the point with the minimum wall shear stress after entering the expanded region. The plot for wall shear stress is shown in Fig. 5.

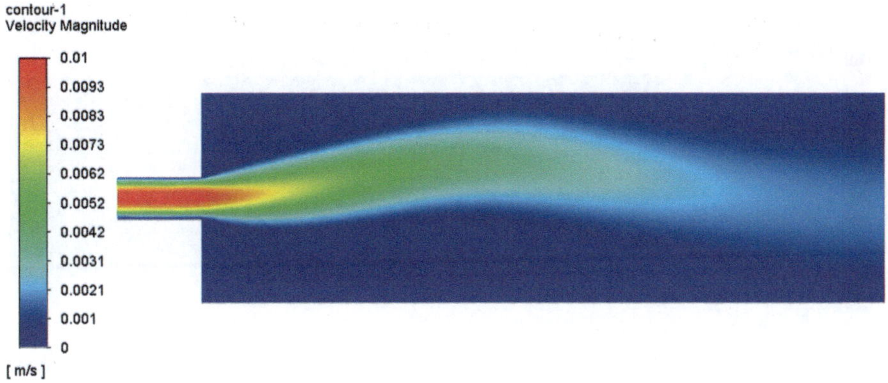

Fig. 4 Contour plot for magnitude of velocity (Reynolds Number = 50)

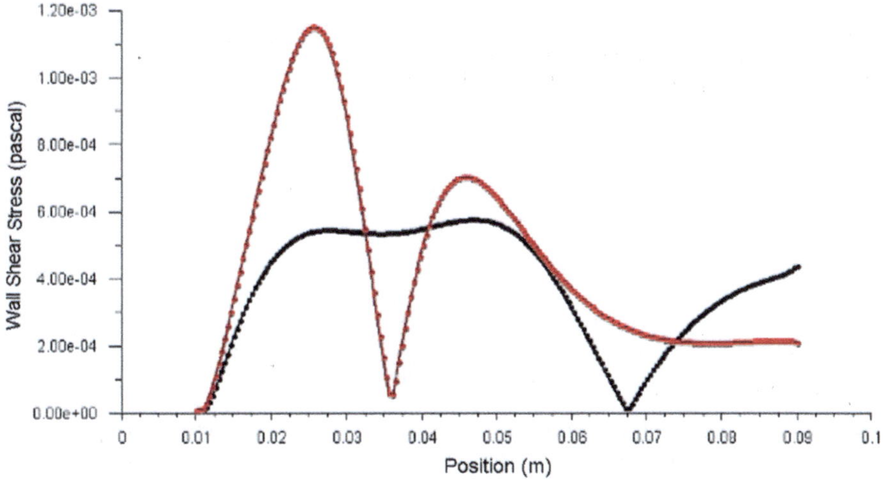

Fig. 5 Plot for wall shear in the expanded region with respect to position in *x*-axis. The red line denotes the curve for values on the upper wall and the black line denotes the curve for values on the lower wall

To get exact values, the solution data for Wall Shear Stress for both upper and lower surfaces is exported, which is read and sorted in MATLAB to find the reattachment point.

The results of flow analysis for other values of Reynolds Number for the particular case of expansion ratio 1:5 are shown in Figs. 6 and 7 to visualise the bifurcation phenomena.

As shown in Fig. 6, for Reynolds Number = 30, bifurcation phenomenon is not observed as the value is below the critical Reynolds Number. The curves of wall

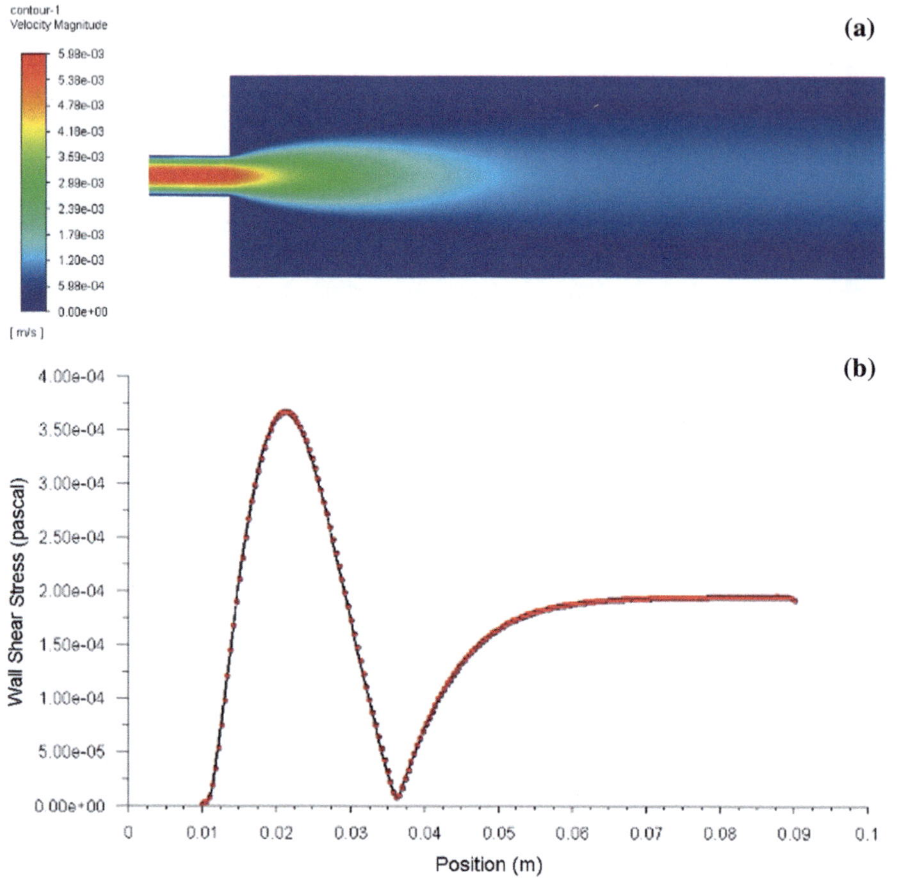

Fig. 6 For Reynolds Number = 30, **a** Contour plot for magnitude of velocity, **b** Plot for wall shear in the expanded region with respect to position in *x*-axis. The red line denotes the curve for values on the upper wall and the black line denotes the curve for values on the lower wall

shear stress for upper wall and lower wall coincide with each other, indicating that the flow is symmetric.

As shown in Fig. 7, for Reynolds Number = 60, bifurcation phenomenon is being clearly observed as the value is above the critical Reynolds Number. The curves of wall shear stress for upper wall and lower wall are distinct and indicate the difference in the location of reattachment point and presence of asymmetric flow.

Results of training the neural network with the data from flow analysis for expansion ratio 1:5 are given in Table 1. Here **XR1** denotes the vortex length in the upper section and **XR2** denotes the vortex length in the lower section.

The loss for each observation is calculated as Mean Square Error:

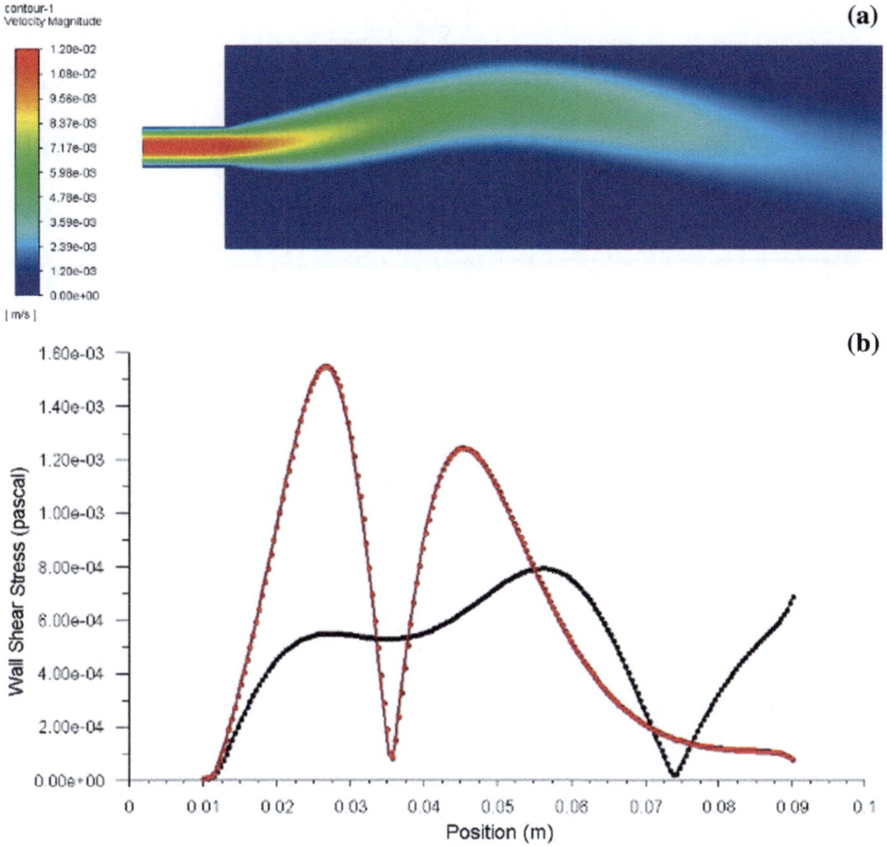

Fig. 7 For Reynolds Number = 60, **a** Contour plot for magnitude of velocity, **b** Plot for wall shear in the expanded region with respect to position in x-axis. The red line denotes the curve for values on the upper wall and the black line denotes the curve for values on the lower wall

Table 1 Results of training the neural network with the data from flow analysis for expansion ratio 1:5

Loss	0.1942809820175171
XR1_Loss	0.04706711694598198
XR1_RMSE	0.21694956719875336
XR2_Loss	0.015353627502918243
XR2_RMSE	0.12390975654125214

$$\text{Loss}^i = \frac{\sum_{i=1}^{m}\left(\frac{x_t^i - x_p^i}{x_t^i}\right)^2}{m} \tag{3}$$

where x_t^i is the true value and x_p^i is the predicted value.

Root Mean Square Error or RMSE is calculated as (Fig. 8):

$$\mathrm{RMSE} = \sqrt{\frac{\sum_{i=1}^{m}\left(\frac{x_t^i - x_p^i}{x_t^i}\right)^2}{m}} \tag{4}$$

The dataset obtained from flow analysis in different geometries was also used to train the neural network. The results are shown in Table 2.

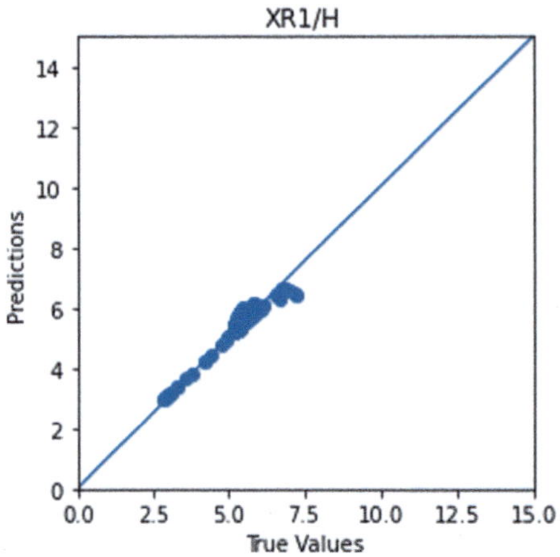

Fig. 8 Plots for predictions versus true values as predicted by the neural network for vortex lengths in upper section and lower section for expansion ratio 1:5

Table 2 Results of training the neural network with the data from flow analysis for varying expansion ratios

Expansion ratio	Critical Reynolds number	Loss (%)	XR1		XR2	
			loss (%)	Root mean square error (%)	loss (%)	Root mean square error
1:2	223	19.5937	4.5330	21.2910	2.5171	15.8655
1:2.5	120.5	21.4720	5.4451	23.3347	3.3575	18.3236
1:3	80.75	30.8973	10.7211	32.7431	7.2005	26.8338
1:3.5	64.5	25.4990	9.0317	30.0529	3.6221	19.0319
1.4	55	37.5469	9.7023	31.1485	14.6957	38.3350
1:4.5	48	21.6608	6.7439	25.9689	2.2632	15.0441
1:5	42	19.4281	4.7067	21.6950	1.5354	12.3910
1:6	34.5	15.5439	2.3840	15.4401	0.3695	6.0785

Table 3 Values of Critical Reynolds Number for different expansion ratios, as predicted by the neural network and comparing the values for Critical Reynolds Number as found by Drikakis et al. [3]

Expansion ratio	Critical Reynolds number	Critical Reynolds number as found by Drikakis et al. [3]
1:2	223	216
1:2.5	120.5	–
1:3	80.75	80
1:3.5	64.5	–
1:4	55	53
1:4.5	48	–
1:5	42	41
1:6	34.5	33

4 Discussion

While performing numerical simulations, it was observed that for low Reynolds Number, the flow remained symmetric and the vortices were of equal lengths. After a certain critical Reynolds Number, the flow became asymmetrical with unequal vortex lengths on upper and lower walls. It was also observed that when expansion ratio was reduced, the height of the expanded section was increased, the critical Reynolds Number seemed to decrease exponentially.

The disparity between the vortices was observed to be not restricted to only configuration. When the simulation was repeated, it was observed that larger vortices were formed on the opposite side. Secondary vortices may also be generated. This explains that bifurcation is a purely random phenomenon generated from the instability of a vortex core along the centreline. Table 3 shows the critical Reynolds Number for different expansion ratios, as predicted by the neural network.

Due to the presence of an open question about the reliability of numerical simulations to mimic experimental results, there is a degree of uncertainty upon results from numerical simulations. Above results seem to align with the results as found by Drikakis et al. [3]

The error from numerical solutions and error from training the neural network add up to a total error in predicting the lengths of vortices. But, the ratio of the lengths of vortices to the thinner inlet section height gives a more precise result which may align with experimental results.

While training the neural network, it was observed that training the model for more number of epochs produced better results. But, the accuracy reaches a constant maximum with further increase in epochs. The number of epochs was decided by trials. The above results infer that we can predict the vortex lengths for other values of Reynolds Number with approximate errors as stated from training and testing the dataset on the model.

Acknowledgements During my study, I have utilised the resources provided by Centre for Computational and Data Sciences, IIT Kharagpur, funded by National Supercomputing Mission. I was granted access to the resources while I was pursuing my summer internship at CCDS, IIT Kharagpur. I also utilised High-performance Computing Facility PARAM Shakti which allowed to reduce the computational time by an admirable factor.

I also acknowledge the cooperation, assistance and encouragement provided by my parents and the students working in the same laboratory as mine throughout my study.

References

1. Mullin T, Seddon JR, Mantle MD, Sederman AJ (2009) Bifurcation phenomena in the flow through a sudden expansion in a circular pipe. Phys Fluids 21(1):014110
2. Battaglia F, Papadopoulos G (2005) Bifurcation characteristics of flows in rectangular sudden expansion channels. J Fluids Eng 128(4):671–679
3. Drikakis D (1997) Bifurcation phenomena in incompressible sudden expansion flows. Phys Fluids 9(1):76–87
4. Grossi E, Buscema M (2008) Introduction to artificial neural networks. Eur J Gastroenterol Hepatol 19(12):1046
5. Lee H, Song J (2019) Introduction to convolutional neural network using Keras; an understanding from a statistician. Commun Stat Appl Methods. 26:591–610
6. Zhi S (2005) Application of artificial intelligence CFD based on neural network in vapor–water two-phase flow. Eng Appl Artif Intell 18(6):663–671

Influence of Porocoat Beaded Coating Structure on Bone Ingrowth Around the Porous Coated Implant: A Two-Dimensional Finite Element Study

Minku and Rajesh Ghosh

Abstract The long-term biomechanical functionality of uncemented orthopaedic implants relies on bone ingrowth around porous coated implants. The present study aims to numerically predict the spatial–temporal distribution of tissue differentiation around the Porocoat coated implant surface by employing the mechanoregulatory tissue differentiation algorithm using 2D Finite Element Analysis. Besides this, the influence of the porosity gradient from the substrate surface to the bone interface on bone ingrowth is also investigated. Two distinct 2D microscale models were developed and designated MODEL A and MODEL B. Both models have a three-layer beaded coating, with MODEL B incorporating the effect of a porosity gradient. Both models were assessed for bone ingrowth distribution. The percentage of bone ingrowth in the inter-bead spacing was predicted to be 66% for MODEL A and 45% for MODEL B after the attainment of equilibrium. The results of the simulation showed that a higher amount of bone ingrowth was predicted in MODEL A, which indicates a strong implant-bone interface. The presence of a porosity gradient was found to weaken the interface through reduced bone formation. The findings of the present study are well in corroboration with relevant studies which suggested that predicted results could be employed to predict bone ingrowth.

Keywords Uncemented implants · Bone ingrowth · Porocoat coated implant · Mechanoregulatory algorithm · Tissue differentiation

1 Introduction

To achieve stable long-term fixation sufficient bone ingrowth is required around the prosthesis [1]. The uncemented porous coated implants have evolved in the orthopaedic field to enhance biological fixation in contrast to cemented implants [2].

Minku · R. Ghosh (✉)
Biomechanics Research Laboratory, School of Mechanical and Materials Engineering, Indian Institute of Technology, Mandi, Himachal Pradesh 175075, India
e-mail: rajesh@iitmandi.ac.in

R. K. Ray et al. (eds.), *Advances in Theoretical and Applied Mechanics*, Lecture Notes in Mechanical Engineering, https://doi.org/10.1007/978-981-97-0418-7_4

Uncemented prostheses enhance initial stability and osseointegration by increasing the prosthesis' surface area and surface roughness [3]. It is well recognized that bone ingrowth is governed by various factors like the design of porous coating, pore size, friction coefficient, and elastic modulus of coating [4]. The different types of porous coatings are hydroxyapatite coating, porocoat coating, gription coating, and titanium plasma spray [5]. The present study emphasizes classic standard porocoat coating, which consists of sintered titanium beads arranged in a random manner to create interconnected voids that facilitate bone growth [6]. The features of this porous coating employed a bead size of 200–250 μm, a pore size in the range of 100–300 μm, and average porosity of 45% [7].

Numerous in silico pre-clinical studies based on Finite Element (FE) analysis were performed to predict the amount of bone ingrowth around uncemented implants. The mechanoregulatory models along with tissue differentiation theories were able to successfully evaluate bone ingrowth around various porous coated implants, including an acetabular cup, a multi-layered porocoat beaded coating, and a femoral implant of the hip stem [6, 8, 9]. Three distinct layers of graded porous coatings are made up of: a dense inner interface layer, a middle layer combination of macro and micropores, and a macropores layer on outer [10]. Rapid bone formation assists in limiting the micromotion at bone-implant interface and encourages higher bone formation for the subsequent iterations [6]. However, till now, none of the researchers explored the influence of the porosity gradient from the implant surface to promote rapid osseointegration for beaded coating.

The aim of the present study is to construct a two-dimensional microscale model of the interface between bone and implant in order to obtain the spatial-temporal distribution of tissue differentiation within the porous region of the bone-implant interface. Furthermore, the effect of the graded porous coating on distribution of bone ingrowth is also being investigated. These two models (MODEL A and MODEL B) were quantitatively compared based on the patterns of tissue differentiation in the inter-bead spacing, as well as the amount of each tissue formation under the identical applied boundary conditions.

2 Materials and Methods

2.1 Development of a Two-Dimensional Microscale Model of the Bone-Implant Interface

Two 2D microscale models (MODEL A and MODEL B) of the Porocoat beaded coated implant adjacent to the bone filled with granulation tissues were developed in ANSYS FE software (ANSYS, Inc., PA, USA) based on μCT scan images of Porocoat coated samples [7]. These two models were constructed loosely based on porosity and represented the cross-sections of a 3D porocoat coated substrate. A three- layer beaded coating was applied to MODEL A, as shown in Fig. 1a. The bead diameter was assumed to be 200 μm similar to the previously established literature [1, 7]. The microgeometric features, which include bead height and inter-bead spacing, were considered to be 540 and 100 μm, respectively. In contrary to MODEL B, the arrangement of beads in a particular manner that porosity incrementally increases from the substrate surface toward the bone layer is shown in Fig. 1b. The diameter of the bead at the bottom layer, middle layer, and top layer was assumed to be 200, 150, and 100 μm, respectively. The bead height for MODEL B was 440 μm. As bead diameter varies along the substrate surface, which in turn forces to vary the inter-bead spacing from the substrate surface, which is 100 μm at the bottom layer, 150 μm at the middle layer, and 200 μm at the topmost layer. The average porosity for MODEL A and MODEL B was estimated to be approximately 42% and 51% of the pore space region, respectively. A polar gap of 50 μm was considered to exist between the bone and the upper surface of the implanted bead, which is initially filled with granulation tissues [11]. Each of the adjacent beads was assumed to penetrate each other by 20 μm to simulate the initial fixation in the microscale model [1]. The interfaces of both the microscale models were considered to be perfectly bonded [11].

The discretization of both models was performed using a plane strain eight-noded quadrilateral element in ANSYS FE software (ANSYS, Inc., PA, USA), with a maximum edge length of 10 μm. The optimum size of the element was taken after performing the mesh convergence study. It was assumed that all the materials possessed the properties of linearity, elasticity, homogeneity, and isotropy (Table 1).

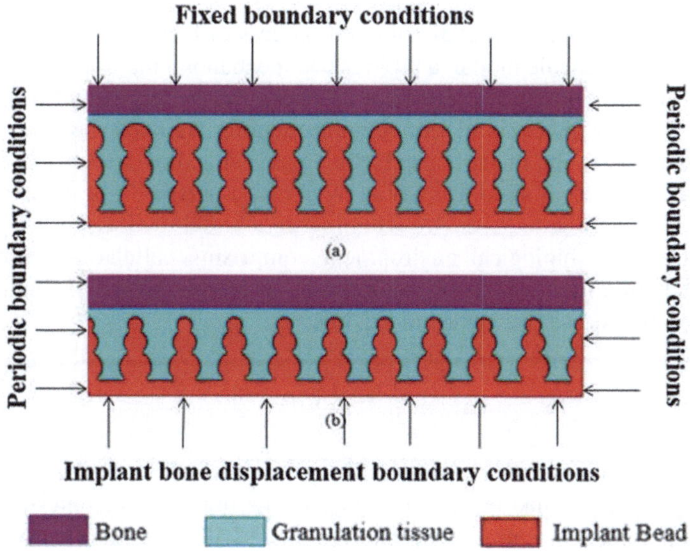

Fig. 1 2D microscale model of the implant-bone interface **a** MODEL A, **b** MODEL B

Table 1 Materials properties applied to bone, tissue, and implant [11]

Materials	Young's Modulus (MPa)	Poisson's ratio
Cancellous bone	500	0.3
Granulation tissue	1	0.167
Titanium	110,000	0.3

2.2 Applied Loading and Boundary Conditions

To evaluate the spatial–temporal process of evolutionary bone ingrowth, the following boundary conditions were employed, as shown in [11]. On both microscale models [11], a tangential displacement of 20 μm was prescribed on the lower surface of the implant along the positive x-direction. To avoid the occurrence of initial debonding of the implant from the granulation tissues, a displacement of 20 μm in the positive y-direction was applied at the lower surface of the implant [11]. The displacement of the top boundary of the bone was considered to be entirely constrained in all directions [11]. Periodic boundary conditions were implemented on the lateral surfaces to simulate the unit of an infinitely long microscale model [11, 12].

2.3 Mechanoregulatory Tissue Differentiation Algorithm

To numerically predict the spatial-temporal process of bone ingrowth into the pore space of each microscale model, a progressive mechanoregulatory tissue differentiation algorithm is employed, similar to an earlier published study by Mukherjee and Gupta [8]. This computer simulation scheme contains two main segments: mechanical and biological. The model's mechanical environment is responsible for calculating local mechanical stimuli, such as hydrostatic pressure and deviatoric strain (Fig. 2).

In contrast, the biological environment comprising cellular activities, which include cell migration and cell proliferation, was simulated by employing the principle of diffusion as described in Eq. (1).

$$\frac{\mathrm{d}\rho}{\mathrm{d}t} = k\nabla^2\rho \tag{1}$$

where k is a diffusion constant taken as 2.37 mm^2/day, which resulted in the maximum cell density being reached in the pore space of the implant-bone interface after 16 weeks and ρ is the concentration of undifferentiated MSC. A rule of mixture, using Eq. (2), was incorporated to determine each element's elastic properties, such as Young's modulus and Poisson's ratio.

When the biophysical stimulus calculation indicated that a new cell phenotype would be favoured, the material properties of an element were altered. Convergence in

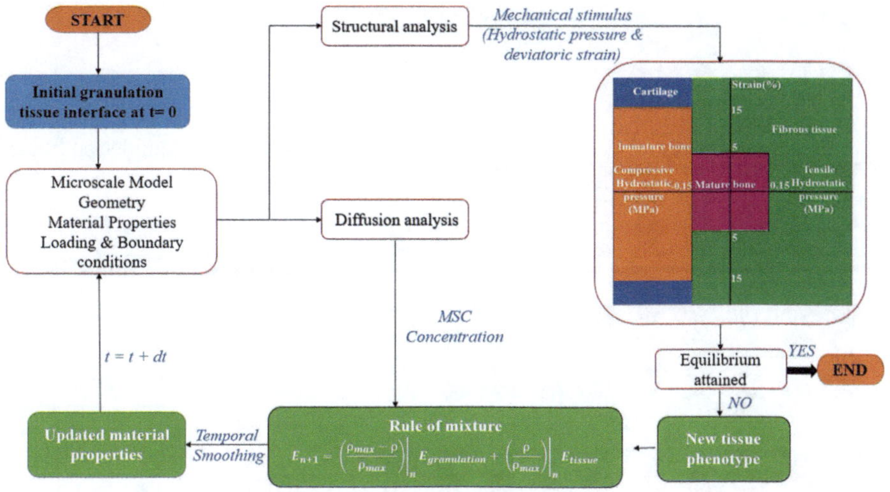

Fig. 2 Schematic overview of the iterative mechanoregulatory tissue differentiation algorithm

tissue development was assumed to have occurred when the variation in the phenotype of the formed tissue between two consecutive iterations was < 1% [8].

$$E_{n+1} = \left(\frac{\rho_{\max} - \rho}{\rho_{\max}}\right)\bigg|_n E_{\text{granulation}} + \left(\frac{\rho}{\rho_{\max}}\right)\bigg|_n E_{\text{tissue}} \quad (2)$$

3 Results

The emphasis of the research was focused on how the initial granulation tissues in the pores transformed into either fibrous, cartilage, or bone, using bone ingrowth simulation corresponding to each microscale model. Both these models were directly evaluated based on the spatial-temporal patterns of tissue differentiation in the inter-bead region and the percentage of each tissue phenotype formation after the attainment of equilibrium. The quantification of tissue differentiation was done by assessing the percentage of bone, cartilage, and fibrous tissue that developed within the interstitial space between the beads following each iteration.

3.1 Bone Ingrowth Prediction

Both models exhibited substantially different bone ingrowth patterns. At the beginning of the algorithm for MODEL A, the majority of cartilage tissue (65%) and

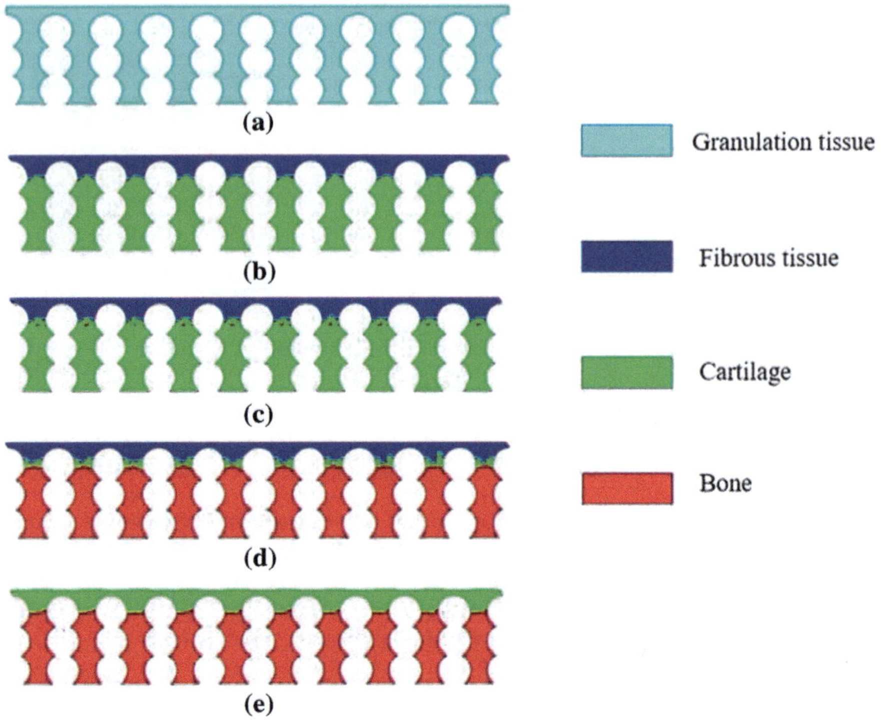

Fig. 3 Predicted bone ingrowth patterns for MODEL A **a** at 0th iteration, **b** after 5th iteration, **c** after 10th iteration, **d** after 15th iteration, **e** after attainment of equilibrium

fibrous tissue (35%) formed, which shows a fragile bone-implant interface. No sign of bone development occurred until around the 10th iteration. Between the 10th to 15th iterations, the region of cartilage tissue was gradually replaced by bone, enabling stronger tissues to form. On the other hand, no change was found in the fibrous tissue region. During the stabilization phase, it was observed that ~66% of the inter-bead spacing was occupied by bone ingrowth, while cartilage tissue accounted for 34%, and no trace of fibrous tissue was observed, as indicated in Fig. 3.

3.2 Effect of Porosity Gradient

During the initial stage of tissue differentiation for MODEL B, the cartilage tissue predominates and accounts for ~95% with some amount of fibrous tissue ~5% produced. In intermediate iterations, the cartilage tissue was transformed into the bone and a portion of the fibrous tissue region remained unchanged. However, upon completion of the process, bone ingrowth was limited to 45% in the inter-bead region, cartilage tissue was prominent at 50.3% in the pore space, and a small amount of

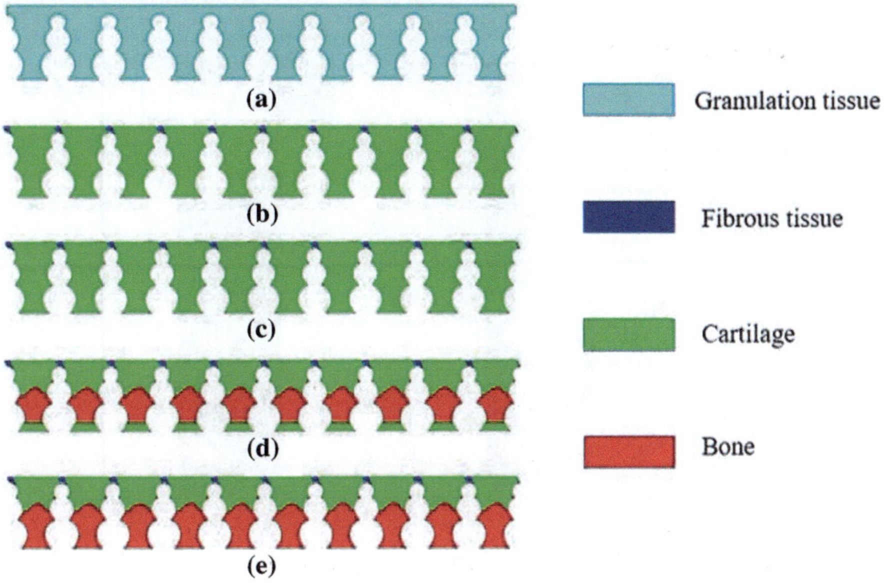

Fig. 4 Predicted bone ingrowth patterns for MODEL B **a** at 0th iteration, **b** after 5th iteration, **c** after 10th iteration, **d** after 15th iteration, **e** after attainment of equilibrium

fibrous tissue formed at 4.5%, as shown in Fig. 4. A comparative graph revealed the amount of each tissue formed once the convergence was attained, as given in Fig. 5. It can be noted that MODEL B has less bone formation, and throughout the process, the presence of cartilage was more prominent (Fig. 5).

4 Discussion

The presence of a porous coating on the surface of the implant facilitates bone ingrowth, thereby leading to enhanced long-term survivorship of orthopaedic implants. In this study, the primary aim was to construct a two-dimensional microscale model of the bone-implant interface of porocoat coated implants and to gain insight into the spatial-temporal distribution of tissue differentiation around the pore space of the bone-implant interface. Additionally, the effect of the graded porous coating on bone ingrowth patterns is also being investigated. Bone ingrowth was simulated using 20 μm micromotion in both cases. A mechanoregulatory tissue differentiation algorithm was developed to predict such distributions.

During the early stage of tissue differentiation, some portions of the pore space were filled with fibrous tissues, while a significant portion was observed to be occupied by cartilage in both models. As the algorithm progresses into the intermediate phase, a significant portion of the cartilage tissue undergoes replacement by bone,

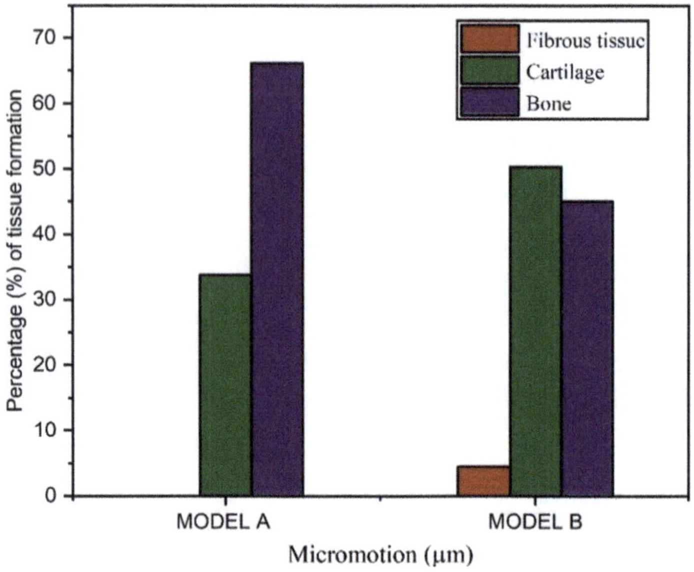

Fig. 5 Amount of different tissue types formed after attainment of equilibrium

which enables a strong interface to form and most of these fibrous tissues are differentiated into the cartilage at a later stage of the bone regeneration process [8, 12, 13].

The predictions of the present study were compared to previously investigated numerical schemes for relevance and corroborn [6, 8, 9, 11]. The percentage of bone ingrowth (66%) predicted by the current study (considering the same level of micromotion) provides agreement with bone ingrowth (70%) predicted by Mukherjee and Gupta [11]. However, Puthumanapully [6] observed 16% of bone ingrowth, which is contrary due to the micromotion applied to the bone layer by keeping the implant boundary fixed. Furthermore, it is worth noting that bone tissue formation was seen within the time period of 10–15 days postoperatively, which was found to be aligned with the results reported in prior studies [8, 9].

It is worth mentioning that the layer of novel bone tissue has the propensity to form further deeper within the inter-bead spacing, sticking to the implant's surface rather than moving toward the host bone, and this process of bone formation from the implant side is known as contact osteogenesis [8].

This study has some noteworthy limitations. The employed geometry was simplified in order to depict an idealistic interface between the bone and porous coating. This is done for the ease of simulation and computational requirements. In reality, the actual geometry is complex, and the arrangement of sintered beads is randomized on the implant surface.

5 Conclusion

Based on the observation and corroboration with other relevant studies, it can be concluded that tissue differentiation patterns in this study were well in consistent with other studies. A higher amount of bone ingrowth was observed in MODEL A than in MODEL B for the same level of micromotion. The presence of a porosity gradient made the implant-bone interface weaker through reduced bone formation and increased cartilage tissue formation. This study will be useful for implant manufacturers as well as the orthopaedic surgeons while choosing appropriate porous coated implants.

Acknowledgements The authors would like to acknowledge the Indian Institute of Technology, Mandi for providing computational facilities for supporting this study.

References

1. Liu X, Niebur GL (2008) Bone ingrowth into a porous coated implant predicted by a mechano-regulatory tissue differentiation algorithm. Biomech Model Mechanobiol 7(4):335–344
2. Frank RM, Fabi D, Levine BR (2013) Modern porous coatings in orthopedic applications. Thin films and coatings in biology. Springer, Dordrecht, pp 69–103
3. Yang X, Gao F, Sun W, Li Z (2022) Clinical application and biological functionalization of different surface coatings in artificial joint prosthesis: a comprehensive research review. Coatings 12(2):117
4. Bobyn JD, Pilliar RM, Cameron HU, Weatherly GC (1980) The optimum pore size for the fixation of porous-surfaced metal implants by the ingrowth of bone. Clin Orthop Related Res 150:263–270
5. Marin E, Fedrizzi L, Zagra L (2010) Porous metallic structures for orthopaedic applications: a short review of materials and technologies. Eur Orthop Traumatol 1(3):103–109
6. Puthumanapully PK (2010) Simulation of tissue differentiation in uncemented hip implants based on a mechanoregulatory hypothesis. Doctoral dissertation, University of Southampton
7. Damm NB, Morlock MM, Bishop NE (2017) Influence of trabecular bone quality and implantation direction on press-fit mechanics. J Orthop Res 35(2):224–233
8. Mukherjee K, Gupta S (2016) Bone ingrowth around porous-coated acetabular implant: a three-dimensional finite element study using mechanoregulatory algorithm. Biomechan Model Mechanobiol 15(2):389–403
9. Mathai B, Gupta S (2022) Bone ingrowth around an uncemented femoral implant using mechanoregulatory algorithm: a multiscale finite element analysis. J Biomech Eng 144(2):021004
10. Yang YZ, Tian JM, Tian JT, Chen ZQ, Deng XJ, Zhang DH (2000) Preparation of graded porous titanium coatings on titanium implant materials by plasma spraying. J Biomed Mater Res 52(2):333–337
11. Mukherjee K, Gupta S (2014) Simulation of tissue differentiation around acetabular cups: the effects of implant-bone relative displacement and polar gap. Adv Biomech Appl 1(2):95–109

12. Ghosh R, Chanda S, Chakraborty D (2020) The influence of macro-textural designs over implant surface on bone on-growth: a computational mechanobiology based study. Comput Biol Med 124:103937
13. Puthumanapully PK, Browne M (2011) Tissue differentiation around a short stemmed metaphyseal loading implant employing a modified mechanoregulatory algorithm: a finite element study. J Orthop Res 29(5):787–794

Magnetohydrodynamic Effects Due to Graphene Oxide Nanofluid Flow in Concentric Cylinders

Jagadeeshwar Pashikanti and D. R. Susmitha Priyadharshini

Abstract Aggregated studies on graphene nanoparticles and magnetohydrodynamic (MHD) flow of nanofluids are important for the effective utilization of their striking thermophysical properties in extensive industrial and medical applications. This investigation is one such study to explore the MHD flow of water dispersed with graphene oxide nanoparticles in concentric cylinders. The model is developed by adapting Buongiorno nanofluid model including viscous dissipation effects and the impacts of nanoparticle attributes namely thermophoretic and Brownian diffusion. Spectral quasilinearization method with Chebyshev's polynomials is adapted to solve the differential equations under convective conditions. On studying the effects of implanted parameters, it is found that the diffusive mass transfer enhancement by Hartmann and Eckert numbers is remarked, while Eckert number improves convective heat transfer and Hartmann number improves conductive heat transfer. Also, special cases of the results are quantitatively analysed and are found to agree with the literature. Furthermore, entropy generation analysis is presented.

Keywords Buongiorno model · Entropy generation · Graphene nanofluids

1 Introduction

Nanofluids are preferred to other conventional viscous and microfluids for their effective heat transfer properties. Additionally, they keep the flow channels from obstruction, deposition and erosion. Literature suggests that the sole significant slip mechanisms in nanofluid flows are effected by thermophoresis and Brownian motion [1]. The performance of nanofluids varies depending on the volume fraction, choice of the geometry, base fluids and hybridized nanoparticles and the needs in demand [2–4]. However, the obtained results from computational studies conclude that graphene

J. Pashikanti (✉) · D. R. S. Priyadharshini
Department of Science and Humanities, Indian Institute of Information Technology, Tiruchirappalli, Tamil Nadu 620012, India
e-mail: jagadeeshwarp@iiitt.ac.in

R. K. Ray et al. (eds.), *Advances in Theoretical and Applied Mechanics*, Lecture Notes in Mechanical Engineering, https://doi.org/10.1007/978-981-97-0418-7_5

oxide (GO) aids in maximum heat transfer because of the excellent thermal conductivity [5–8]. Some of the applications of graphene based nanofluids include their uses in lithium-ion batteries, biosensors, supercapacitors, medical suspensions etc.

Some of the notable numerical studies of nanofluid flow in concentric cylinders include the studies of silver, copper, alumina and titania nanoparticles in water [9, 10]. From the conclusions, it is established for high Rayleigh numbers, the nanofluids enhance convective heat transfer. The study conducted by Yu et al. [11] investigates the aqueous nanofluids with copper oxide particles for four different volume fractions for natural convection and concludes that the time-avergaed Nusselt number reduces with nanoparticle volume fraction. Interpretations on the numerical studies of forced convection in two porous geometries containing water based copper nanofluid [12] reveal that the Nusselt number increases with porosity.

Moreover, graphene has the largest thermal conductivity of all known materials. For its advantageous thermophysical, mechanical and chemical properties, its immense use as coolant is raising aloft. The contribution to renewable energy reassures economically large scale practical applications in view of power storage and capacity. Such significance interests us in computationally studying graphene based nanofluids flow in several geometries. This paper aims at bridging the gap of computationally studying the magnetohydrodynamic (MHD) flow of graphene oxide nanoparticles dispersed in water in concentric cylinders. The flow is modelled and the equations are numerically solved to graph the results.

2 Mathematical Formulation

A vertical cylinder with radius b rotates in anticlockwise direction with the angular velocity Ω. A fixed cylinder with radius a, such that $b > a$, is placed inside the rotating cylinder. We assume that the nanofluid of graphene particles in water is flowing between the two cylinders with the velocity (v, u, w). The tangential pressure gradient of the fluid $\partial P/\partial \Theta$ is maintained at a constant value with the application of radial magnetic field B_0. The fully developed flow is presented using cylindrical coordinates (r, Θ, z). We also assume that the fluid's concentration and temperature changes radially. Thus, the representational diagram is portrayed in Fig. 1.

Considering the body force caused by gravity, the effects of chemical reaction, radiative heat flux, the effects caused by characteristic nanoparticles phenomena namely Brownian motion and thermophoresis and adapting the nanofluid model [1], the flow is modelled as follows:

$$\frac{\partial u}{\partial \Theta} = 0 \tag{1}$$

$$\frac{\rho_{\mathrm{nf}} u^2}{r} = \frac{\partial p}{\partial r} \tag{2}$$

Fig. 1 Representational diagram of the flow

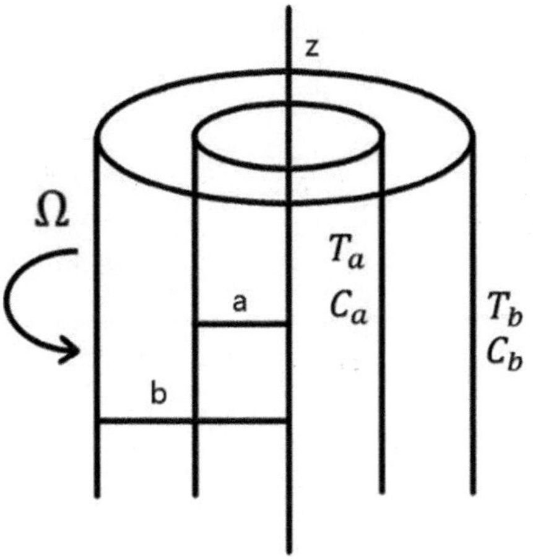

$$\mu_{\text{nf}}\left(\frac{\partial^2 u}{\partial r^2}+\frac{1}{r}\frac{\partial u}{\partial r}-\frac{u}{r^2}\right)+(1-C_a)(T-T_a)g(\rho\beta)_{\text{nf}}$$
$$-(\rho_{\text{sp}}-\rho_{\text{bf}})(C-C_a)g-\sigma_{\text{nf}}\text{B}_0^2 u-\frac{1}{r}\frac{\partial p}{\partial\Theta}=0 \tag{3}$$

$$\frac{\kappa_{\text{nf}}}{(\rho C_p)_{\text{nf}}}\left(\frac{\partial^2 T}{\partial r^2}+\frac{1}{r}\frac{\partial T}{\partial r}\right)+\tau\left(D_B\left(\frac{\partial C}{\partial r}\frac{\partial T}{\partial r}\right)+\frac{D_T}{T_\infty}\left(\frac{\partial T}{\partial r}\right)^2\right)$$
$$+\frac{\mu_{\text{nf}}}{(\rho C_p)_{\text{nf}}}\left(\frac{\partial u}{\partial r}-\frac{u}{r}\right)^2=0 \tag{4}$$

$$D_B\left(\frac{\partial^2 C}{\partial r^2}+\frac{1}{r}\frac{\partial C}{\partial r}\right)+\frac{D_T}{T_\infty}\left(\frac{\partial^2 T}{\partial r^2}+\frac{1}{r}\frac{\partial T}{\partial r}\right)=0 \tag{5}$$

The corresponding convective no-slip boundary conditions are

$$\begin{aligned}
&u=0,\ -\kappa_{nf}\frac{\partial T}{\partial r}=h(T_a-T),\quad -D_m\frac{\partial C}{\partial r}=k_m(C_a-C),\quad \text{at } r=a\\
&u=b\Omega,\quad -\kappa_{nf}\frac{\partial T}{\partial r}=h(T-T_b),\quad -D_m\frac{\partial C}{\partial r}=k_m(C-C_b),\quad \text{at } r=b
\end{aligned} \tag{6}$$

The notations T_a, T_b, C_a and C_b respectively denote the temperatures and concentrations at the inner and outer cylindrical surfaces, h and k_m denote the coefficients of heat and mass transfer and κ and D_m denote thermal conductivity and mass diffusivity.

Table 1 presents the values of thermophysical properties and the definitions of nanofluid properties [13, 14] are as follows:

$$\mu_{\text{nf}}=\frac{\mu_{\text{bf}}}{(1-\Phi)^{2.5}},\quad \rho_{\text{nf}}=(1-\Phi)\rho_{\text{bf}}+\Phi\rho_{\text{sp}},\quad \alpha_{\text{nf}}=\frac{\kappa_{\text{nf}}}{(\rho C_p)_{nf}},$$

Table 1 Thermophysical properties [18–21]

Property (Units)	Water	Graphene oxide
ρ (kg/m^3)	997.1	1800
C_p (J/kg K)	4179	717
κ (W/mK)	0.613	5000
β (10^{-5}/K)	21	28.4
σ (S/m)	0.005	10^7

$$(\rho C_p)_{\mathrm{nf}} = (1-\Phi)(\rho C_p)_{\mathrm{bf}} + \Phi(\rho C_p)_{\mathrm{sp}}, \quad (\rho\beta)_{\mathrm{nf}} = (1-\Phi)(\rho\beta)_{\mathrm{bf}} + \Phi(\rho\beta)_{\mathrm{sp}},$$
$$\frac{\kappa_{\mathrm{nf}}}{\kappa_{\mathrm{bf}}} = \frac{\kappa_{\mathrm{sp}} + 2\kappa_{\mathrm{bf}} + 2\Phi(\kappa_{\mathrm{bf}} - \kappa_{\mathrm{sp}})}{\kappa_{\mathrm{sp}} + 2\kappa_{\mathrm{bf}} - \Phi(\kappa_{\mathrm{bf}} - \kappa_{\mathrm{sp}})}, \quad \frac{\sigma_{\mathrm{nf}}}{\sigma_{\mathrm{bf}}} = \frac{\sigma_{\mathrm{sp}} + 2\sigma_{\mathrm{bf}} + 2\Phi(\sigma_{\mathrm{bf}} - \sigma_{\mathrm{sp}})}{\sigma_{\mathrm{sp}} + 2\sigma_{\mathrm{bf}} - \Phi(\sigma_{\mathrm{bf}} - \sigma_{\mathrm{sp}})}$$

The subscripts here namely nf, sp and bf indicate nanofluid, solid particle and base fluid, whereas, the quantities C_p, κ, μ, β and ρ respectively denote specific heat capacity, thermal conductivity, dynamic viscosity, thermal expansion coefficient and density.

The following are the similarity variables used to transform the modelled pde's (1)–(6):

$$\eta = \frac{r^2}{b^2}, \quad u = \frac{\Omega}{\sqrt{\eta}} f(\eta), \quad \theta(\eta) = \frac{T - T_a}{T_b - T_a}, \quad \phi(\eta) = \frac{C - C_a}{C_b - C_a}, \quad P = \frac{pb}{\Omega \mu_{\mathrm{bf}}} \tag{7}$$

The transformed ODEs with the boundary conditions are given as

$$4\eta f'' + A_1 \frac{\mathrm{Gr}}{\mathrm{Re}} \sqrt{\eta}(A_2\theta - Nr\phi) - A_1 A_3 Ha\, f - A_1 P_1 = 0 \tag{8}$$

$$\eta^3\theta'' + \eta^2\theta' + A_4\eta^3\theta'(N_b\phi' + N_t\theta') + A_5 Ec Pr(\eta f' - f)^2 = 0 \tag{9}$$

$$\eta\phi'' + \phi' + \frac{N_t}{N_b}(\eta\theta'' + \theta') = 0 \tag{10}$$

such that

$$\begin{aligned} &\text{at } \eta = \eta_0 = \frac{a^2}{b^2}, \quad f = 0, \quad \frac{2a}{b}\theta' = A_6 Bi_t \theta, \quad \frac{2a}{b}\phi' = Bi_c \phi \\ &\text{at } \eta = 1, \quad f = 0, \quad 2\theta' = A_6 Bi_t(1-\theta), \quad 2\phi' = Bi_c(1-\phi) \end{aligned} \tag{11}$$

The constants and the dimensionless parameters used namely the Reynolds number Re, constant pressure gradient P_1, Grashof number Gr, buoyancy ratio N_r, Eckert number Ec, Prandtl number Pr, Brownian motion parameter N_b, Hartmann number Ha, thermophoresis parameter N_t, thermal and concentration Biot numbers

Bi_t and Bi_c and heat capacity ratio τ are given as

$$A_1 = (1-\Phi)^{2.5},\ A_2 = (1-\Phi) + \Phi\frac{(\rho\beta)_{\mathrm{sp}}}{(\rho\beta)_{\mathrm{bf}}},\ A_3 = \frac{\sigma_{\mathrm{nf}}}{\sigma_{\mathrm{bf}}},$$

$$A_4 = \left((1-\Phi) + \Phi\frac{(\rho C_p)_{\mathrm{sp}}}{(\rho C_p)_{\mathrm{bf}}}\right)\frac{\kappa_{\mathrm{bf}}}{\kappa_{\mathrm{nf}}},$$

$$\tau = \frac{(\rho C_p)_{\mathrm{sp}}}{(\rho C_p)_{\mathrm{nf}}},\ \mathrm{Pr} = \frac{\mu_{\mathrm{bf}} C_{p_{\mathrm{bf}}}}{\kappa_{\mathrm{bf}}},\ A_5 = (1-\Phi)^{-2.5}\frac{\kappa_{\mathrm{bf}}}{\kappa_{\mathrm{nf}}},\ A_6 = \frac{\kappa_{\mathrm{bf}}}{\kappa_{\mathrm{nf}}},$$

$$P_1 = \frac{\partial p}{\partial \Theta},\ N_r = \frac{(\rho_p - \rho_{\mathrm{bf}})(C_b - C_a)}{(\rho\beta)_{\mathrm{bf}}(1-C_a)(T_b - T_a)},\ \mathrm{Gr} = \frac{\rho_{\mathrm{bf}}^2 g \beta_{\mathrm{bf}}(T_b - T_a)(1-C_a)b^3}{\mu_{\mathrm{bf}}^2},$$

$$\mathrm{Ha} = \frac{\sigma_{\mathrm{bf}} B_0^2 b^2}{\mu_{\mathrm{bf}}},\ \mathrm{Re} = \frac{\rho_{\mathrm{bf}}\Omega b}{\mu_{\mathrm{bf}}},\ \mathrm{Bi}_t = \frac{h}{\kappa_{\mathrm{bf}}},\ \mathrm{Bi}_c = \frac{k_m}{D_m},$$

$$\mathrm{Ec} = \frac{\Omega^2}{C_{p_{\mathrm{bf}}}(T_b - T_a)},\ N_b = \frac{\tau D_B (C_b - C_a)(\rho C_p)_{\mathrm{bf}}}{\kappa_{\mathrm{bf}}},\ N_t = \frac{\tau D_T (T_b - T_a)(\rho C_p)_{\mathrm{bf}}}{\kappa_{\mathrm{bf}} T_a}$$

Practically significant values such as Sherwood number Sh_z, skin friction C_f and Nusselt number Nu_z are derived as

$$\frac{\mathrm{Nu}_\eta}{2\sqrt{\eta}} = -\theta',\quad \frac{\mathrm{Sh}_\eta}{2\sqrt{\eta}} = -\phi',\quad \eta C_f \mathrm{Re}_\eta = C_2(\eta f' - f) \tag{12}$$

where the constant is given by $C_2 = 4(1-\Phi)^{-2.5}((1-\Phi) + \Phi\rho_{\mathrm{sp}}/\rho_{\mathrm{bf}})^{-1}$. In the next section, a derivation and an analysis of the entropy are presented.

3 Entropy Analysis

The generated entropy is contributed from temperature, viscous dissipation and concentration. Thus, the entropy generation rate, S_G is written as [15]

$$S_G = \frac{\kappa_{\mathrm{nf}}}{T_a^2}\left(\frac{\partial T}{\partial r}\right)^2 + \frac{\mu_{\mathrm{nf}}}{T_a}\left(\frac{\partial u}{\partial r} - \frac{u}{r}\right)^2 + \mathrm{RD}_B\left(\frac{\partial C}{\partial r}\right)\left[\frac{1}{C_a}\left(\frac{\partial C}{\partial r}\right) + \frac{1}{T_a}\left(\frac{\partial T}{\partial r}\right)\right] \tag{13}$$

Here, the expressions on the right hand side are ascribed to thermodynamic irreversibilities caused due to temperature, fluid friction and combined mass and heat transfer. From S_G and characteristic entropy generation rate, $S_{G0} = \kappa_{\mathrm{nf}}(T_b - T_a)^2/(T_a L)^2$, we write the nondimensional entropy generation number as

$$N_S = S_G / S_{G_0} \tag{14}$$

From (7), N_S is given by

$$\frac{\eta^2}{4}N_S = \frac{1}{\chi}\left(\eta^3\theta'^2 + A_5\frac{\mathrm{EcPr}}{\Omega_T}(\eta f' - f)^2 + A_6 M_m \frac{\Omega_C}{\Omega_T}\eta^3\phi'\left(\frac{\Omega_C}{\Omega_T}\phi' + \theta'\right)\right) \quad (15)$$

$$= N_{S_h} + N_{S_f} + N_{S_m}, \quad (16)$$

where the subscripts *h, f, m* respectively correspond to the irreversibilities brought about by heat transfer, fluid friction and combined heat and mass transfer. The parameters in (15) are given by Eckert number constant parameter $\chi = d^2/L^2$, temperature parameter $\Omega_T = (T_b - T_a)/T_a$, concentration parameter $\Omega_C = (C_b - C_a)/C_a$ and the combined mass and heat transfer parameter $M_m = \mathrm{RD_B}C_a/\kappa_{bf}$.

Bejan number, $\mathrm{Be} = N_{Sh}/N_S$ determines the principal source for entropy generation [16]. From this ratio, it is dictated that heat transfer mainly influences the entropy if $\mathrm{Be} > 0.5$, while, fluid friction and mass and heat transfer mainly contribute to the entropy if $\mathrm{Be} < 0.5$ and all the three irreversibilities contribute equally if $\mathrm{Be} = 0.5$ [17].

4 Numerical Solution

The ODEs (8)–(11) are solved by adapting the Spectral quasilinearization method (SQLM) [22–24]. We apply the collocation method with the Chebyshev polynomials and solve the linearized equations by approximating at the Gauss–Lobatto points. The obtained matrix equation is then solved by choosing initial approximations satisfying the suitable boundary conditions.

5 Results

In this section, the matrix equation is solved to graphically depict the results with interpretations. The impacts of the embedded parameters are studied by varying the parameter values in the practical range [25, 26]. Since, the Newtonian behaviour of water based nanofluid is considered, fixed values of $\Phi = 0.01$ and $\mathrm{Pr} = 6.5$ are taken. The other parameter values are taken to be $N_b = 2 \times 10^{-4}$, $N_t = 3 \times 10^{-4}$, $\mathrm{Gr} = 2 \times 10^5$, $R_d = 5$, $C_R = 0.08$, $N_r = 2$, $\mathrm{Re} = 300$ and $\mathrm{Sc} = 170$ unless mentioned otherwise. The order of SLM approximation is taken to be $N = 100$ and the convergence of results is obtained at the third iteration. The results for the case of $S = \mathrm{Gr} = P_1 = R_d = C_R = 0$ presented in Table 3 accord well with the results from Sinha and Chaudhary [27].

The effects of Bi_t and Bi_c on N_S and Be are depicted in Fig. 2. When Bi_t increases, there is an increased entropy generation (Fig. 2a). This causes the Bejan number to increase, which implies the impacts of heat transfer on the increasing entropy (Fig. 2b). In the case of Bi_c, entropy number decreases, near the outer cylinder, increases for $\eta < 0.55$, decreases for $\eta > 0.55$ and $\eta < 0.75$ and increases for $\eta > 0.75$ (Fig. 2c). While, Be increases near the surface, decreases for $\eta < 0.55$, increases for $\eta > 0.55$ and $\eta < 0.75$ and decreases for $\eta > 0.75$ (Fig. 2d). Thus, mass transfer and fluid friction influence the entropy generated througout the channel, in the case of Bi_c. From Fig. 2e, it is observed that Ec causes an increase in N_S. The increase in viscous dissipation explains the raised entropy. Whereas, Ec causes a depletion in Be, thus confirming the evident influence of mass transfer and fluid friction on increasing entropy (Fig. 2f). Whereas, enhancing the magnetic field strength, depletes N_S, in the proximity of the inner cylinder, outer cylinder and throughout the channel, with an exception in the middle of the flow (Fig. 2g). Thus, Be increases and decreases respectively (Fig. 2h). Hence, when magnetic field is strengthened, mass transfer and fluid friction dissipates the thermal energy.

Figure 3 represents the influence of Bi_t, Bi_c, Ec and Ha on velocity profiles. When Bi_t increases, the boundary layer increases and hence the fluid velocity decreases (Fig. 3a). Whereas, when Bi_c increases, f increases (Fig. 3b) inspite of an increase in concentration, unlike the case of N_b and N_t. Figure 3c shows the variations with respect to Ec, where an increasing Eckert number implies an increased kinetic energy of the nanofluid causing enhanced f values. Similarly, with the Hartmann number, contradicting the anticipated reduction in velocity due to Lorentz force, the nanofluid velocity is improved (Table 2). This is due to the graphene oxide nanoparticles which results in the depletion of skin friction drag (Fig. 3d).

The effects of Bi_t, Ec and Ha on temperature profiles are presented in Fig. 4. Since, Biot number is directly proportional to convection, a raise in Bi_t is supposed to raise the temperature in turn. But there is an increased entropy generation dissipating the thermal energy, causing θ to reduce (Fig. 4a). Similarly, an enhancement in Eckert number transforms the nanofluid's kinetic energy into thermal energy, thus increasing the fluid temperature (Fig. 4b). Also, with Ha, the effect of Lorentz force reduces the fluid temperature (Fig. 4c).

Figure 5 shows the influence of N_b, N_t, Bi_c, Ec and Ha on the concentration of the nanofluid. As mentioned in the description of Fig. 3, an increase in N_b increases Brownian diffusion, thus enhancing ϕ (Fig. 5a). Similarly, when N_t, increases, nanoparticles diffuse from the outer cylinder to inner cylinder and hence the concentration decreases (Fig. 5b). When Bi_c is raised, convective mass transfer increases and the raised temperature causes the nanoparticles to move from outer cylinder to inner cylinder. Hence, ϕ increases for $\eta > 0.55$ and decreases otherwise (Fig. 5c). Whereas, the Eckert number negatively impacts the concentration profiles (Fig. 5d). Strengthening the magnetic field, improves the concentration of the nanofluid (Fig. 5e).

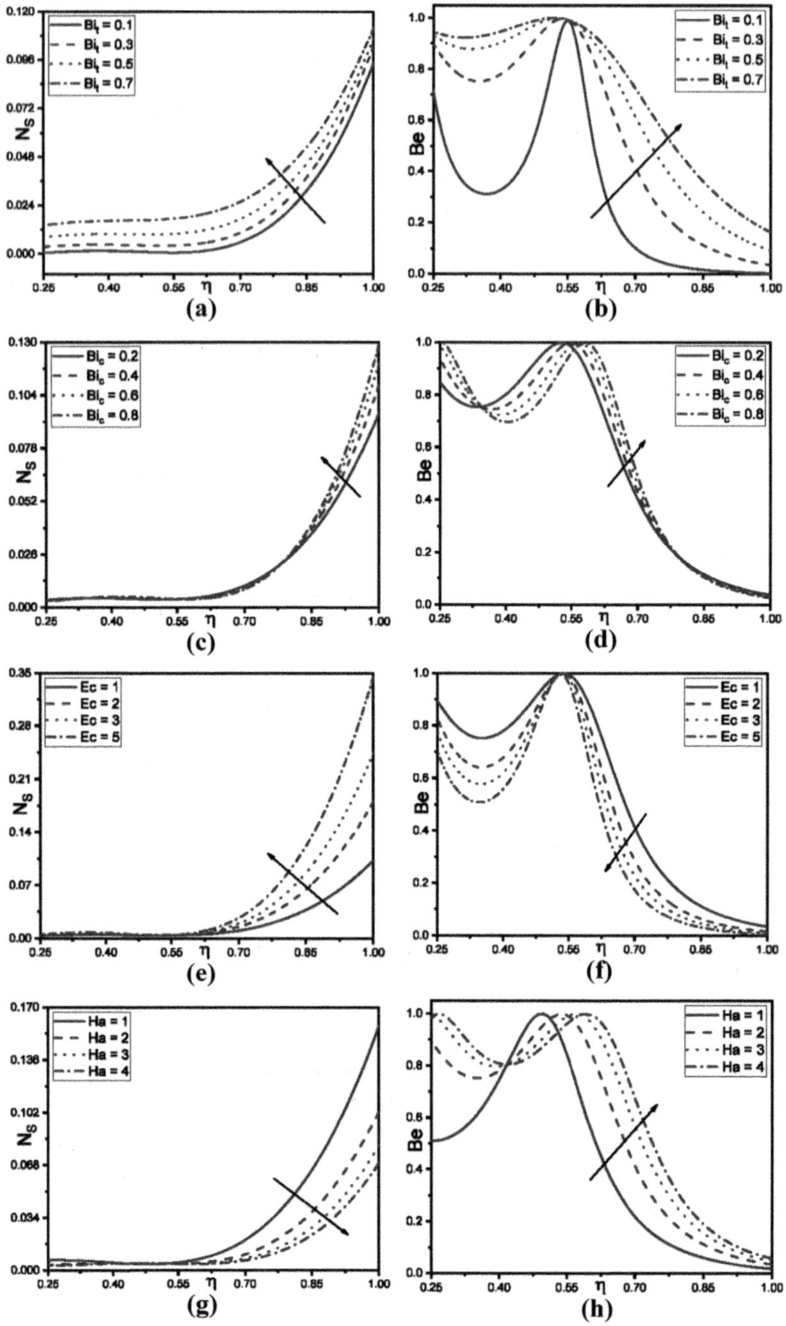

Fig. 2 Influence of **a** Bi_t on $\mathrm{N_S}$, **b** Bi_t on Be, **c** $\mathrm{Bi_c}$ on N_S, **d** $\mathrm{Bi_c}$ on Be, **e** Ec on $\mathrm{N_S}$, **f** Ec on Be, **g** Ha on $\mathrm{N_S}$ and **h** Ha on Be

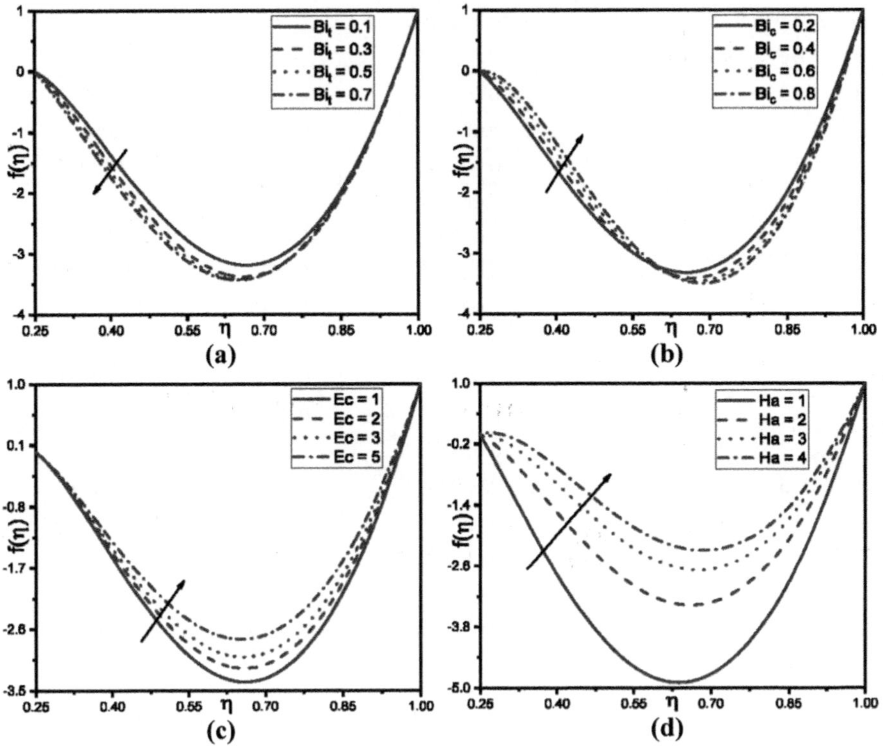

Fig. 3 Varying effects of f with respect to **a** Bi_t, **b** Bi_c, **c** Ec and **d** Ha

Table 2 Comparison of $f(\eta)$ for $\mathrm{Gr} = \mathrm{P}_1 = \mathrm{Ec} = \mathrm{Ha} = 0$

η	Present study	Sinha and Chaudhary [17]
1	1	1
0.981646194	0.975528	0.97546
0.928381373	0.904508	0.90453
0.84541947	0.793893	0.79386
0.740881373	0.654508	0.65453
0.625	0.5	0.5
0.509118627	0.345492	0.34546
0.40458053	0.206107	0.20613
0.321618627	0.095492	0.09546
0.268353806	0.024472	0.02453
0.25	0	0

Table 3 Nusselt, Sherwood and skin friction values

Bi_t	Bi_c	Ec (10^{-5})	Ha	$-\theta'(1)$	$-\phi'(1)$	$f'(1) - f(1)$
0.1	0.3	1	2	– 0.008190	– 0.035051	76.09075
0.3	0.3	1	2	– 0.029879	– 0.035225	78.08673
0.5	0.3	1	2	– 0.049644	– 0.035245	77.12948
0.7	0.3	1	2	– 0.067544	– 0.035245	75.83458
0.3	0.2	1	2	– 0.030046	– 0.024679	75.08822
0.3	0.4	1	2	– 0.029757	– 0.045221	80.47699
0.3	0.6	1	2	– 0.029554	– 0.063760	84.44778
0.3	0.8	1	2	– 0.029368	– 0.080598	87.81777
0.3	0.3	1	2	– 0.029879	– 0.035225	78.08673
0.3	0.3	2	2	– 0.026539	– 0.036895	73.93239
0.3	0.3	3	2	– 0.023773	– 0.038277	70.49317
0.3	0.3	4	2	– 0.019389	– 0.040469	65.04071
0.3	0.3	5	1	– 0.026255	– 0.037037	97.71582
0.3	0.3	1	2	– 0.029879	– 0.035225	78.08673
0.3	0.3	1	3	– 0.031109	– 0.034610	68.77080
0.3	0.3	1	4	– 0.031681	– 0.034324	63.41776

Table 3 presents the values of Nu, Sh and C_f at the surface of the outer cylinder. By definition, Nusselt number is the ratio between convection to conduction at the surface, and at the outer cylinder, Nu increases with increasing Bi_c and Ec. Whereas, increasing Bi_t and Ha decreases Nu, thus enhancing the heat transfer by conduction at the surface. Similary, increasing Bi_t, Bi_c, Ec and Ha decreases Sh, thus denoting an increased diffusive mass transfer. Furthermore, C_f increases with Bi_c, while depleting with respect to Bi_t, Ec and Ha.

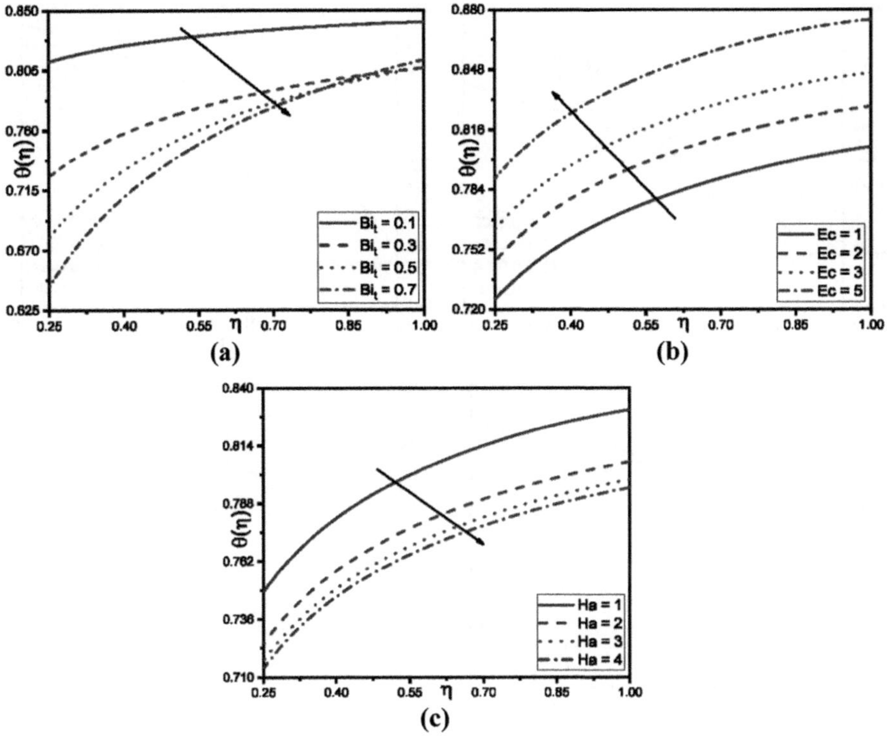

Fig. 4 Varying effects of θ with respect to **a** Bi_t, **b** Ec and **c** Ha

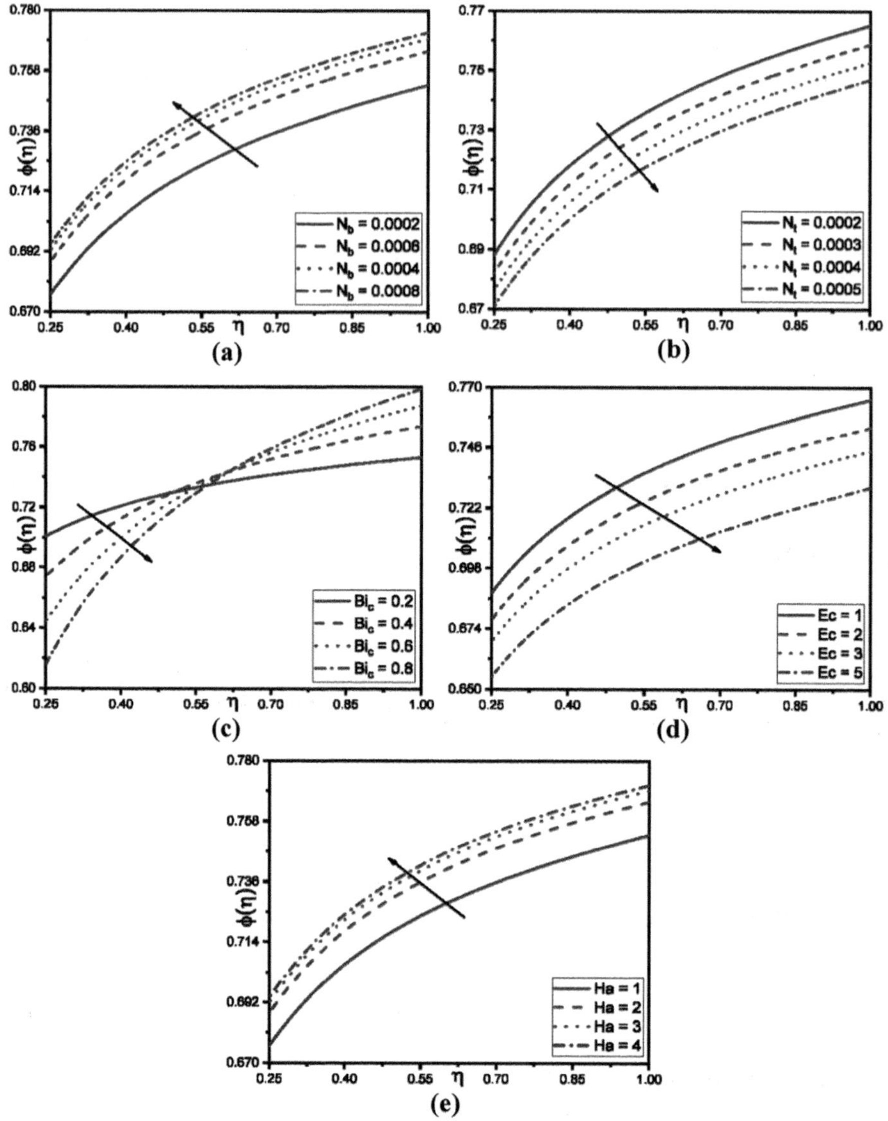

Fig. 5 Varying effects of ϕ in view of **a** N_b, **b** N_t, **c** $\mathrm{Bi_c}$, **d** Ec and **e** Ha

6 Conclusion

On studying the flow of water dispersed with the nanoparticles of graphene oxide in concentric cylinders including the effects of magnetic field and viscous dissipation, we infer the following:

- Velocity is enhanced by enhancing Hartmann number, Eckert number, thermophoresis parameter and thermal Biot number.
- Heat transfer is improved by enhancing the Brownian motion parameter, Eckert number and concentration Biot number.
- Mass transfer is improved by improving Brownian motion parameter and Hartmann number.
- For $\eta > 0.67$, Be < 0.5. Hence, fluid friction and mass transfer is the primary cause for entropy generation at the outer cylinder and viscous dissipation and mass transfer contribute to the generated entropy at the inner cylinder.
- Both the Hartmann number and Eckert number enhance mass transfer through diffusion and deplete skin friction. While, Hartmann number improves conductive heat transfer, Eckert number aids in convective heat transfer.

References

1. Buongiorno J (2006) Convective transport in nanofluids. J Heat Transfer 128:240–250. https://doi.org/10.1115/1.2150834
2. Abro KA, Abdon A (2022) A computational technique for thermal analysis in coaxial cylinder of onedimensional flow of fractional Oldroyd-B nanofluid. Int J Ambient Energy 43:5357–5365. https://doi.org/10.1080/01430750.2021.1939157
3. Akram M, Jamshed W, Goud BS, Pasha AA, Sajid T, Rahman MM, Arshad M, Weera W (2022) Irregular heat source impact on carreau nanofluid flowing via exponential expanding cylinder: a thermal case study. Case Stud Therm Eng 36:102171. https://doi.org/10.1016/j.csite.2022.102171
4. Shah RA, Ullah H, Khan MS, Khan A (2021) Parametric analysis of the heat transfer behavior of the nano-particle ionic-liquid flow between concentric cylinders. Adv Mech Eng 13:168781402110240. https://doi.org/10.1177/16878140211024009
5. Javanmard M, Salmani H, Taheri MH, Askari N, Kazemi MA (2021) Heat transfer of a radial, nanofluid water-graphene oxide hydromagnetic flow between coaxial pipes with a variable radius ratio. Proc Inst Mech Eng Part E J Process Mech Eng 235:124–133. https://doi.org/10.1177/0954408920948194
6. Masood S, Farooq M (2021) Influence of thermal stratification and thermal radiation on graphene oxide-Ag/H_2O hybrid nanofluid. J Therm Anal Calorim 143:1361–1370. https://doi.org/10.1007/s10973-020-10227-7
7. Ramesh GK, Shehzad SA, Izadi M (2022) Falkner-Skan flow of aqueous magnetite–graphene oxide nanoliquid driven by a wedge. Chinese J Phys 77:733–746. https://doi.org/10.1016/j.cjph.2021.07.023
8. Bhattacharyya A, Sharma R, Hussain SM, Chamkha AJ, Mamatha E (2022) A numerical and statistical approach to capture the flow characteristics of Maxwell hybrid nanofluid containing copper and graphene nanoparticles. Chinese J Phys 77:1278–1290. https://doi.org/10.1016/j.cjph.2021.09.015
9. Abu-Nada E, Masoud Z, Hijazi A (2008) Natural convection heat transfer enhancement in horizontal concentric annuli using nanofluids. Int Commun Heat Mass Transf 35:657–665. https://doi.org/10.1016/j.icheatmasstransfer.2007.11.004
10. Abedini A, Emadoddin S, Armaghani T (2019) Numerical analysis of mixed convection of different nanofluids in concentric annulus. Int. J. Numer. Methods Heat Fluid Flow. 29:1506–1525. https://doi.org/10.1108/HFF-06-2018-0337

11. Yu Z-T, Xu X, Hu Y-C, Fan L-W, Cen K-F (2012) A numerical investigation of transient natural convection heat transfer of aqueous nanofluids in a horizontal concentric annulus. Int J Heat Mass Transf 55:1141–1148. https://doi.org/10.1016/j.ijheatmasstransfer.2011.09.058
12. Liu X, Toghraie D, Hekmatifar M, Akbari OA, Karimipour A, Afrand M (2020) Numerical investigation of nanofluid laminar forced convection heat transfer between two horizontal concentric cylinders in the presence of porous medium. J Therm Anal Calorim 141:2095–2108. https://doi.org/10.1007/s10973-020-09406-3
13. Gul T, Ullah MZ, Alzahrani AK, Amiri IS (2019) Thermal performance of the graphene oxide nanofluids flow in an upright channel through a permeable medium. IEEE Access 7:102345–102355. https://doi.org/10.1109/ACCESS.2019.2927787
14. Elsaid K, Abdelkareem MA, Maghrabie HM, Sayed ET, Wilberforce T, Baroutaji A, Olabi AG (2021) Thermophysical properties of graphene-based nanofluids. Int J Thermofluids 10:100073. https://doi.org/10.1016/j.ijft.2021.100073
15. Bejan A (1996) Entropy generation minimization: the new thermodynamics of finite-size devices and finite-time processes. J Appl Phys 79:1191–1218. https://doi.org/10.1063/1.362674
16. Bejan A, Kestin J (1983) Entropy generation through heat and fluid flow. J Appl Mech 50:475–475. https://doi.org/10.1115/1.3167072
17. Paoletti S, Rispoli F, Sciubba E (1989) Calculation of exergetic losses in compact heat exchanger passages. In: ASME AES, pp 21–29
18. Lide DR (ed) (2004) CRC handbook of chemistry and physics. CRC Press, Boca Raton
19. Ghadikolaei SS, Hosseinzadeh K, Hatami M, Ganji DD, Armin M (2018) Investigation for squeezing flow of ethylene glycol ($C_2H_6O_2$) carbon nanotubes (CNTs) in rotating stretching channel with nonlinear thermal radiation. J Mol Liq 263:10–21
20. Chu YM, Nisar KS, Khan U, DaeiKasmaei H, Malaver M, Zaib A, Khan I (2020) Mixed convection in MHD water-based molybdenum disulfide-graphene oxide hybrid nanofluid through an upright cylinder with shape factor. Water 12(6):1723
21. Al-Sankoor K, Al-Gayyim H, Al-Musaedi S, Asadi Z, Ganji DD (2021) Analytically investigating of heat transfer parameters with presence of graphene oxide nanoparticles in Williamson-magnetic fluid by AGM and HPM methods. Case Stud Thermal Engi 27:101236
22. Srinivasacharya D, HimaBindu K (2015) Entropy generation in a micropolar fluid flow through an inclined channel with slip and convective boundary conditions. Energy 91:72–83. https://doi.org/10.1016/j.energy.2015.08.014
23. Bellman RE, Kalaba RE (1967) Quasilinearization and nonlinear boundary-value problems. Am Math Mon 74:1157. https://doi.org/10.2307/2313669
24. Canuto C et al (2006) Spectral methods. Berlin, Heidelberg: Springer (Scientific Computation)
25. Malashetty MS, Umavathi JC, Prathap Kumar J (2001) Convective magnetohydrodynamic two fluid flow and heat transfer in an inclined channel. Heat Mass Transf 37:259–264. https://doi.org/10.1007/s002310000134
26. Behseresht A, Noghrehabadi A, Ghalambaz M (2014) Natural-convection heat and mass transfer from a vertical cone in porous media filled with nanofluids using the practical ranges of nanofluids thermo-physical properties. Chem Eng Res Des 92:447–452. https://doi.org/10.1016/j.cherd.2013.08.028
27. Sinha KD, Chaudhary RC (1966) Viscous incompressible flow between two coaxial rotating porous cylinders. Proc Nat Inst Sci India 32:81

Study of Parameters Influencing Aerodynamics of Train-Tunnel System

M. Harikrishnan and K. M. Pervaiz Fathima

Abstract Various aerodynamic effects are produced as the train passes through a tunnel, especially when the train speed is high. These effects depend on a wide range of variables such as train speed, tunnel geometry, and train area. For the safety of the train and the passengers, it is crucial to understand the relationship between these variables and the aerodynamic effects. The goal of the current study is to determine which combination of these factors has the highest impact on the train-tunnel aerodynamics by analysing the pressure variation of these combinations. For different set of parameters, the pressure-time history curves are plotted and analysed to obtain the critical combination of parameters. The aerodynamic loads coming on to the train and the passenger comfort criteria for this critical combination are also discussed.

Keywords Aerodynamic effects · Pressure-time history · Maximum pressure · Aerodynamical phenomena

1 Introduction

The train-tunnel system experiences several aerodynamic influences as train enters the tunnel. Increased rail speed can meet higher traffic capacity requirements. As the speed increases, the aerodynamic effects become more pronounced. This has a direct impact on the tunnel and train safety, as well as the safety of passengers. As a result, research into the parameters that influence train-tunnel aerodynamics becomes critical. Several studies in this area are available in the literature, and some of them are briefly mentioned. Tunnel portal has an effect on the pressure changes inside and outside of the train, and also the pressure-time history curve is affected by train direction [1]. Effect of airshaft [1, 2], train speed [1, 3], tunnel cross-section [4, 5] and ambient

M. Harikrishnan (✉) · K. M. Pervaiz Fathima
Indian Institute of Technology Jammu, Jagti, Nagrota, Jammu, Jammu and Kashmir, India
e-mail: 2021pce1003@iitjammu.ac.in

K. M. Pervaiz Fathima
e-mail: pervaiz.khatoon@iitjammu.ac.in

R. K. Ray et al. (eds.), *Advances in Theoretical and Applied Mechanics*, Lecture Notes in Mechanical Engineering, https://doi.org/10.1007/978-981-97-0418-7_6

pressure [6] on the pressure variations was studied and reported by using full-scale model tests [4], moving model tests and numerical simulations [7]. According to existing research, tunnel length, tunnel portal, tunnel cross-section, train nose shape, train speed, ambient pressure, train length, crosswind, temperature, blockage ratio, tunnel shape, tunnel auxiliary facilities, train nose length, train surface roughness, sealing of trains, running direction of trains and airflow velocity influence the aerodynamics of trains inside tunnels. However, there has been relatively little research on the combination of factors that lead to the highest aerodynamic effects. So, a 1-D analysis is done, considering the flow to be uniform in each cross-section. Two theoretical models are used for simulating the tunnel airflows. The first model solves full equations of continuity, momentum and energy, and in the second model, energy equations are replaced with an assumption of homentropic flow [8]. Tunnel length, train length, train speed, train area and tunnel area are studied using the pressure-time history curves obtained by ThermoTun software. Initially, the pressure-time history curve of each parameter is analysed, and after that, some critical combination of parameters are selected and their effects analysed. The pressure-time history curves are drawn for points in the tunnel as well as inside the train. The conclusions are drawn, and the permissible pressure values suggested by codes of different countries are compared for both the safety of train and the comfort of passengers.

2 Influence of Various Train-Tunnel Parameters on the Pressure Variation

The factors considered for the study are

- Tunnel length
- Train length
- Train area
- Tunnel area
- Train speed.

In the parametric study performed, one parameter is varied to study its effect on the aerodynamic properties while keeping all the other parameters at their base values shown in Table 1.

2.1 Tunnel Length

Two tunnel lengths of 500 m (L500) and 2000 m (L2000) are considered in the analysis. Figure 1 shows the pressure variation for 500 m and 2000 m tunnel at position 125 m and 500 m, respectively from the tunnel entrance. After the train passes the tunnel portal, larger pressure fluctuations are seen when the tunnel length is higher.

Table 1 Parameters and their base values used in the study

Parameters	Base values
Tunnel length	2000 m
Train length	200 m
Train area	8 m^2
Tunnel area	50 m^2
Train speed	200 kmph

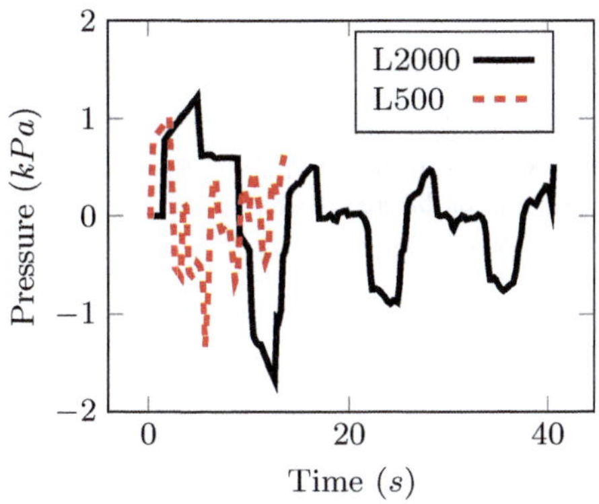

Fig. 1 Pressure-time history variation for different tunnel lengths.

Due to the excessive pressure fluctuations that happen for a longer period in a longer tunnel cause an increased pressure variation [9]. As suggested by Ko et al. [4], the maximum negative and positive peak pressures are used to compare the results. The maximum positive and negative pressure peaks are 25.7% and 25.56% higher, respectively, for 2000 m compared to 500 m as shown in Fig. 1.

2.2 *Train Length*

Two train lengths of 200 m (TL200) and 400 m (TL400) are considered in the analysis. Figure 2 shows the pressure variation for 200 m and 400 m trains, respectively. As the train having a higher train length enters the tunnel, the maximum positive peak pressure is produced when the train enters into the portal. The highest value of positive and negative peak pressure is about 38% and 31.5% more for 400 m train compared to 200 m as shown in Fig. 2 (pressure variation on the tunnel at 500 m from the portal). The higher pressure is due to the difference in the development of the boundary layer caused due to train length, which in turn leads to different degrees of aerodynamic interaction with the tunnel inner wall [10]. Once the long and short trains enter the tunnel, then the pressure peaks vary in the similar ranges.

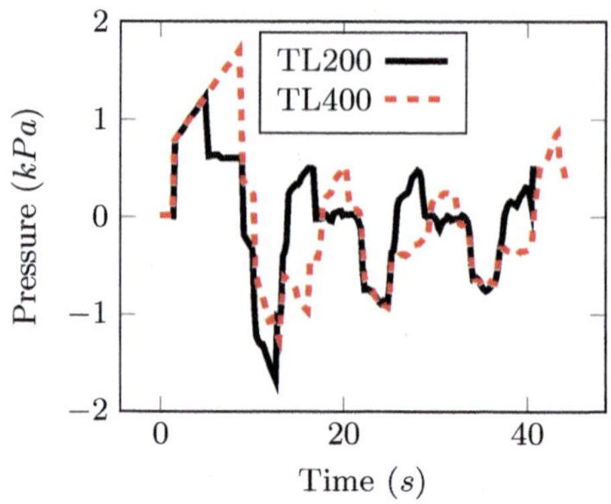

Fig. 2 Pressure-time history variation for different train lengths

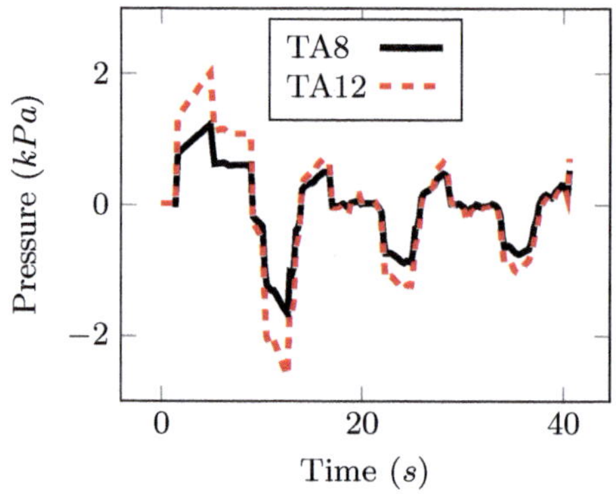

Fig. 3 Pressure-time history variation for different train areas

2.3 *Train Area*

The train areas considered in the analysis are 8 m^2 (TA8) and 12 m^2 (TA12). The pressure variations for 8 and 12 m^2 are shown in Fig. 3 (pressure variation on the tunnel at 500 m from the portal). In the case of the train having higher area, it can be observed that the pressure waveforms are similar to that of train with smaller area. The difference is that there is an increase in pressure when the trains enter the portal, and after that, the negative peaks shows a higher value for the case of higher train area. When the train with larger front area enters into tunnel, it causes more compression of air at the portal resulting in a higher pressure increase compared to the tunnel with a smaller area. Further, the higher negative peaks shown by train with larger area could be because of the more expansion waves generated, resulting in a higher negative pressure. As the train area is increased from 8 to 12 m^2 the maximum positive and negative peak gets increased by 66.67 and 60.4%.

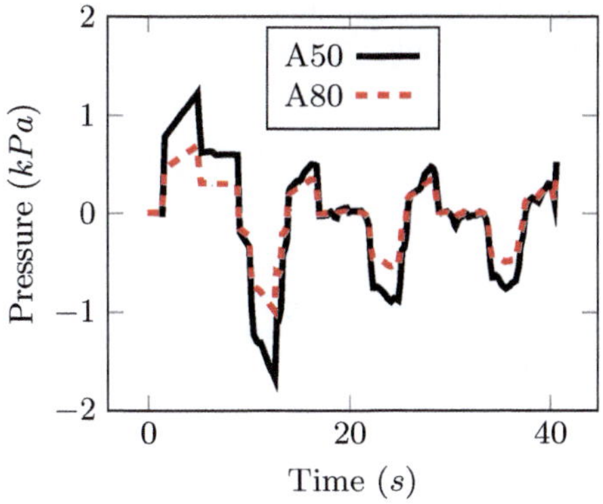

Fig. 4 Pressure-time history variation for different tunnel areas

2.4 *Tunnel Area*

Two tunnel cross-section areas considered in the analysis are 50 m^2 (A50) and 80 m^2 (A80). The pressure variations for 50 m^2 and 80 m^2 are shown in Fig. 4 (pressure variation on the tunnel at 500 m from the portal). It can be seen from the graph that the case with larger tunnel area shows lower pressure variations compared to that with smaller tunnel area. Increase in tunnel area can be perceived as decrease in train area or vice-versa. So, for higher tunnel area, it increases gap between the train and tunnel, thus reducing the compression of air caused due to the entry of train into the tunnel portal. Also, further pressure fluctuations occur at lower values compared to that in the case with smaller tunnel area. From the graph, the tunnel area is very critical parameter as the increase in positive and negative peak pressure for low area tunnel is about 79 and 68.6% more compared to higher area tunnel. The area of tunnel and train are important parameters, and the ratio of train area to tunnel area is known as blockage ratio [11]. It is seen from the literature that this value should be kept lower to keep the pressure fluctuations to a minimum value.

2.5 *Train Speed*

Two train speeds of 200 kmph (V200) and 300 kmph (V400) are considered in the analysis. Figure 5 shows the pressure variation at 500 m from tunnel portal for 200 kmph and 300 kmph train speeds, respectively. When the train speed is increased from 200 to 300 kmph, the maximum positive and negative peak pressures are increased by 122 and 55% shown in Fig. 5 (pressure variation on the tunnel at 500 m from the portal). So at higher train speed, higher pressure variations occur due to stronger pressure waves that are formed inside the tunnel. Thus train speed is a very important factor affecting aerodynamics of train-tunnel system.

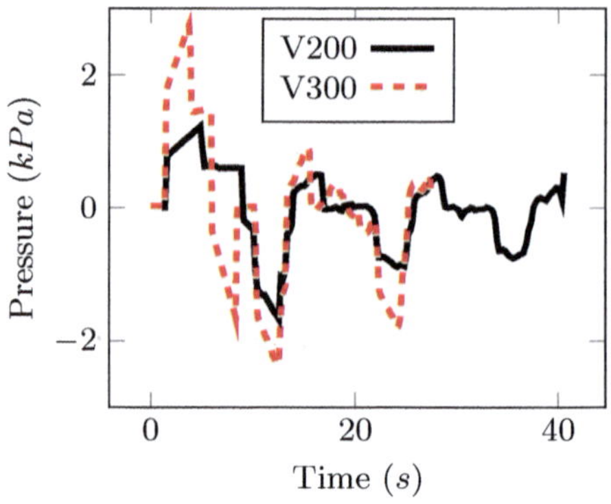

Fig. 5 Pressure-time history variation for different train speed

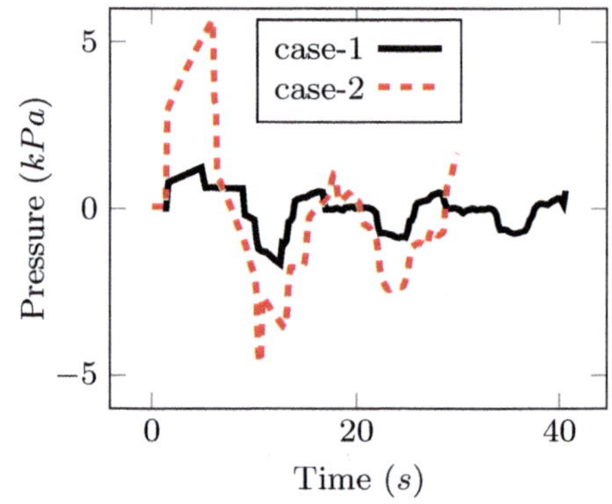

Fig. 6 Pressure-time history variation by considering train speed, train length and train area. Case-1: Tunnel length (2000 m), tunnel area (50 m^2), train speed (200 kmph), train length (200 m), train area (8 m^2)(All parameters are set at base values). Case-2: Train length (400 m), train speed (300 kmph), tunnel length (2000 m), train area (12 m^2), tunnel area (50 m^2)

3 Critical Combination of Parameters

The combination of parameters considered are train length, train area and train speed. The effect of combination of these three parameters is shown in Fig. 6. Here, case-1 represents the case where all parameters are at their base values (Table 1). The case-2, on the other hand, represents all parameters set to the values that result in the highest pressures. When higher values of train speed, train length and train area are compared with lower values of train speed, train length and train area, the maximum positive pressure peak is about 7 times. And the maximum negative peak pressure varies about 3 times, as shown in Fig. 6 (pressure variation on the tunnel at 500 m from the portal).

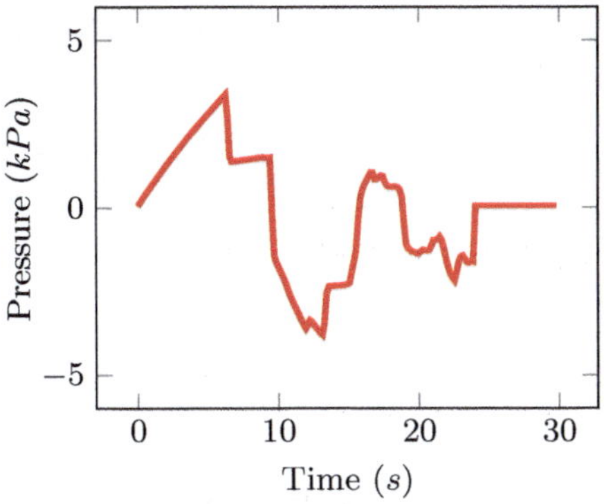

Fig. 7 Pressure-time history variation by considering train speed, tunnel length, tunnel area, train length, train area on the front coach of train

Table 2 Aerodynamic loads in German DB AG standard [12]

Train speed (kmph)	Maximum positive pressure (Pa)	Maximum negative pressure (Pa)
280 enter or 280 leave	3000	– 5100
300 enter or 300 leave	3300	– 5400
330	1000–1800	– 3000 to – 4000
330 enter or 330 leave	1950–2600	– 3000 to – 4000

4 Safety of Trains in Tunnels

4.1 Aerodynamic Loads

Various codes from different countries have given limits for the aerodynamic pressure that acts on the train. Figure 7 shows the pressure-time variation for the critical combination (higher value of train length (400 m), train speed (300 kmph), tunnel length (2000 m), train area (12 m^2) and lower value of tunnel area (50 m^2)) at the front coach of the train (outside). The maximum positive and negative pressure that are permissible for various speeds are shown in Table 2 given below as per German DB AG standard. As per Table 2, the allowable value of pressure for 300 kmph speed is 3300 Pa for maximum positive pressure and – 5400 Pa for maximum negative pressure. From Fig. 7, it can be seen that the maximum positive pressure is about 3316 Pa which exceeds the limit and not permissible. However, the maximum negative pressure is – 3851 Pa which is within permissible limit.

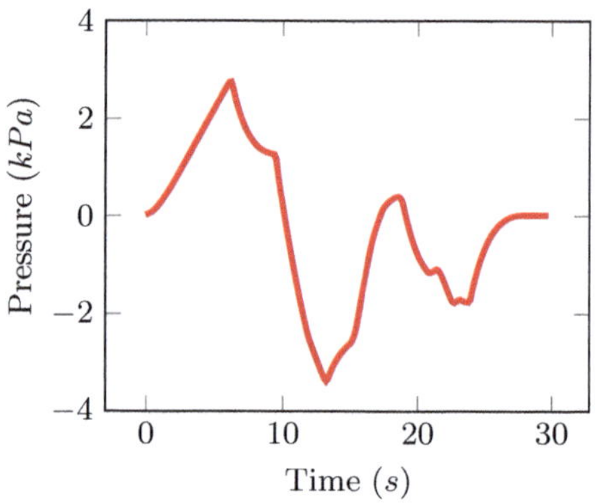

Fig. 8 Pressure-time history variation by considering train speed, tunnel length, tunnel area, train length, train area inside the train

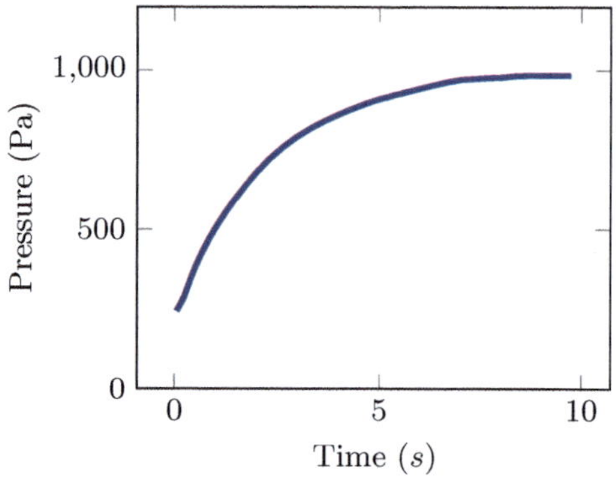

Fig. 9 Allowable pressure increase based on Japanese standards [12]

4.2 Passenger Comfort

Figure 8 shows the pressure-time history variation at the front coach inside the train for the combination of higher value of train length (400 m), train speed (300 kmph), tunnel length (2000 m), train area (12 m^2) and lower value of tunnel area (50 m^2). Figure 9 gives the permissible increase in pressure for different time intervals that are used in Japan. From the graph, the increase in pressure for 5 s is 2202 kPa. When we compare this with increase in pressure for 5 s shown in Fig. 9, the value is well above the permissible limit. So it comes under non-comfort zone. For the combinations of parameters considered, the pressure variation will cause discomfort to passengers.

5 Conclusions

The following conclusions can be drawn from this study

- Train length, train area, tunnel area, tunnel length and train speed all these parameters significantly affect the aerodynamics of train-tunnel system.
- The train area, tunnel area and train speed show large increases in the pressure. So these factors have larger effect on the aerodynamics of train-tunnel system.
- The effect of tunnel area and train area can be accounted for by adjusting the blockage ratio.
- The combination of parameters considered with values of train length 400 m, tunnel length 2000 m, train speed 300 m, tunnel area 50 m^2, train area 12 m^2 result in aerodynamic loads on the train (maximum positive pressure) that exceeds the permissible limits given as per German DB AG standard.
- Similarly, the pressure increase inside the train falls in the non-comfort zone (as per Japanese standards). So this combination of parameters with the given values will cause discomfort to the passengers travelling in it.

The reported pressure variations are at 500 m from the tunnel portal. However, the maximum pressure could occur at a different location, this analysis was not carried out in the present study. In this study, we have considered five parameters, there are other parameters, such as tunnel portal, airshaft, ambient pressure, which affect the aerodynamics of train-tunnel system. So, newer models have to be developed by including all these factors to accurately predict the pressure variation.

References

1. Xionga X, Zhu L, Zhanga J, Lia A, Lia X, Tang M (2020) Field measurements of the interior and exterior aerodynamic pressure induced by a metro train passing through a tunnel. Sustain Cities Soc 53:101928
2. Zhou WB (2009) Study on the influence of vertical shafts on the aerodynamic effect of double-line tunnel
3. Liu TH, Tian HQ, Liang XF (2010) Aerodynamic effects caused by trains entering tunnels. J Trans Eng 136(9):846–853
4. Ko YY, Chen CH, Hoe T, Wang ST (2012) Field measurements of aerodynamic pressures in tunnels induced by high speed trains. J Wind Eng Indus Aerodyn 100(1):19–29
5. Huang S, Che ZX, Li ZW, Jiang YN, Wang ZG (2020) Influence of tunnel cross sectional shape on surface pressure change induced by passing metro trains. Tunn Undergr Space Technol 106:103611
6. Liu F, Wang F, Han J, Zhao S, Weng M (2022) Effects of ambient pressure on aerodynamic pressures induced by passing metro trains in tunnels. Tunn Undergr Space Technol 126:104540
7. Liu TH, Chen ZW, Chen XD, Xie TZ, Zhang J (2017) Transient loads and their influence on the dynamic responses of trains in a tunnel. Tunn Undergr Space Technol 66:121–133
8. ThermoTun-Online Homepage. https://www.thermotun-online.com/
9. Liu TH, Chen XD, Li WH, Xie TZ, Chen ZW (2017) Field study on the interior pressure variations in high-speed trains passing through tunnels of different lengths. J Wind Eng Indus Aerodyn 169:54–66

10. Guo Z, Liu T, Deng E, Yang W (2022) Influence of the length configuration of trains on the aerodynamic environment in a metro system. Tunn Undergr Space Technol 123:104413
11. Cross D, Hughes B, Ingham D, Ma L (2015) A validated numerical investigation of the effects of high blockage ratio and train and tunnel length upon underground railway aerodynamics. J Wind Eng Indus Aerodyn 146:195–206
12. Niu J, Sui Y, Yu Q, Cao X, Yuan Y (2020) Aerodynamics of railway train/tunnel system: a review of recent research. Energy Built Environ 1(4):351–375

Wake Modelling of Horizontal-Axis Wind Turbines Using Sparse Identification of Non-linear Dynamics (SINDy)

Suchita C. Barhate, Ojing Siram, and Niranjan Sahoo

Abstract Performing accurate modelling of measurement data from a non-linear dynamic system is a challenging task prone to many idealized assumptions and sensitivity to noise resulting in non-reliable and inaccurate models. Several data-driven approaches, such as genetic programming, artificial neural network, and others, are capable of producing dynamic models with extended relations. However, such models often carry long and non-intuitive Equations, it thus becomes difficult to derive a linear representation of the non-linear dynamics. The paper proposes an innovative application of the SINDy algorithm for deriving a dynamical model from the measurement data obtained using an optical method called Particle Image Velocimetry (PIV). The methodology has been used to identify and model the non-linear dynamics involved in wind turbine wake. The application of the model analogy will have a great implication in wind farm modelling, where economic constraints related to available space and the turbine installation is a prime concern. The study helps in determining the optimal distance of placement of turbines to minimize the effect of wake interference and enhancement of performance.

Keywords Sparse identification · Wake modelling · Particle image velocimetry · Coherent structures · Velocity of the flow field · Vorticity vectors · Proper orthogonal decomposition (POD) · POD modes

S. C. Barhate (✉) · O. Siram
School of Energy Science and Engineering, Indian Institute of Technology Guwahati, Guwahati 781039, India
e-mail: suchita_chandrakant@iitg.ac.in

N. Sahoo
Department of Mechanical Engineering, Indian Institute of Technology Guwahati, Guwahati 781039, India

R. K. Ray et al. (eds.), *Advances in Theoretical and Applied Mechanics*, Lecture Notes in Mechanical Engineering, https://doi.org/10.1007/978-981-97-0418-7_7

1 Introduction

Wind energy is one of the key renewable energy sources in the energy sector, and has shown rapid growth in recent years and ongoing expansion can be observed owing to the climate change mitigation steps that are been taken all over the globe to attain energy sustainability. Wind energy's contribution to the world's electricity generation in the coming three decades is predicted to increase by up to 18% [1]. In wind farms, excessive losses in total power output have been observed due to wake interactions between the upstream and downstream turbines. The wake impact from upstream turbines results in significant power losses for the downstream turbines decreasing the overall electricity generated by the turbines while causing fatigue loading condition on the turbine components. A thorough comprehensive and precise turbine wake prediction may dramatically increase the effectiveness of converting wind energy in a large-scale wind farm and device better turbine spacing plans for the farm [2, 3]. The wake is characterized by a mean velocity decrease and turbulence increase in the fluid behind the rotor. For a deployment to be effective, prediction of wake is crucial as maintaining optimum efficiency of the wind farm is one such motive that is lately gaining attention from the community of wind farm layout developers and researchers.

Dynamical systems are the models of reality, they describe the world around us, especially how the different things change and co-evolve in time. Systems that observe variation in physical properties with respect to change in spatial and temporal coordinates can be modelled to derive principles and governing Equations that are capable of predicting the system behavior in near future. ML and AI are finding its applications in a wide range of areas of research, one such recent area of research is the automation of discovery of physical Equations or models that govern a dynamical system. These Equations and coordinate systems are derived from the data obtained from measurement alone [1]. Although supporting an Equation only on the basis of data is challenging, the model obtained must also account for the inevitable misalliance of theory and measurement data. The application of data-driven modelling and control for complex systems is a vastly interesting and potent field that is capable of transforming engineering, and physical sciences. The availability of data from experiments using high-precision instruments and numerical simulations in abundance has formed one of the prominent reasons for the rapid evolvement of data-driven techniques. A fluid flow system showcases immensely complex and multi-dimensional dynamics but there are methods that extract related features in the form of datasets from such systems. These methods observe dominant patterns in the flow system via measurement data that give rise to coherent structures capable of defining the system sparsely. Efficient analysis of the coherent structures observable from the high- fidelity data has given rise to the data-driven modelling methods like Koopman theory [4], dynamic mode decomposition (DMD), non-linear Laplacian Spectral Analysis, Genetic Programming [5], and ANN (Artificial Neural Networks), DNN (Deep Neural Networks) and others, have been found useful for such complex flows. The coherent structure analysis in the field of fluid dynamics is evolving therefore, it does not take into consideration all the elusive details of the flow for analysis of Equations of motion, instead fluid interactions are described in the form of representable structural hierarchy synthesized from relevant configurations of the flow also known as decomposition techniques. Proper orthogonal decomposition is easier to interpret and has wide applicability to data obtained from experiments as well as numerical simulations [2]. Such dimensionality reduction techniques can be thought of as the primary use of data science in modelling complex dynamics of a physical system.

1.1 Wake Modelling

Numerous wake models were developed and have evolved over a period of time. The wake models consider the various aerodynamic loadings that act upon a wind turbine. The broad categories of wake modelling include analytical modelling, computational modelling, and experimental modelling which is also known as the data-driven approach. Characterization of wake propagation through a set of analytical and empirical Equations is carried out in analytical models like the Jensen, Larsen, Bastankhah and Porte-Agel (B.P), Frandsen, Ainslie, Lissaman and others [6, 7]. These models are based on the physical principles like conservation of mass, and momentum conservation using empirical relations of wake decay to model the velocity deficit behind the turbine rotor. The power loss of the modelled wind farm with the wake effect is analyzed under the ramifications of the wind speed time delay and the wind direction. Analytical models are typically applied for wake modelling in micro-siting scenarios and wind farm output predictions. Such models tend to overlook the actual nature of flow and characterize on the basis of the energy content in the flow, neglecting the ambient velocity deficit perturbation. Such models have many idealistic assumptions and fail to capture and predict the real-world fluid flow regimes.

1.2 Drawbacks in the Existing Wake Models

When compared with advanced CFD simulations, analytical wake models have been observed to be physically deficient in predicting the wake velocities given the need of calibrating a wake expansion factor. Analytical models fall short in capturing the complex geometry induced effects specific for turbine blades on the wake [6, 7]. Apart from using an undesirable top-hat wake profile for the velocity, Jensen's model tends to underestimate the wake velocity in the modelling process. The other major limitation of the Jensen wake model is an improper constant wake decay parameter expressed by 'k' through the entire wake flow field. Similarly, models like the Lissaman wake model tend to neglect the pressure differential [6, 7]. Also, the real-time data for wind turbine wake is 3-D and therefore it becomes very computationally expensive to model using prevalent methods like the CFD-RANS approach.

1.3 Introduction to Sparse Identification of Non-linear Dynamics (SINDy) Algorithm

In the backdrop of the limitations encountered by the analytical model and to reduce the computational inefficiency, data-driven modelling for identification of the system and producing governing Equations for the estimation of the parameters from the data is a solution. The major setback in modelling that needs consideration is the possibility of approximating the nonlinear dynamics in the system as sparsely as possible, instead of using a set of complicated basis of variables and expansion terms that not only increase the complexity of the solution but are observed to contribute exiguously in understanding the dynamics of the system in the first place. Such setbacks can be overcome using sparse regression models like the **S**parse **I**dentification of **N**on-linear **Dy**namics (SINDy) [8, 9]. Identification of the system in terms of its features, complexity, control variables and the trend in which they influence the system behavior is of utmost important and improving the efficiency of discovery of these terms or variables can be achieved with the help of methods discussed in the previous sections. Although regression methods, ANNs, and other prevalent evolutionary algorithms in computational science are capable of producing models from high-precision data. They fall short to predict the fault occurrences precisely [10] and therefore the models derived often lack flexibility thereby leading to overfitting and high model complexity. Hence, SINDy offers the means to generate a robust, parsimonious dynamical model with a descriptive ability and exceeding capability of balancing complexity and reducing the chances of overfitting. The SINDy algorithm has been applied to various applications including space weather modelling, flow turbulence modelling, atmospheric convection modelling, spread of disease modelling, financial market stability prediction, and others [8–10].

1.4 SINDy Algorithm Theory

The objective here is to extract sparse model equations representing the dynamics of the time-series data obtained from PIV measurements. The initial process will include extraction of velocity vectors from the images captured by PIV and representing them in temporal and space varying data matrix $[X]$ matrix such that a $[\dot{X}]$ matrix denotes the time derivative terms of the initial matrix [8]. SINDy is a way to translate the measurement data into a plausible dynamical system governing the time evolution of the particle system, where it provides with a means to represent the data into an interpretable and generalized system. The next step includes the formation of a library matrix $\theta[X]$ containing all the candidate terms that are capable of describing the dynamics of the system under consideration. Finally, a matrix $\in$ consisting of sparse coefficient vectors is formed. The columns of which can be calculated using the Equation $[\dot{X}] \approx \Xi * \theta[X]$. The method assumes that every dynamical system in existence can be represented in the form of the Equation which represents the time evolution of $x(t)$ in terms of a nonlinear function f narrowing down to the Equation $dx/dt = f(x, t, u; \beta)$, where f is a vector valued function that demonstrates the state change in the next Δt time period, given the current state x, where $x(t) \in R^n$ are the quantifiable state variables at time $t \in [0, T]$. β represents the parameters that the system offers explicit control over. For example, in a pendulum system mounted on a moving cart, β will comprise of the length of the pendulum string, the mass of the pendulum, velocity of the cart, and acceleration of the system as whole. The β parameters help us to understand the explicit dependence of the dynamics f on changes in β. The application of SINDy in recent time has gained interest in modelling and representing a dynamical fluid flow problem. The system follows a set of non-linear partial differential equation and the SINDy algorithm has been used to exemplify the dynamics involved in this system as sparsely as possible and be generalized in the form of intuitive partial differential Equations. The system of a fluid flow past a two-dimensional cylinder is an eminent choice of many aerodynamics researchers, since it is relevant to numerous engineering problems that we observe related to fluid flow past an obstacle. Such studies signify how the fluid effects the obstacle and its behavior around different geometries of obstacles for engineering applications. Brunton et al. [9] considered a low Reynolds number ($\approx$100) flow across a circular cylinder that has impact over the flow pattern and the fluid surface forces acting on the solid body, such that the system has rich dynamics of the fluid properties. The measurement data for the system was taken using Navier–Stokes direct numerical simulations. The vortex shedding past a cylinder data includes the numeric matrices containing the velocities in x and y directions and the vorticity matrix containing ω_z for all the temporal frames varying spatially. Since the data obtained for the system is high-dimensional, POD is performed on the dataset for the development of a mean-field model for the cylinder wake which facilitates in providing an optimal low-rank basis system in the Frobenius norm space (Fig. 1). POD performed on the fluid velocity fields presents a hierarchy of orthonormal modes which tend to encapsulate majority of the energy for a given rank truncation [8]. Figure 1, show the POD modes for the data such that the POD basis tends to reduce the dimensions of the fluid system which are successful in representing the actual quadratic non-linearities and time scales that the Navier–Stokes Equations are associated with [9]. The first and second POD modes tend to capture substantial portion of the energy from the system and hence define the coordinate system for the flow.

2 Research Objective

The operation of wind turbines is non-autonomous (meaning time dependent), in terms of the magnitude of 2-D velocity coordinates, and vorticity. The data used for the analysis has been extracted using particle image velocimetry (PIV). A system that showcases pattern formation or follows a trend when plotted has the ability to automatically learn and improve performance and stability via accurate modelling.

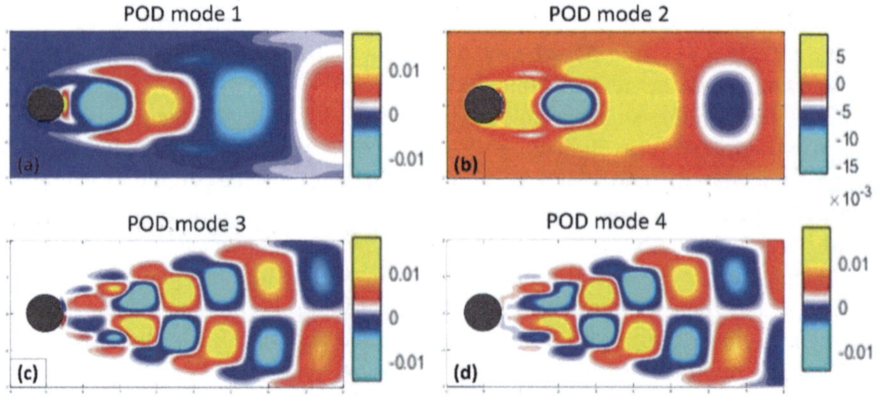

Fig. 1 POD Modes of the vorticity data for flow past a cylinder. The modes have been produced on publicly available data in Brunton et al. [8]

Therefore, the primary objective includes the formulation of a model that is capable of precise wake prediction, and for attaining it the secondary objectives consist of identification and extraction of high-energy coherent structures by applying decomposition methods like proper orthogonal decomposition (POD). The choice of basic functions to be added in the sparse matrix for realistic systems must be made such that the elements of the matrix should be capable of clearly representing the dynamics of the system. The basis used for the problem statement elaborated in Fig. 1 are polynomial basis with degree up to fifth order.

3 Experimental and Numerical Setup

3.1 Wind Tunnel and Acquisition

The assessment of the performance for the modelled wind turbines was carried out in a low turbulence level wind tunnel. The wind tunnel used is a subsonic wind tunnel, with a 3-phase 30 HP induction motor powered tunnel fan, the dimensional layout of the tunnel is given in Fig. 2a. The wind tunnel is an open-circuit type and can produce a maximum wind speed of 50 m/s at the test section. The investigation has been carried out in a subsonic low turbulence wind tunnel of 600 mm × 600 mm × 2000 mm test section (Fig. 2a). The wind turbine model used has a rotor hub height of 300 mm, and a radius of 65 mm, and further details on the design and development of model rotor has been mentioned in articles [11, 12]. A downscale HAWT model with a blockage factor of about 6% is taken for the case study. The turbulence level in the tunnel is estimated to be < 0.3% using the hot-wire anemometry system. The time-resolved instantaneous velocity extraction from the PIV images produces an average

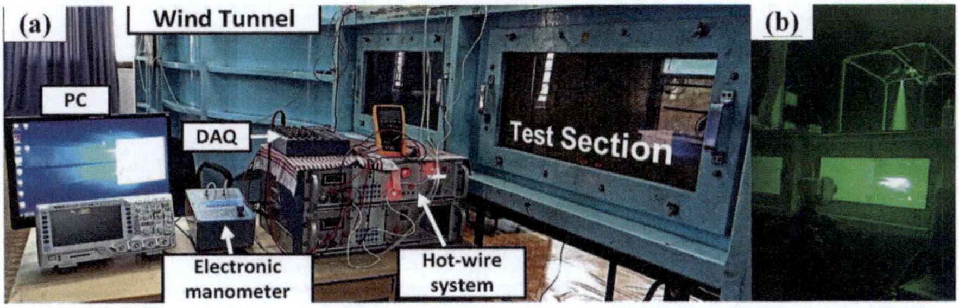

Fig. 2 Illustration of **a** subsonic wind tunnel at IITG, and **b** stereo PIV setup

SD of about 2%. The wind speed inside the tunnel was measured through a pitot-static probe, and the differential pressure is measured through a single channel electronic manometer, and the resultant pressure difference is called the dynamic pressure. The electronic manometer has a sampling frequency of 100 Hz. The measurement data for the SINDy algorithm are taken from the PIV system, for the present study a stereo PIV system has been used to map the rotor wake field. Thus, time-resolved x, y, and z-component of velocity can be obtained. The PIV system (Fig. 2b) is supported by high-speed phantom camera and the laser sheet is produced by the Litron system, which is able to generate a double pulse green light of wavelength 527 nm. The wake field of the turbine has been assessed within the field of view (FOV) of 2500 × 1600 pixels, at the rate of 0.8 kHz and with an interrogation window of 24 × 24 pixels. This range of acquisition rate is noted as well as enough to dictate the underlying low and high frequency fluctuations. For the assessment of PIV frames, a vector field correlation algorithm is used, for which DaV of 10.0 is employed. The time-resolved instantaneous velocity extraction from the PIV images produces an average SD of about 2%. The measurements of the system can be represented as $[y = g(x, t)]$ ideally if no noise is considered, but realistic data has a noise component to it which can be shown as $y = g(x, t) + n$, where 'n' represents the noise component.

3.2 *Numerical Setup and Methodology*

The numerical setup of the experiment contains the matrices vel_u, vel_v, vort, and vel_res which stand for velocity u, velocity v, vorticity, and resultant velocity respectively to be used as the theta matrix, also the code development for visualizing the coherent high energy patterns from the data was done using MATLAB R2021a. The post-processing and cleaning of the PIV data was performed using Python 3.9 and the Pandas library. The matrix formation of the velocity vector magnitudes in x and y direction will look like Eq. (1) where the column vectors will be similar to matrix relation Eq. (2), through (4) [13].

$$V = \begin{bmatrix} | & | & \cdots & | \\ V_1 & V_2 & \cdots & V_n \\ | & | & \cdots & | \end{bmatrix} \tag{1}$$

$$V_1 = \left[\, v_1(t_1)\ v_1(t_2)\ v_1(t_3)\ v_1(t_4)\ \cdots\ v_1(t_m) \,\right]^{\mathrm{T}} \tag{2}$$

$$V_2 = \left[\, v_2(t_1)\ v_2(t_2)\ v_2(t_3)\ v_2(t_4)\ \cdots\ v_2(t_m) \,\right]^{\mathrm{T}} \tag{3}$$

$$V_n = \left[\, v_n(t_1)\ v_n(t_2)\ v_n(t_3)\ v_n(t_4)\ \cdots\ v_n(t_m) \,\right]^{\mathrm{T}} \tag{4}$$

whereas, the method followed for the decomposition consisting of the self-transposed matrix C (Eq. 5) formed using large number of snapshots of the field, such that the complete data is reshaped into a single numeric symmetric and square matrix $C \in R^{n\times m}$, where each column of $[V]$ represents a single instantaneous velocity field vector.

$$C = \frac{1}{M} V \cdot V^{\mathrm{T}} \tag{5}$$

The eigen-decomposition by performing singular value decomposition (SVD) on matrix C (Eq. 6), such that $[U]$ is an orthogonal matrix that contains the left eigenvectors of C, W^{T} is a transpose of an orthogonal matrix that contains the right eigenvectors of C. Whereas, S is a diagonal matrix with eigenvalues as the diagonal elements

$$C = U\ S\ W^{\mathrm{T}} \tag{6}$$

Here, the columns of the U matrix represent the POD modes (Eq. 7) such that the ith snapshot is capable of being represented as a linear combination of the columns of U where, u_k represents the columns of matrix U.

$$q^i(t) = \sum_{K=1}^{M} a_k(t_i)\, u_k \tag{7}$$

As for the matrix formation in SINDy, measurements from $y = g\ (x, t) + n$, will form the $[V]$ matrix where, the columns of $[V]$ are essentially the snapshots of the data from system reshaped into very tall column vectors that are constantly evolving in time and with the dynamics of the system. So, V_1 is a column vector, V_2 from the next frame of the snapshot and so on. The library matrix, $\theta_1(V)$ (Eq. 7) is a candidate non-linear function of $[V]$ columns (Eq. 9), and it may denote polynomial, trigonometric, hyperbolic, exponential, or logarithmic non-linearities. The problem statements in fluid mechanics are known to obey by the Navier Stokes equations which are inherently first and second order polynomial equations, therefore when modelling dynamical fluid flows the selection of polynomials as the basis functions is suggested.

$$\theta(V) = \begin{bmatrix} | & | & | & \cdots\cdots & | & | & | \\ 1 & V & V^{p_2} & \cdots\cdots & \sin v & \cos v & \sin 2v \\ | & | & | & \cdots\cdots & | & | & | \end{bmatrix} \tag{8}$$

$$V^{P_{2(t)}} = \begin{bmatrix} v_1^2(t_1) & v_1 v_2(t_1) & \cdots & v_2^2(t_1) & \cdots & v_n^2(t_1) \\ v_1^2(t_2) & v_1 v_2(t_2) & \cdots & v_2^2(t_2) & \cdots & v_n^2(t_2) \\ . & . & \cdots & . & \cdots & . \\ . & . & \cdots & . & \cdots & . \\ . & . & \cdots & . & \cdots & . \\ v_1^2(t_m) & v_1 v_2(t_m) & \cdots & v_2^2(t_m) & \cdots & v_n^2(t_m) \end{bmatrix} \tag{9}$$

The sparse matrix (Eq. 10) for the algorithm contains the column vectors of coefficients which are representatives of the active terms from $f(x)$ which will be present in the final output.

$$\Xi = \begin{bmatrix} | & | & | & \cdots & | \\ \xi_1 & \xi_2 & \xi_3 & \cdots & \xi_n \\ | & | & | & \cdots & | \end{bmatrix} \tag{10}$$

While processing and feeding the data to the algorithm, even noise in trivial magnitude can in turn increase the noise in the derivative matrix, and to account for this issue it is suggested to normalize the data and calibrate the term $[\dot{V}] \approx \Xi * \theta[V]$ by adding a noise term '$\eta\ Z$' where 'Z' is a matrix of independent identically distributed Gaussian entries with zero mean and noise magnitude η, making it $[\dot{V}] \approx \Xi * \theta[V] + \eta\ Z$ [14].

4 Results and Discussion

As shown below, Fig. 3 represents the plots that were extracted using particle image velocimetry (PIV). Figure 3a showcases the distribution of the magnitude of instantaneous velocity in the x-direction in the masked domain from the frame, Fig. 3b displays the instantaneous velocity in the y-direction in the masked domain from the frame, Fig. 3c showcases the vortex formations in the domain, and Fig. 3d displays the mean intensity of velocities across the domain.

The dimensions of the data matrix were high, implying that decomposition methods will be needed to realize the prominent characteristics and their direction of high impact in the considered flow. The data matrix $[V]_{m\times n}$ used has the dimensions of 105230 × 500, where 'm' represents the total number of spatial coordinates in the form of (x, y), whereas 'n' represents the magnitude of the quantity in varying time. The numeric matrices were processed using MATLABR2021a, as per the layout

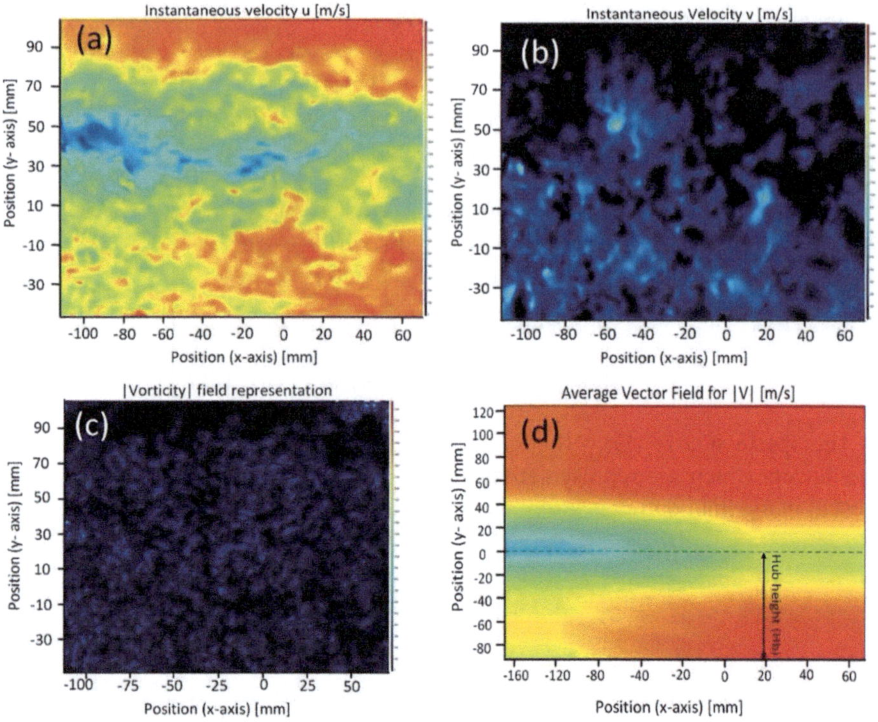

Fig. 3 Domain plots for **a**, **b** instantaneous velocities, **c** vorticity, and **d** mean flow

algorithm discussed in the numerical setup section. Figure 4 shows the scatter plots for the instantaneous velocity values in x, y, and z direction respectively at inflow wind velocity 1.8 m/s and tip speed ratio 2.5, these values form the matrices for the decomposition method further. The modes for the vorticity data matrix were plotted, and the results are shown in Fig. 5, which shows the different decomposition mode plots that were obtained using the PIV data. Although the plots are not very intuitive and clear, small patterns of high-intensity energy field can be observed in the contours which showcase the distribution of energy in the flow. The results obtained are unclear and the reason may lie in the noise present in the data obtained. The recognition of coherent structures in a particular coordinate axis might look confusing but the repetitive pattern fading along (x/R) shows the energy structures' decrease and dissociation. These POD modes, which showcase vorticity fluctuations in a particular direction, can be used as a reference to select our variable of interest for further analysis. Further study and scope of the project include the refinement of the obtained results and applying the SINDy algorithm on the selected coherent structures along the coordinates attained using decomposition of the numeric matrices. The results shown in Fig. 5 were expected to give a clear idea and intuition showcasing the vorticity streaks distribution in the fluid flow, suggesting coherence in the wake span region.

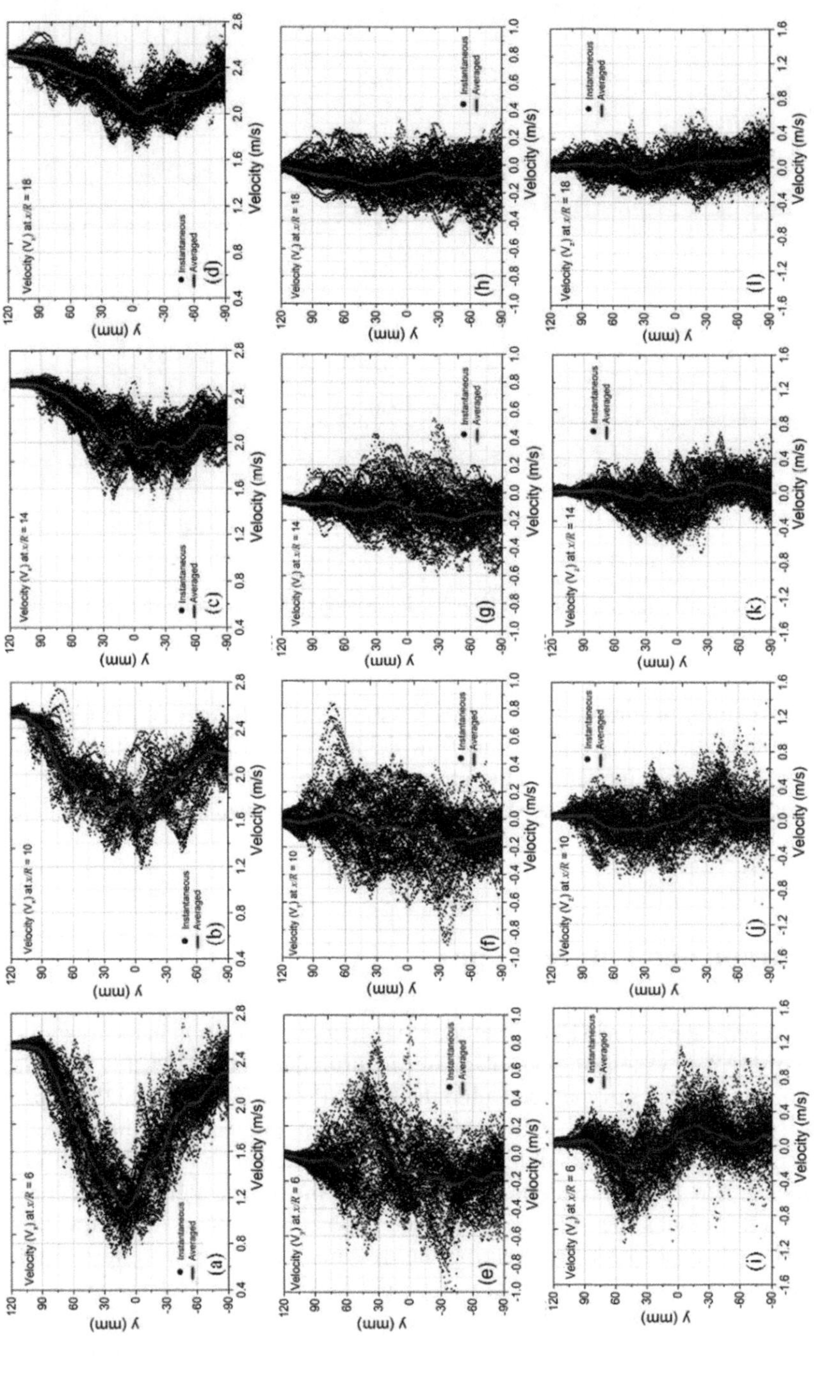

Fig. 4 Scatter plots for—V_x velocity distribution (**a–d**), V_y velocity distribution (**e–h**), V_z velocity distribution (**i–l**). At inflow wind velocity 1.8 m/s and tip speed ratio of 2.5

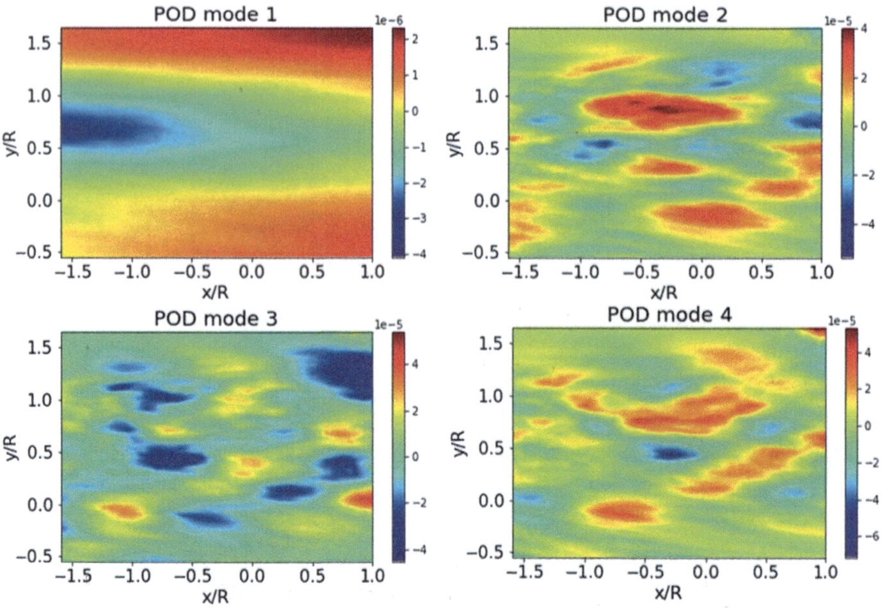

Fig. 5 Proper orthogonal decomposition (POD) modes contour plots based on the data derived from the experimental

The investigation of turbine wake characteristics was performed to identify the embedded coherent structures in the fluid flow field, to be considered for the further development of the wake model, and therefore the PIV data obtained must be considered for heavy post-processing since the presence of missing data at some coordinate data points led to distorted mode generation.

5 Conclusion and Future Scope

The paper presents the use of SINDy methodology and its application in the dynamic modelling of HAWT wake using out-of-phase/ free run data retrieved using Particle Image Velocimetry (PIV). A method called the POD (Proper Orthogonal Decomposition) has been used to extract spatial-temporal coherent points from fluid datasets by collecting multiple snapshots of high-dimensional fluid data in time. The present study applies dimensionality reduction techniques like the proper orthogonal decomposition which includes the singular value decomposition method to obtain low-dimensional basis capable of representing the dynamics of the flow system. Such low dimensional basis of the system helps in proper as well as efficient processing of the otherwise high dimensional data generated from the PIV system. The selection of the best fit terms in the sparse vector matrix will be taken on the basis of prior knowledge and classical intuition, recalling that SINDy is a type of guided model. Based on the Navier Stokes equations it is expected that the model will be sparse with terms containing the first and second order polynomial terms. The application of this model analogy can be found to be effective and intuitive in presenting the non-linear dynamics of wind turbine wake, and will assertively have a significant implication in wind farm modeling.

Acknowledgements During the period of this study, the financial assistantship extended to the first author by the Indian Institute of Technology Guwahati, India, is gratefully acknowledged. Authors are very much thankful to Department of Science and Technology (DST)-New Delhi and Technology Innovation Hub (TIH)-Indian Institute of Technology Guwahati for providing the necessary infrastructure and financial support.

Funding The research work reported did not receive any specific grant from funding agencies.

Conflict of Interest There are no conflicts of interest.

References

1. Kutz JN, Brunton SL, Brunton BW, Proctor JL (2016) Dynamic mode decomposition. Society for Industrial and Applied Mathematics, Philadelphia, PA. https://doi.org/10.1137/1.9781611974508
2. Berkooz G, Holmes P, Lumley JL (1993) The proper orthogonal decomposition in the analysis of turbulent flows. Annu Rev Fluid Mech 25(1):539–575. https://doi.org/10.1146/annurev.fl.25.010193.002543
3. Harlow FH, Welch JE (1965) Numerical calculation of time-dependent viscous incompressible flow of fluid with free surface. Phys Fluids 8:2182–2189. https://doi.org/10.1063/1.1761178
4. Mezic I (2013) Analysis of fluid flows via spectral properties of the Koopman operator. Annu Rev Fluid Mech 45:357–378. https://doi.org/10.1146/annurev-fluid-011212-140652
5. Bongard J, Lipson H (2007) Automated reverse engineering of nonlinear dynamical systems. Proc Natl Acad Sci USA 104:9943–9948. https://doi.org/10.1073/pnas.0609476104
6. Siram O, Kumar R, Saha UK, Sahoo N (2022) A comprehensive review on analytical formulations of wind turbine wake models and future scope in the development of wind farms of assorted configurations. J Energy Resour Technol 144:110801. https://doi.org/10.1115/1.4054175
7. Andersen SJ, Sørensen JN, Ivanell S, Mikkelsen RF (2014) Comparison of engineering wake models with CFD simulations. J Phys Conf Ser 524:393–399. https://doi.org/10.1088/1742-6596/524/1/012161
8. Brunton SL, Proctor JL, Kutz JN (2016) Discovering governing equations from data by sparse identification of nonlinear dynamical systems. Proc Natl Acad Sci USA 113:3932–3937. https://doi.org/10.1073/pnas.1517384113
9. Cortiella A, Park KC, Doostan A (2021) Sparse identification of nonlinear dynamical systems via reweighted l1-regularized least squares. Comput Methods Appl Mech Eng 376:113620. https://doi.org/10.1016/j.cma.2020.113620
10. Mangan NM, Kutz JN, Brunton SL, Proctor JL (2017) Model selection for dynamical systems via sparse regression and information criteria. Proc R Soc A Math Phys Eng Sci 473:20170009. https://doi.org/10.1098/rspa.2017.0009

11. Siram O, Saha UK, Sahoo N (2022) Blade design considerations of small wind turbines: from classical to emerging bio-inspired profiles/shapes. J Renew Sustain Energy. https://doi.org/10.1063/5.0096402
12. Siram O, Kesharwani N, Sahoo N, Saha UK (2022) Aerodynamic design and wind tunnel tests of small-scale horizontal-axis wind turbines for low tip speed ratio applications. J Solar Energy Eng Trans ASME 144:1–34. https://doi.org/10.1115/1.4053453
13. Champion KP, Brunton SL, Kutz JN (2019) Discovery of nonlinear multiscale systems: Sampling strategies and embeddings. SIAM J Appl Dyn Syst 18:312–333. https://doi.org/10.1137/18M1188227
14. Quade M, Abel M, NathanKutz J, Brunton SL (2018) Sparse identification of nonlinear dynamics for rapid model recovery. Chaos. https://doi.org/10.1063/1.5027470

Surface Wave Profile Due to Oscillatory Motion of an Asymmetric Block of Ocean Floor

Ritika Das and Santu Das

Abstract The study of waves generated due to ocean bottom displacements is of great importance. In an underwater earthquake, the ocean bed undergoes movement for a short duration which induces wave motion in the seawater that travels along the surface for a significant amount of time. More than often, such wave propagation creates a tsunami at the shoreline. We here consider a rather simplified problem of the horizontal and rigid ocean floor, a finite chunk of which oscillates in the vertical direction at a rate that varies with respect to space. Such a variation creates an asymmetric floor surface profile of the oscillating block. The impact of ocean water compressibility is introduced by generating acoustic-gravity waves that significantly alter the ocean surface profile. We show the variation in surface profile elevation due to changes in ocean depth, the width of the oscillating ocean floor, and its slope.

Keywords Acoustic-gravity waves · Ocean bottom displacement · Fourier transformation · Surface waves

1 Introduction

A majority of studies on the modeling of tsunamis consider the incompressibility of ocean water. However, contemporary studies signified the importance of water compressibility in tsunami generation [1–5] following the fundamental works done by Sells [6] and Yamamoto [7]. Without compressibility, the pure gravity case is the only progressive wave for a given frequency of oscillation that corresponds to a single propagating mode and infinitely many evanescent modes [8]. However, with the inclusion of water compressibility, it has been shown that multiple propagating modes will exist as an outcome of the conversion of evanescent modes from the corresponding incompressible case [7]. Besides the pure gravity one, these extra

R. Das (✉) · S. Das
Mathematical and Computational Sciences (Physical Sciences Division), Institute of Advanced Study in Science and Technology, Guwahati 781035, India
e-mail: ritikadas1004@gmail.com

Academy of Scientific and Innovative Research (AcSIR), Ghaziabad 201002, India

R. K. Ray et al. (eds.), *Advances in Theoretical and Applied Mechanics*, Lecture Notes in Mechanical Engineering, https://doi.org/10.1007/978-981-97-0418-7_8

propagating modes are called acoustic-gravity waves (AGW). Thus, inclusion of compressibility is of utmost significance.

We consider a moving and vibrating large portion of the ocean floor holding the possibility of triggering a tsunami. The vertical motion of the portion is considered to be oscillatory at a rate that varies with space. It leads to an asymmetrically raised block of the ocean floor. We aim to study the propagation of AGW and pure gravity waves and the impact of different parameters of interest, such as ocean depth, width, and slope of the block.

2 Mathematical Formulation

A Cartesian coordinate system is utilized, where the origin is positioned at the center of the oscillating ocean floor block. The x-axis is parallel to the sea floor horizontally, and the y-axis is vertically oriented upward, as depicted in Fig. 1. As illustrated, the block has a total width of $2b$ and an assumed infinite length, while the ocean maintains a constant depth of h. It is also assumed that the block is undergoing vertical oscillations with a small amplitude a_0 at $x = -b$ and a_1 at $x = b$ with an angular frequency α, resulting in the displacement of the ocean floor given by

$$\xi(x, t) = H(b^2 - x^2)[a_0 + M(x + b)] \exp(\mathrm{i}\alpha t), \tag{1}$$

where $M = \dfrac{a_1 - a_0}{2b}$ is the slope of the line joining the points $(-b, a_0)$ and (b, a_1) as shown in Fig. 1 and H is the Heaviside step function.

Considering the problem as time-harmonic and taking into account the standard assumptions of linearized water wave theory, the displacement potential $\phi(x, y, t)$ satisfies the following governing equation:

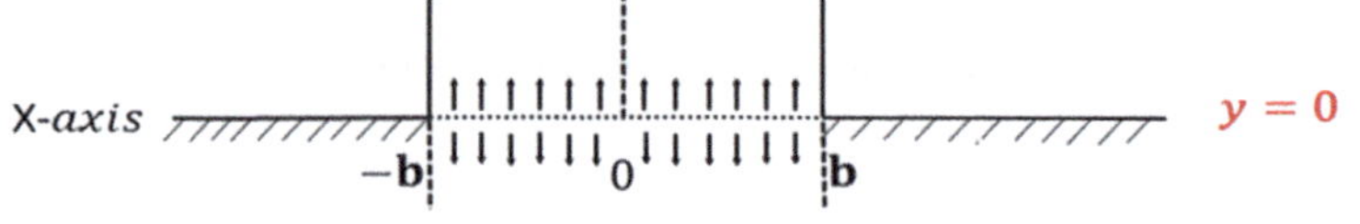

Fig. 1 Schematic diagram of the physical problem depicting the asymmetric motion of a portion of the ocean block

$$\nabla^2 f = -\frac{\alpha^2}{c^2} f, \tag{2a}$$

where f is the spatial part of $\phi(x, y, t)$ and c is the speed of sound in the undisturbed ocean medium. The matching of spatial vertical displacement at the ocean floor and the linearized free surface boundary condition are given by

$$\frac{\partial f}{\partial y} = H(b^2 - x^2)[a_0 + M(x + b)] \qquad (y = 0), \tag{2b}$$

$$\frac{\partial f}{\partial y} = \frac{\alpha^2}{g} f \qquad (y = h), \tag{2c}$$

where g is the acceleration due to gravity. Equations (2a)–(2c) collectively form a *Boundary Value Problem* (BVP).

Solution Process We apply the method of Yamamoto [3] to analytically solve the BVP. Solving the problem in the space domain is challenging due to the inhomogeneous nature of the bottom boundary condition (2b). However, by utilizing the Fourier wave number transformation, as defined below,

$$F(w, y) = \int_{-\infty}^{\infty} \exp(-iwx) f(x, y) dx, \tag{3a}$$

$$f(x, y) = \frac{1}{2\pi} \int_{-\infty}^{\infty} \exp(iwx) F(w, y) dw, \tag{3b}$$

the boundary conditions can be made homogeneous in the wave number domain, making it easier to solve the problem.

The transformed BVP in terms of $F(w, y)$ is given as

$$\left(\frac{\alpha^2}{c^2} - w^2\right) F + \frac{\partial^2 F}{\partial y^2} = 0 \tag{4a}$$

with boundary conditions

$$\frac{\partial F}{\partial y} = \frac{a_0 \exp(iwb)}{iw} - \frac{a_1 \exp(-iwb)}{iw} - \frac{2M \sin wb}{iw^2} \qquad (y = 0), \tag{4b}$$

$$\frac{\partial F}{\partial y} = \frac{\alpha^2}{g} F \qquad (y = h). \tag{4c}$$

The general solution for Eq. (4a) using separation of variables is given by:

$$F(w, y) = c_1(w) \exp(\mu y) + c_2(w) \exp(-\mu y), \tag{5}$$

where

$$\mu^2 = w^2 - \frac{\alpha^2}{c^2}; \ (\mathrm{Re}(\mu) \geq 0). \tag{6}$$

Thus from Eqs. (4b), (4c), and (5), we obtain $F(w, y)$ as

$$F(w, y) = -\frac{[a_0 \exp(iwb) - a_1 \exp(-iwb) + \frac{2M}{w} \sin wb]}{2\, i\mu w[\mu \sinh \mu h - \frac{\alpha^2}{g} \cosh \mu h]} \Upsilon(\mu, y), \tag{7}$$

where

$$\Upsilon(\mu, y) = (\mu + \frac{\alpha^2}{g}) \exp(\mu(y - h)) + (\mu - \frac{\alpha^2}{g}) \exp(-\mu(y - h)). \tag{8}$$

We now perform inverse Fourier transform on Eq. (7) to get

$$\begin{aligned} f(x, y) =& \frac{1}{2\pi} \int_{-\infty}^{\infty} \frac{\mu \cosh \mu(y - h) + \frac{\alpha^2}{g} \sinh \mu(y - h)}{\mu w^2(\mu \sinh \mu h - \frac{\alpha^2}{g} \cosh \mu h)} \\ & \times \{(iwa_0 + M) \exp[iw(x + b)] - (iwa_1 + M) \exp[iw(x - b)]\}\, dw. \end{aligned} \tag{9}$$

We introduce the function $N(\xi, y)$ by

$$N(\xi, y) = \int_{-\infty}^{\infty} \frac{\mu \cosh \mu(y - h) + \frac{\alpha^2}{g} \sinh \mu(y - h)}{\mu w^2(\mu \sinh \mu h - \frac{\alpha^2}{g} \cosh \mu h)} T(\xi, w) \exp(iw\xi) dw, \tag{10}$$

where

$$T(\xi, w) = \begin{cases} iwa_0 + M & \xi = x + b, \\ iwa_1 + M & \xi = x - b. \end{cases} \tag{11}$$

such that $f(x, y)$ can be written explicitly as

$$f(x, y) = \frac{1}{2\pi} [N(x + b, y) - N(x - b, y)]. \tag{12}$$

It is easier to realize from Eq. (9) the equivalence between the flow generated by the motion of the asymmetric block of the ocean floor and that by the two line sources located at the edges of the block $x = \pm b$. Around the singularities, we now apply the contour integration (Fig. 2) method and use residue calculus to obtain the solution in the two regions $|x| < b$ and $|x| > b$, respectively, as

$$
\begin{aligned}
f(x,y)_{|x|>b} = \pm 2\Big\{ & \Big[\mu_0 \cosh \mu_0 y\Big(k_0[a_0 \cos k_0(x+b) - a_1 \cos k_0(x-b)] \\
& + M[\sin k_0(x+b) - \sin k_0(x-b)]\Big)\Big] \Big/ k_0^3[2\mu_0 h + \sinh 2\mu_0 h] \\
& + \sum_{n=1}^{N} \Big\{\mu_n \cos \mu_n y\Big(k_n[a_0 \cos k_n(x+b) - a_1 \cos k_n(x-b)] \\
& + M[\sin k_n(x+b) - \sin k_n(x-b)]\Big)\Big\} \Big/ k_n^3[2\mu_n h + \sin 2\mu_n h] \\
& - \sum_{n=N+1}^{\infty} \Big\{\mu_n \cos \mu_n y \exp(\pm\lambda_n x)\Big(\lambda_n(a_0 \exp(\pm\lambda_n b) - a_1 \exp(\mp\lambda_n b))\Big) \\
& + 2M \sinh \lambda_n b\Big\} \Big/ \lambda_n^3[2\mu_n h + \sin 2\mu_n h]\Big\},
\end{aligned} \tag{13a}
$$

$$
\begin{aligned}
f(x,y)_{|x|<b} = 2\Big\{ & \Big[- \mu_0 \cosh \mu_0 y\Big(k_0[a_0 \cos k_0(x+b) + a_1 \cos k_0(x-b)] \\
& + M[\sin k_0(x+b) + \sin k_0(x-b)]\Big)\Big] \Big/ k_0^3[2\mu_0 h + \sinh 2\mu_0 h] \\
& - \sum_{n=1}^{N} \Big\{\mu_n \cos \mu_n y\Big(k_n[a_0 \cos k_n(x+b) + a_1 \cos k_n(x-b)] \\
& + M[\sin k_n(x+b) + \sin k_n(x-b)]\Big)\Big\} \Big/ k_n^3[2\mu_n h + \sin 2\mu_n h] \\
& + \sum_{n=N+1}^{\infty} \Big\{\mu_n \cos \mu_n y \exp(-\lambda_n b)\Big(\lambda_n(a_0 \exp(-\lambda_n x) + a_1 \exp(\lambda_n x))\Big) \\
& + 2M \sinh \lambda_n b\Big\} \Big/ \lambda_n^3[2\mu_n h + \sin 2\mu_n h] \\
& + \frac{k_s \cos k_s(y-h) + \frac{\alpha^2}{g} \sin k_s(y-h)}{4k_s[k_s \sin k_s h + \frac{\alpha^2}{g} \cos k_s h]}\Big(2Mx + a_0 + a_1\Big)\Big\},
\end{aligned} \tag{13b}
$$

where μ_0, μ_n are the wave numbers associated with the propagating mode and the attenuation coefficients of the evanescent modes exhibited by water waves in an incompressible fluid, satisfying the dispersion relations

$$
\alpha^2 = \mu_0 g \tanh \mu_0 h \quad \text{and} \quad \alpha^2 = -\mu_n g \tan \mu_n h, \tag{14}
$$

where k_s, k_0, k_n, λ_n are the wave numbers of the sound wave in an unbounded medium, propagating gravity wave, propagating acoustic waves, and attenuation coefficients of evanescent modes in a compressible fluid, respectively. These are positive real-valued and given by

$$
\begin{aligned}
& k_s = \alpha/c\,; \quad k_0 = (\mu_0^2 + k_s^2)^{1/2}; \quad k_n = (k_s^2 - \mu_n^2)^{1/2} \quad \text{for} \quad k_s > \mu_n, \\
& \text{and} \quad \lambda_n = (\mu_n^2 - k_s^2)^{1/2} \text{ for } k_s < \mu_n.
\end{aligned} \tag{15}
$$

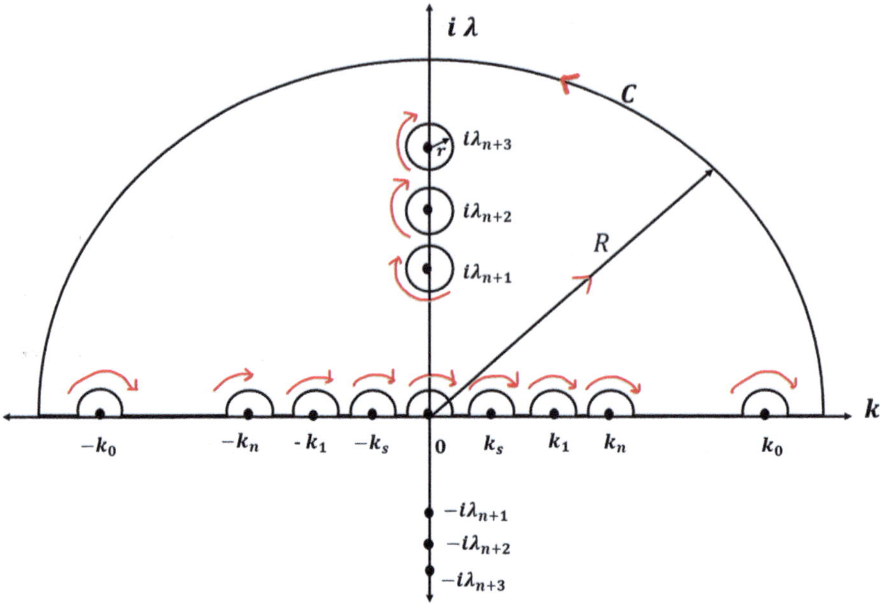

Fig. 2 Singularities and contours to evaluate the integral (9)

The positive and negative signs in Eq. (13a) represent the region $x < -b$ and $x > b$, respectively. Thus, we get the required solution for the BVP.

3 Results and Discussion

When the slope is zero, the potential function matches with the one obtained by Yamamoto [3]. This confirms the correctness of the solution process. This paper aims to determine the surface wave profile by plotting $\frac{\partial f}{\partial y}$, representing the vertical displacement. The behavior of vertical displacement of the surface wave against the horizontal distance due to the oscillation of the ocean block with angular velocity $\alpha = 2\pi$ has been studied.

In Fig. 3, we have observed the surface wave profile for different values of the slope M, keeping ocean depth and width of the block constant at $h = 5000$ m, $b = 2000$ m, amplitude $a_0 = 1$ m. Figure 4 demonstrates the surface wave profile for fixed width $b = 5000$ m, slope $M = 0.0001$ w.r.t varying ocean depth. In Fig. 5, the width of the oscillating block is changing while keeping the other parameters fixed as $h = 6500$ m, $M = 0.00025$, $a_0 = 1$ m.

Near the region of bottom fluctuation $|x| < b$, the amplitude of the surface waves is proportional to the slope. With the varying water depth, we observe that the amplitude

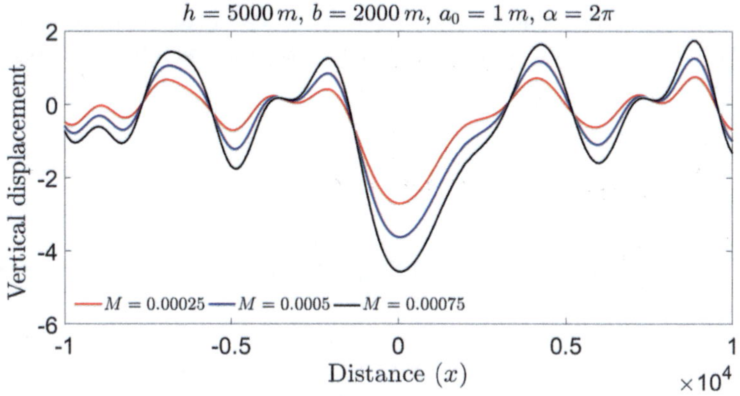

Fig. 3 Surface wave profile against horizontal distance for different slope values

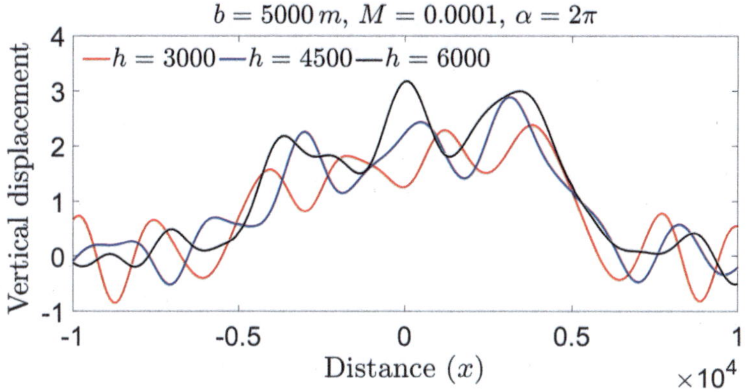

Fig. 4 Surface wave profile against horizontal distance for different ocean depths

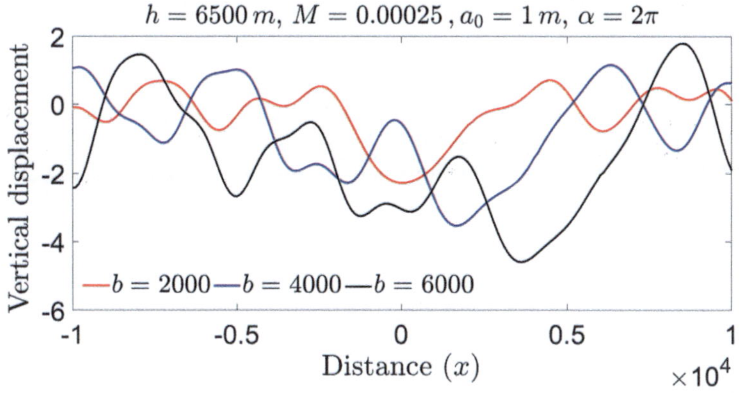

Fig. 5 Surface wave profile w.r.t varying width of the ocean block

increases inside the fluctuation region with increased water depth. In the case of varying width of bottom fluctuation, we see that the maximum amplitude is proportional to the width inside the fluctuation region.

4 Conclusion

An analytical solution to the problem is obtained using the Fourier transformation technique. All the linearized water wave theory assumptions have been used for the analytical approach to the problem. We can conclude from these observations that the variations in the vertical displacement of the water surface induced by the acoustic-gravity waves increase, with an increase in the ocean depth, slope, and width of the oscillating ocean block.

Acknowledgements The author would like to acknowledge the Institute of Advanced Study in Science and Technology **(IASST)** for giving financial assistance and the Academy of Scientific and Innovative Research **(AcSIR)** for enrollment in the Ph.D. program.

References

1. Nosov MA (1999) Tsunami generation in a compressible ocean. Phys Chem Earth B(5):437–441
2. Nosov MA (2000) Tsunami generation in a compressible ocean by vertical bottom motions. Izy Atmos Ocean Phys 36:661–669
3. Stiassnie M (2010) Tsunamis and acoustic-gravity waves from underwater earthquakes. J Eng Math 67:23–32
4. Abdolali A, Kirby JT (2017) Role of compressibility on tsunami propagation. J Geophys Res Ocean 122(12):9780–9794
5. Das S, Meylan MH (2023) Time-domain wave response of a compressible ocean due to an arbitrary ocean bottom motion. Appl Math Model 118:832–852
6. Sells CCL (1965) The effect of a sudden change of shape of the bottom of a slightly compressible ocean. Philos Tran Roy Soc A 258(1092):495–528
7. Yamamoto T (1982) Gravity waves and acoustic waves generated by submarine earthquakes. Int J Soil Dyn Earthquake Eng 1:75–82
8. Kadri U (2014) Deep ocean water transport by acoustic-gravity waves. J Geophys Res Ocean 119:7925–7930

Scalar and Directional Localized Artificial Diffusivity Methods to Capture Shock–Turbulence Interaction

Ritu Raj Kumar and Nagabhushana Rao Vadlamani

Abstract The local artificial diffusivity (LAD)-based methods rely on adding a grid-dependent artificial fluid transport coefficient (shear viscosity, bulk viscosity, and thermal diffusivity) to the physical fluid transport coefficients to capture the shock discontinuity in high-speed flows. In this paper, we demonstrate the performance of LAD when coupled with a high-order in-house solver COMPSQUARE, which uses central difference schemes (explicit and compact) for spatial discretization and an explicit fourth-order Runge–Kutta scheme for time integration. The efficacy of this method is tested on a variety of test cases which include a 1D stationary normal shock, 2D compression ramp, 2D incident oblique shock-boundary layer interaction, and a 3D compressible turbulent boundary layer. We further examine the efficacy of the scalar and directional forms of the artificial diffusivity method. On highly stretched grids with higher aspect ratios, the scalar formulation of LAD adds excessive artificial dissipation to the governing equations resulting in smaller time steps due to increased numerical stiffness. In contrast, the directional artificial diffusivity approach outperforms the scalar LAD on higher aspect ratio grids, as will be demonstrated for the test cases of the incident oblique shock and supersonic turbulent boundary layer. For the incident oblique shock test case in particular, the computation using directional LAD is 14.3 times faster than the scalar LAD.

Keywords Artificial diffusivity · Shock–turbulence interaction · High-order schemes · Turbulent boundary layer

1 Introduction

The interaction between shock waves and turbulence is a common phenomenon observed in supersonic and hypersonic flows over high-speed air-breathing vehicle intakes, spacecraft reentry vehicles, turbomachinery blades, etc. In high-speed turbu-

R. R. Kumar · N. R. Vadlamani (✉)
Indian Institute of Technology Madras, Chennai, India
e-mail: nrv@iitm.ac.in

R. R. Kumar
e-mail: ae22d016@smail.iitm.ac.in

R. K. Ray et al. (eds.), *Advances in Theoretical and Applied Mechanics*, Lecture Notes in Mechanical Engineering, https://doi.org/10.1007/978-981-97-0418-7_9

lent flows, capturing the sharp discontinuities due to shocks while equally resolving turbulence scales is a challenging task. On the one hand, the numerical approach should be capable of resolving all the turbulent scales limiting the numerical dissipation. On the other hand, shock-capturing schemes tend to introduce artificial numerical dissipation to capture the shock waves and retain the stability of the schemes. In the context of large eddy simulation (LES), several schemes have been proposed in the literature to meet such conflicting requirements. These include WENO, hybrid WENO-central difference, artificial diffusivity, adaptive characteristic-based filter, and shock fitting [1]. Cook et al. [2] proposed a methodology to capture discontinuities in the flows involving shocks, turbulence, and mixing. The 1D formulation of artificial diffusivity coefficients was further extended to multi-dimensional curvilinear coordinates [3]. The method is based on high-wavenumber damping, which adds localized artificial fluid transport coefficients such as dynamic viscosity, bulk viscosity, thermal conductivity, and species diffusion coefficient to the transport fluid coefficients across the discontinuity. In this method, the discontinuity is smeared over a numerically resolvable scale. Further, Olson and Lele [4] developed a directional artificial fluid diffusivity for capturing shocks on the grids with a high aspect ratio (AR).

Although the central difference schemes are preferred over the upwind schemes in smooth regions of the flow, it is well known that such schemes result in spurious oscillations across steep gradients or discontinuities such as shock waves, material discontinuity, or contact surfaces. Kawai et al. [5] have investigated the performance of different local artificial diffusivity (LAD) models for the large eddy simulation (LES) of compressible turbulent flows. The authors proposed a novel switching function (shock sensor), which is a combination of a negative dilatation and Ducros-type sensor [6] to model the artificial bulk viscosity. The key advantages of this method include its ease of implementation, low computational cost, localization of artificial fluid transport coefficients, and lack of a weighting/hybrid scheme. The primary objective of this paper is to assess the performance of the scalar and directional localized artificial diffusivity (LAD) methods proposed by [4, 5] toward capturing compressible turbulent flows with shocks. In particular, we demonstrate the performance of LAD when coupled with a high-order in-house solver COMPSQUARE [7]. The solver utilizes high-order central difference schemes (explicit/compact) with a low-pass filter [8] to suppress the dispersion associated with high frequencies.

The outline of the paper is as follows: Sect. 2 describes the methodology of the scalar and directional LAD. In Sect. 3, the efficacy of the scalar and direction LAD method is demonstrated using a variety of 2D and 3D test cases. The key conclusions from the study are summarized in Sect. 4.

2 Methodology

COMPSQUARE solves the unsteady three-dimensional compressible form of Navier–Stokes equation in generalized coordinates. These equations can be written as:

$$\frac{\partial \rho}{\partial t} + \nabla \cdot (\rho \boldsymbol{u}) = 0, \tag{1}$$

$$\frac{\partial \rho \boldsymbol{u}}{\partial t} + \nabla \cdot (\rho \boldsymbol{u}\boldsymbol{u} + p\underline{\boldsymbol{\delta}} - \underline{\tau}) = 0, \tag{2}$$

$$\frac{\partial E}{\partial t} + \nabla \cdot [E\boldsymbol{u} + (p\underline{\boldsymbol{\delta}} - \underline{\tau}) \cdot \boldsymbol{u} - \kappa \nabla T] = 0, \tag{3}$$

$$E = \frac{p}{\gamma - 1} + \frac{1}{2}\rho \boldsymbol{u} \cdot \boldsymbol{u}, \quad p = \rho R T, \tag{4}$$

where ρ is the density, $\boldsymbol{u}$ is the velocity vector, T is the temperature, p is the static pressure, and E is the total energy. The ratio of specific heat, $\gamma = 1.4$, κ is the thermal conductivity, R is the gas constant, and $\underline{\boldsymbol{\delta}}$ is the unit tensor. The viscous stress tensor, $\underline{\tau}$ is given as

$$\tau = \mu(2\underline{S}) + (\beta - \frac{2}{3})(\nabla \cdot \boldsymbol{u})\underline{\boldsymbol{\delta}}, \quad \underline{S} = \frac{1}{2}(\nabla \boldsymbol{u} + (\nabla \boldsymbol{u})^T) \tag{5}$$

where μ is the dynamic viscosity, β is the bulk viscosity, and $\underline{\boldsymbol{S}}$ is the strain rate tensor. Shock and material discontinuities are captured by adding grid-dependent artificial fluid transport coefficients to the physical fluid transport coefficients following the approach of [5] as:

$$\mu = \mu_f + \mu^*, \quad \beta = \beta_f + \beta^*, \quad \kappa = \kappa_f + \kappa^*. \tag{6}$$

An attractive feature of this model is the automatic deactivation of the artificial transport coefficients in smooth well-resolved flow regions while providing dissipation in the non-smooth regions of the flow.

Scalar LAD: The formulation of artificial diffusivity coefficients for the scalar LAD is given as,

$$\mu^* = C_\mu \rho \overline{\left| \sum_{l=1}^{3} \frac{\partial^4 S}{\partial \xi_l^4} \Delta\xi_l^4 \Delta_{l,\mu}^2 \right|}, \quad \kappa^* = C_\kappa \frac{\rho c_s}{T} \overline{\left| \sum_{l=1}^{3} \frac{\partial^4 e}{\partial \xi_l^4} \Delta\xi_l^4 \Delta_{l,\kappa} \right|} \tag{7}$$

$$\beta^* = C_\beta \rho f_{sw} \overline{\left| \sum_{l=1}^{3} \frac{\partial^4 \nabla \cdot \mathbf{u}}{\partial \xi_l^4} \Delta\xi_l^4 \Delta_{l,\beta}^2 \right|}, \quad f_{sw} = \frac{H(-\nabla \cdot \mathbf{u})(\nabla \cdot \mathbf{u})^2}{(\nabla \cdot \mathbf{u})^2 + (\nabla \times \mathbf{u})^2 + \epsilon}, \tag{8}$$

where $C_\mu = 0.002$, $C_\beta = 1.75$, and $C_\kappa = 0.01$ are the dimensionless user-specified coefficients. The generalized coordinate is denoted by ξ_l, where l = 1, 2, and 3 signifies ξ, η, and ζ directions, respectively. H is a Heaviside function, and ϵ is a small positive constant of magnitude 10^{-32} to prevent division by zero. The f_{sw} is the switching function, a combination of negative dilatation $H(-\nabla \cdot u)$ and Ducros-type sensor [6] responsible for localizing the artificial bulk viscosity only near the shocks waves. The overbar denotes an approximate truncated-Gaussian filter (see [3] for further details on the filter operation). The $\Delta\xi_l$ and $\Delta_l\cdot$ are the grid spacing in computational and physical space, where $\Delta\xi_l = 1$. The grid spacing $\Delta_l\cdot$ is defined as,

$$\Delta_{l,\mu} = |\boldsymbol{\Delta x_l}|, \quad \Delta_{l,\beta} = \left|\boldsymbol{\Delta x_l} \cdot \frac{\nabla \rho}{|\nabla \rho|}\right|, \quad \Delta_{l,\kappa} = \left|\boldsymbol{\Delta x_l} \cdot \frac{\nabla e}{|\nabla e|}\right|. \tag{9}$$

where $\boldsymbol{\Delta x_l} = (\frac{x_{i+1}-x_{i-1}}{2}, \frac{y_{i+1}-y_{i-1}}{2}, \frac{z_{i+1}-z_{i-1}}{2})^T$ is the local displacement vector along the ξ_l direction, with the nodes in ξ_l direction indexed by i.

Directional LAD: As noted earlier, scalar LAD adds excessive dissipation on highly stretched grids as the associated length scales to estimate the artificial dissipation coefficients are independent of the computational direction. In contrast, the directional LAD alleviates this problem by independently adding the coefficients in each computational direction (ξ, η, and ζ) to the Navier–Stokes equations. These are given as,

$$\mu^*_{\xi_i} = C_\mu \rho \overline{\left|\frac{\partial^4 \mathbf{S}}{\partial \xi_i^4}\right| \Delta\xi_i^4 \Delta L^2_{\xi_i}}, \quad \kappa^*_{\xi_i} = C_\kappa \rho \overline{\frac{c_s}{T}\left(\sum_{j=1}^{3}\left|\frac{\partial^4 e}{\partial \xi_j^4}\right| \Delta\xi_j^4\right) \Delta L_{\xi_i}}, \tag{10}$$

$$\beta^*_{\xi_i} = C_\beta \rho f_{sw} \left(\sum_{j=1}^{3} \overline{\left|\frac{\partial^4 (S_{kk})}{\partial \xi_j^4}\right|} \Delta\xi_j^4 \Delta Ls_{\xi_j}\right) \Delta L_{\xi_i}, \quad i = 1, 2, 3 \tag{11}$$

where the computational grid length scale, $\Delta\xi = 1$, ΔL_{ξ_i}, and ΔLs_{ξ_i} are the physical length scales in each computational direction defined as:

$$\Delta L_{\xi_i} = \sqrt{\sum_{j=1}^{3}\left(\frac{\partial x_j}{\partial \xi_i}\Delta\xi_i\right)^2}, \quad \Delta Ls_{\xi_i} = \frac{\frac{\partial \rho}{\partial \xi_i}\Delta L_{\xi_i}}{\sqrt{\sum_j \left(\frac{\partial \rho}{\partial \xi_j}\right)^2}}, \quad i = 1, 2, 3 \tag{12}$$

3 Numerical Test Cases

Both the scalar and directional LAD have been incorporated into the in-house high-order solver COMPSQUARE (see [7] for further details of the solver). The implementation has been ported to the GPUs to accelerate the computations. Subsequently, the efficacy of these methods has been verified using several test cases discussed below.

3.1 Stationary Normal Shock Wave

A Mach 2 stationary normal shock wave with a pressure jump of $p_r/p_l = 4.5$ is simulated on a 2D isotropic mesh. Here, l and r subscripts denote the pre and post-shock conditions. A computational domain of $L_x \times L_y = 1.0 \times 1.0$ is discretized with $N_x \times N_y = 51 \times 51$ grid points. The shock wave is located at $0.5L_x$. Simulations are carried out with a compact sixth-order scheme for spatial discretization and an eighth-order low-pass filter [8] with a filter coefficient of $\alpha_f = 0.495$. A fourth-order Runge–Kutta explicit scheme is used for time integration with a non-dimensional time step of $\Delta t = 5.0 \times 10^{-4}$. Figure 1a shows the contours of normalized scalar artificial bulk viscosity (β^*) distribution across a Mach 2.0 stationary normal shock wave. Figure 1b shows the streamwise variation of pressure across the shock. The results demonstrate a clear suppression of the high-frequency Gibbs oscillations across the normal shock with LAD.

3.2 2D Laminar Compression Ramp

A two-dimensional laminar shock-boundary layer interaction on a compression ramp is simulated. The computational setup is inline with the simulations of

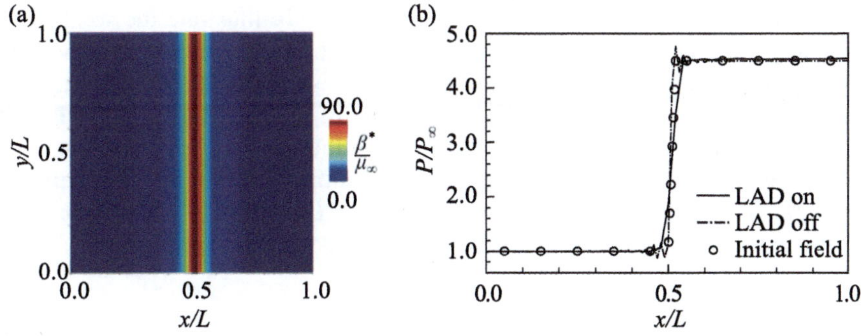

Fig. 1 **a** Normalized β^* distribution across the normal shock, and **b** streamwise pressure variation across the stationary shock along the centerline $L_y = 0.5$ obtained using scalar LAD

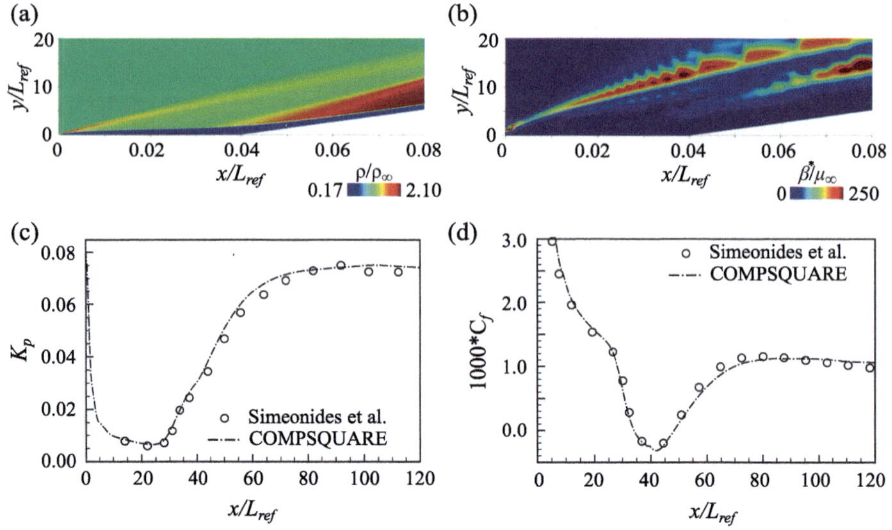

Fig. 2 Contours of **a** Density, **b** scalar β^* across the shocks, streamwise variation of **c** wall-pressure coefficient, and **d** skin-friction coefficient on the compression ramp compared with the reference data [9]

Simeonides et al. [9]. For this test case, a scalar form of LAD is used. The flow conditions considered in this study correspond to a free-stream Mach number of $M_\infty = 6.0$, Reynolds number $\mathrm{Re}_x = 10 \times 10^6$/m, and a ramp angle of $\beta = 7.5°$. The distance from the leading edge to the ramp corner is $0.04L_{\mathrm{ref}}$, wall-normal extent $L_y = 0.04L_{\mathrm{ref}}$, where the reference length scale $L_{\mathrm{ref}} = 1.0$ m. The distance from the compression corner to the outflow plane along the ramp is $0.12L_{\mathrm{ref}}$. The domain is discretized with $N_x \times N_y = 378 \times 81$ grid points in the streamwise and wall-normal direction, respectively.

Figure 2a shows the density contour and the flow features on the compression ramp. At the leading edge of the flat plate, a laminar boundary layer forms and flows over a compression ramp. The findings correctly illustrate the weak shock developing at the leading edge of the plate, the separation shock upstream of the ramp, the interaction between the shock and boundary layer, and the flow reattachment over the ramp. Figure 2b shows the distribution of scalar artificial bulk viscosity (β^*). Higher values of β^* across the shock discontinuities are clearly evident in this figure. Figure 2c, d also shows the streamwise variation of the skin-friction coefficient ($C_f = 2\tau_w/(\rho_\infty u_\infty^2)$) and wall-pressure coefficient ($K_p = (P - P_\infty)/(0.5\rho_\infty u_\infty^2)$) over the plate. The results compare favorably against the reference data [9], further validating the current framework.

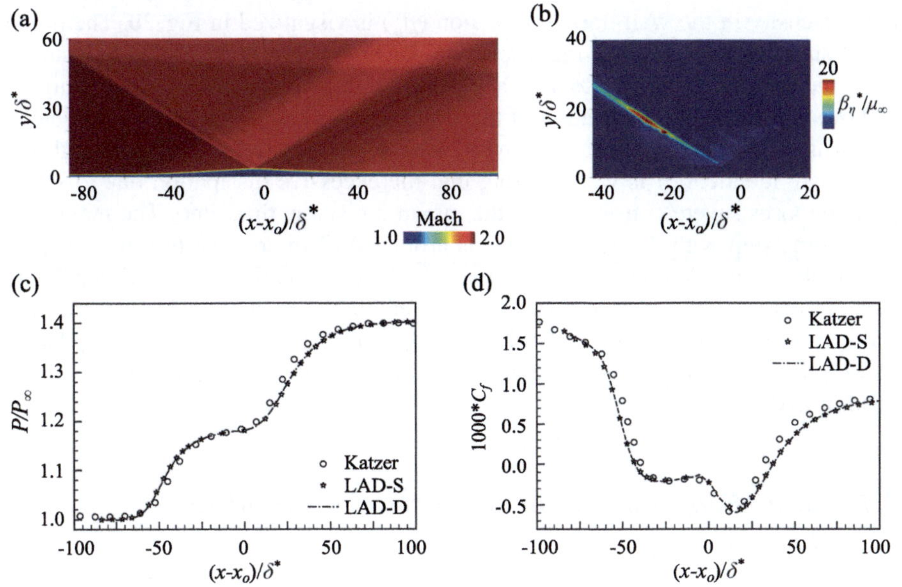

Fig. 3 Contours of **a** Mach number, and **b** wall-normal bulk viscosity (β_η^*) using the directional form of LAD. Streamwise variation of **c** wall-pressure, and **d** skin-friction coefficient obtained using scalar (LAD-S) and directional (LAD-D) artificial bulk viscosity (β^*), compared with reference computation [10]

3.3 2D Incident Oblique Shock-Laminar Boundary Layer Interaction

The interaction of a laminar boundary layer with incident oblique shock on an adiabatic flat plate is studied to examine the efficacy of scalar and directional artificial bulk viscosity methods.

The simulations have been performed using both scalar and directional forms of LAD. The computational setup is based on Katzer [10]. The free-stream Mach number, $M_\infty = 2.0$. The Reynolds number $\mathrm{Re}_{x_o} = 3 \times 10^5$ is based on free-stream properties and reference length scale x_o from the leading edge of the flat plate. The oblique shock impingement position in the inviscid case is $x_o = 169.1\delta^*$, where δ^* is the boundary layer displacement thickness at x_o. The incident shock angle is 32.58° with the shock strength of $p_3/p_1 = 1.40$. The extent of the computational domain is $222\delta^* \times 100\delta^*$ in the streamwise and wall-normal directions. An inflow laminar boundary layer is prescribed at $83.9\delta^*$ upstream of the shock impingement position. Simulations are carried out on a grid comprising of $N_x \times N_y = 300 \times 100$ points, where N_x and N_y represent the number of points in the streamwise and wall-normal directions, respectively.

Figure 3a shows the Mach contour. The shock structures, which include incident shock, reflected shock, and expansion waves, are accurately captured. The directional bulk viscosity in the wall-normal direction (β_η^*) is visualized in Fig. 3b. The initial set of simulations was run using directional LAD, and after converging to a steady state, the flow field was used to initialize the simulation using scalar LAD. Figure 3c, d shows the streamwise variation of the wall-pressure and skin-friction coefficient, which is in good agreement with the reference computations [10]. Although the results are identical, it is worth noting that the excessive dissipation due to scalar LAD imposes severe constraints on the maximum stable time step. The maximum stable time step with directional LAD is around 0.17 in contrast to a time step of 0.01 with scalar LAD. Despite an increase in the computation of direction-specific bulk viscosities with directional LAD, the benefit due to an increase in the maximum stable time-step outweighs the additional computational overhead. A speedup factor of 14.3 times is observed with the directional form of LAD compared to the scalar form of LAD.

3.4 Large Eddy Simulation of Supersonic Turbulent Boundary Layer

The efficacy of scalar and directional LAD is further examined by carrying out the large eddy simulation (LES) of a supersonic turbulent boundary layer. The flow conditions considered in this study correspond to a free-stream Mach number of $M_\infty = 1.3$ and a Reynolds number based on the inlet momentum thickness and free-stream viscosity of $\mathrm{Re}_\theta = 1181$. The computational setup is based on [11]. The computational domain is $L_x = 297\theta_{\mathrm{in}}$ in streamwise direction, $L_y = 79.2\theta_{\mathrm{in}}$ in the wall-normal direction, and $L_z = 63.4\theta_{\mathrm{in}}$ in the spanwise direction, where θ_{in} is the momentum thickness of the inlet boundary layer. Simulations are carried out on two different grids comprising of $N_x \times N_y \times N_z = 768 \times 256 \times 128$ and $512 \times 192 \times 96$ to demonstrate the effect of grid resolution. The coarse and fine meshes comprise 9.4 million and 25.2 million grid points, respectively. To generate inflow turbulence at the inlet, a digital filter approach [12] is incorporated into the in-house solver. A no-slip, adiabatic wall condition is imposed on the bottom wall. Periodicity is imposed in the spanwise direction, while an extrapolation boundary condition is specified at the top and outlet boundaries. A wall damping function $\mathcal{D}_w = 1 - \exp(-y^+/A^+)$ ($A^+ = 26$) is multiplied with the physical length scale Δ_{l^*} to reduce the effect of artificial shear viscosity near the wall in the boundary layer. With a single NVIDIA V100 GPU card, the computational time required to simulate 15 flow-throughs on the finer mesh approximately varied between 21 and 25 h (depending on the variants of the LAD scheme).

Figure 4a shows the instantaneous contours of density on xy plane and streamwise velocity on xz plane at a wall-normal height of $y^+ = 18.5$. Figure 5a shows the van Driest transformed mean velocity with the directional LAD. All the profiles have been extracted at $\mathrm{Re}_{\delta_2} = 1327$, where Re_{δ_2} is the Reynolds number based on the momentum thickness and the wall-viscosity. The fine mesh results compare favorably against the reference data [11]. Figure 5b compares the density-scaled Reynolds stresses with

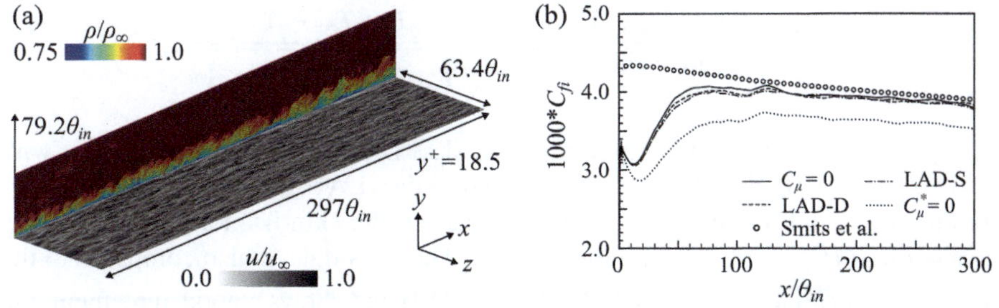

Fig. 4 **a** Instantaneous density contours visualized in xy plane and streamwise instantaneous velocity contours in xz plane, and **b** streamwise variation of van Driest II scaled skin-friction coefficient with LAD turned off ($C_\mu = 0$), directional (LAD-D), and scalar (LAD-S) artificial viscosity. Legends with $C_\mu = 0$, LAD-S and LAD-D have been computed on a fine grid, while $C_\mu^* = 0$ on the coarse mesh

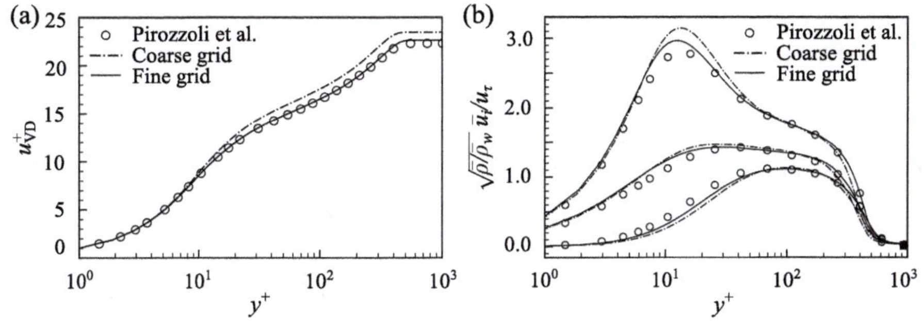

Fig. 5 Profiles of **a** Van Driest transformed mean velocity in wall units, and **b** density-scaled Reynolds stress components in wall units with reference DNS data [11]. The profiles are obtained using the directional form of LAD (μ_i^*), extracted at $\mathrm{Re}_{\delta_2} = 1327$

the DNS (Direct Numerical Simulation) data of Pirozzoli et al. [11] who simulated a similar test case with 65 million grid points. While the results are in good agreement with the reference, the streamwise Reynolds stress is marginally over-predicted in the buffer layer. As expected, the discrepancy is higher on the coarser grid. This is attributed to the under-prediction of skin-friction coefficient and turbulent mixing on the LES grids used for the current simulations when compared to DNS [13].

The LES data are generally compared here with pre-existing, well-established incompressible correlations since there are few direct measurements of these features in the supersonic regime. In this regard, we observe that it is typical to utilize appropriate transformations, with the van Driest II transformation possibly being the best known of them, to compare skin-friction coefficient values at different Mach numbers. The transformed incompressible skin-friction coefficient is $C_{fi} = F_c C_f$, and the incompressible Reynolds number based on momentum thickness is $\mathrm{Re}_{\theta i} = (\mu_\infty/\overline{\mu}_w)^* Re_\theta$. With an adiabatic wall, the term F_c can be computed as,

$$F_c = \frac{\overline{T}_w/T_\infty - 1}{\arcsin^2\alpha}, \quad \alpha = \frac{\overline{T}_w/T_\infty - 1}{\sqrt{\overline{T}_w/T_\infty(\overline{T}_w/T_\infty - 1)}} \tag{13}$$

Fig. 4b shows the profiles of van Driest II scaled skin-friction coefficient with LAD off ($C_\mu = 0$), directional (LAD-D) and scalar (LAD-S) artificial shear viscosity on coarse and fine mesh. As discussed earlier, the skin friction predicted using coarse mesh ($C_\mu^* = 0$) is consistently lower. The LES data is also compared to the theoretical correlations ($C_{fi} = 0.024 Re_{\theta i}^{-1/4}$) [14] and shows a good agreement. On the fine mesh, simulation with LAD turned off performs better than with LAD turned on. Further, the prediction of the incompressible skin-friction coefficient using directional LAD is shown to outperform the scalar LAD, particularly on coarser grids. However, on a finer mesh, the improvements due to directional LAD are marginal.

4 Conclusions

We have demonstrated the performance of LAD when coupled with a high-order in-house solver, which uses central difference schemes (explicit and compact) for spatial discretization. The efficacy of this method is tested on a variety of test cases which include a 1D stationary normal shock, 2D compression ramp, 2D incident oblique shock-boundary layer interaction, and a 3D compressible turbulent boundary layer. We further studied the efficacy of the scalar and directional forms of the artificial diffusivity method. On highly stretched grids with higher aspect ratios, the scalar formulation of LAD adds excessive artificial dissipation to the governing equations resulting in smaller time steps due to increased numerical stiffness. In contrast, the directional artificial diffusivity approach outperforms the scalar LAD on such grids allowing for larger time steps. For the incident oblique shock test case in particular, the computation using directional LAD is 14.3 times faster than the scalar LAD.

Acknowledgements The authors wish to acknowledge NVIDIA for generously donating A100 GPU card used for these simulations under the Academic Hardware Grant Program. We thank P.G. Senapathy Centre, IIT Madras, and the National PARAM Supercomputing Facility (NPSF) for providing the computing resources on the PARAM Siddhi cluster under the National Supercomputing Mission.

References

1. Johnsen E et al (2010) Assessment of high-resolution methods for numerical simulations of compressible turbulence with shock waves. J Comput Phys 229(4):1213–1237
2. Cook AW ((2007) Artificial fluid properties for large-eddy simulation of compressible turbulent mixing. Phys Fluids 19(5):055103
3. Kawai Soshi et al (2008) Localized artificial diffusivity scheme for discontinuity capturing on curvilinear meshes. J Comput Phys 227(22):9498–9526

4. Olson BJ, Lele SK (2013) Directional artificial fluid properties for compressible large-eddy simulation. J Comput Phys 246:207–220
5. Kawai S, Shankar SK, Lele SK (2010) Assessment of localized artificial diffusivity scheme for large-eddy simulation of compressible turbulent flows. J Comput Phys 229(5):1739–1762
6. Ducros F et al (1999) Large-eddy simulation of the shock/turbulence interaction. J Comput Phys 152(2):517–549
7. Vadlamani et al (2018) Distributed roughness effects on transitional and turbulent boundary layers. Flow Turbul Combustion 100:627–649
8. Gaitonde DV et al (1998) High-order schemes for Navier-Stokes equations: algorithm and implementation into FDL3DI. Air Vehicles Directorte, Air Force Research Laboratory, Air Force Materiel
9. Simeonides G et al (1994) Experimental, analytical, and computational methods applied to hypersonic compression ramp flows. AIAA J 32(2):301–310
10. Katzer E (1989) On the length scales of laminar shock/boundary-layer interaction. J Fluid Mech 206:477–496
11. Pirozzoli S et al (2010) Direct numerical simulation of transonic shock/boundary layer interaction under conditions of incipient separation. J Fluid Mech 657:361–393
12. Touber E et al (2009) Large-eddy simulation of low-frequency unsteadiness in a turbulent shock-induced separation bubble. Theor Comput Fluid Dyn 23:79–107
13. Poggie J et al (2015) Resolution effects in compressible, turbulent boundary layer simulations. Comput Fluids 120:57–69
14. Smits AJ et al (1983) Low-Reynolds-number turbulent boundary layers in zero and favorable pressure gradients. J Ship Res 27(03):147–157

Wave Diffraction in a Two-Layer Fluid by the Submerged Horizontal Circular Cylindrical Pipe in Front of a Cliff as a Vertical Wall

Minakshi Ghosh and Dilip Das

Abstract The submerged cylindrical structure in front of a cliff gets affected by the waves reflected from the cliff which behaves as a vertical wall of infinite length. This makes the hydrodynamic results different from those for the cylinder in a horizontally unbounded fluid field. We investigated the wave exciting forces on the circular cylinder when it is submerged below the interface in a density-stratified two-layer fluid, in front of the vertical wall. Employing the image principle, the present cylinder-wall problem is transformed into an equivalent imaginary physical problem of wave diffraction in an unbounded fluid domain by two identical submerged horizontal cylinders placed symmetrically concerning the vertical wall in the original problem. We solved this using the multipole expansion method and obtained infinite systems of linear algebraic equations, which are solved numerically for the unknowns by truncation. Then, the wave exciting forces on the cylinder (sway and heave forces) are solved analytically and calculated numerically. They are depicted graphically against wave numbers in several figures and have been analyzed to provide valuable results for engineering purposes.

Keywords Wave diffraction · Submerged cylinder · Vertical wall · Two-layer fluid

1 Introduction

Many submerged machines or offshore constructions such as pipe bridges or suspended tunnels have cylindrical structures. The submerged cylindrical structure in front of a cliff gets affected by the waves reflected from the cliff as a vertical wall of infinite length. This makes the wave behavior and the corresponding hydrodynamic coefficients different from those for the cylinder submerged in a horizontally unrestrained fluid field. This research aims to study the effect on the hydrodynamic

M. Ghosh · D. Das (✉)
Department of Mathematics, Diamond Harbour Women's University, Sarisha, West Bengal 743368, India
e-mail: dilipdas@dhwu.ac.in

R. K. Ray et al. (eds.), *Advances in Theoretical and Applied Mechanics*, Lecture Notes in Mechanical Engineering, https://doi.org/10.1007/978-981-97-0418-7_10

quantities of the cylinder by the vertical wall which is placed along the cylinder's axis at a particular distance in front of it.

Various studies based on problems of scattering waves by the obstruction of a cylinder in an unbounded infinitely deep constant density fluid domain were studied (cf. Dean [1], Ursell [2], and others). The multipole expansion method gives swift convergence. The suggestion of building a pipe bridge to pass a Norwegian fjord by Friis et al. [3] led to such studies in a two-layer fluid with salty water below the fresh river water (cf. Stokes [4], Lamb [5], Linton and McIver [6, 7], Feng et al. [8], and others). Wave interaction problems with various submerged structures have also been studied in an ice-covered two-layer fluid (cf. Das and Mandal [9, 10], Sahu and Das [11], and others). Each such work considers an absence of a barrier in front of the submerged structures (horizontally unbounded fluid domain).

Using the image principle effectively solves wave problems with a submerged body in front of a cliff as an infinitely deep vertical wall. It modifies the model of this cylinder-wall problem into an imaginary physical problem with an unbounded fluid domain where waves are diffracted by two similar cylinders submerged horizontally and placed symmetrically about the vertical wall in the original problem. Recently, many researchers studied various water wave interaction problems with different bodies placed close to a vertical wall (cf. Zheng and Zhang [12], Cong et al. [13, 14], Li and Liu [15], Li et al. [16], and others). All these studies have observed a notable effect of the vertical wall on the hydrodynamic calculations. However, this genre's literature lacks the study of fluids with density inhomogeneity, a common feature of water resources.

Thus, we extended Li and Liu [15] to investigate the wave exciting forces on the circular cylinder (pipe bridge) when it is submerged below the interface in a two-layer fluid, in front of the vertical wall (cliff) of infinite length. Making use of the image principle and multipole expansion method, the infinite systems of equations have been obtained and solved the wave exciting forces on the cylinder (sway and heave forces) analytically. Calculating several different numerical examples, the results have been represented graphically against wave numbers in numerous figures to analyze their effects. Furthermore, this study can be extended to investigate such a bounded fluid model with an interface-piercing cylinder (cf. Sturova and Motygin [17]). Additionally, ongoing research on wave interaction problems with submerged porous structures (cf. Sharma et al. [18], Tabssum et al. [19], and others) could be explored within the framework of the present fluid model. Recent interest in wave interaction with submerged bodies in fluids with three layers of different densities (cf. Chakrabarti et al. [20], Das [21], Das and Sahu [22], Majumder and Das [23], Ghosh et al. [24], Ghosh and Das [25]) suggests further investigation, particularly considering cliff walls acting as vertical barriers in front of submerged structures.

2 Mathematical Formulation

The present wave diffraction problem considers a submerged floating horizontal circular cylinder below the interface of a density-stratified two-layer fluid in front of a vertical wall. Figure 1a shows the ideal outline of the cylinder-wall interaction problem. The two fluids and the coordinate system of the mathematical model are as in Linton and McIver [6] with the upper layer of depth h. The same variables as in Li and Liu [15] represent the dimension of the cylinder and its position concerning the vertical wall. The cylinder is submerged below the interface (in the lower layer), and thus, the transformed model with the application of the image principle is presented in Fig. 1b. It shows two symmetrical horizontal circular cylinders in the lower layer of the fluid model in a horizontally open fluid domain. Their centers and local coordinate system are as defined by Li and Liu [15]. Thus, the fluid motion in the two regions I and II is described as (cf. Linton and McIver [6], Li and Liu [15])

$$\Phi^q(x, z, t) = \mathrm{Re}\left\{-\frac{\mathrm{i}gH}{2\omega}[\phi^q_{\mathrm{inc}}(x, z) + \phi^q_{\mathrm{D}}(x, z)]\mathrm{e}^{-\mathrm{i}\omega t}\right\}, \quad q = \mathrm{I}, \mathrm{II}, \tag{1}$$

where ϕ_{inc} and ϕ_{D} are the velocity potentials of the incident and diffracted waves due to fixed cylinders, respectively. The velocity potentials satisfy Laplace's equation and linearized boundary conditions (cf. Linton and McIver [6])

$$\begin{aligned} &\nabla^2\phi^q = 0, \quad q = \mathrm{I}, \mathrm{II}, \\ &\phi^{\mathrm{I}}_z = \phi^{\mathrm{II}}_z, \quad \rho\left(\phi^{\mathrm{I}}_z - K\phi^{\mathrm{I}}\right) = \phi^{\mathrm{II}}_z - K\phi^{\mathrm{II}} \text{ on } z = 0, \\ &\phi^{\mathrm{I}}_z = K\phi^{\mathrm{I}} \text{ on } z = h \text{ and } \nabla\phi^{\mathrm{II}} \to 0 \text{ as } z \to -\infty. \end{aligned} \tag{2}$$

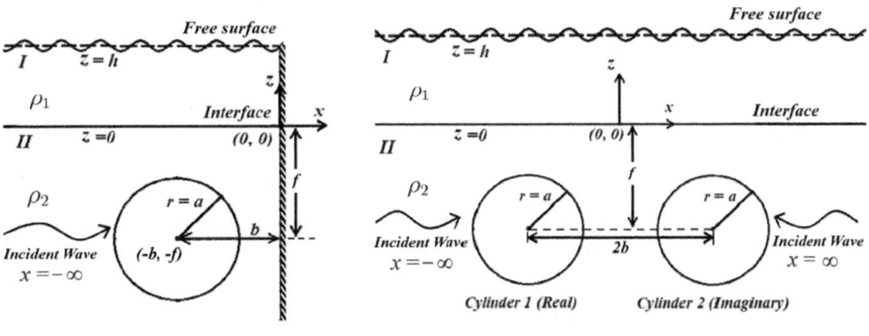

(a) A horizontal circular cylinder in the lower layer in front of a vertical wall.

(b) Two symmetrical horizontal circular cylinders in the lower layer in a horizontally unbounded fluid domain.

Fig. 1 Idealized schematic diagram of the wave-structure interaction problem

In the imaginary system (Fig. 1b), the symmetrical real and imaginary cylinders below the interface are subjected to two incident waves having the properties described in Li and Liu [15]. Thus, the incident waves of distinct wave numbers K and k for fluid of two layers (cf. Linton and McIver [6]) are respectively given as follows:

$$\phi_{\text{inc}}^{\text{II}(K,k)} = \mathrm{e}^{(\mathrm{i}x+z)(K,k)} + \mathrm{e}^{(-\mathrm{i}x+z)(K,k)} \tag{3}$$

The structural boundary conditions are (cf. Li and Liu [15], (8(a)))

$$\frac{\partial \phi_{\text{D}}^{q(K,k)}}{\partial r_l} = -\frac{\partial \phi_{\text{inc}}^{q(K,k)}}{\partial r_l}, \quad r_l = a, q = \text{I}, \text{II}. \tag{4}$$

3 Method of Solution

3.1 Multipole Expansions

The symmetric (+) and anti-symmetric (−) multipoles which are singular at the center of cylinder l for the imaginary model can be written as (cf. Linton and McIver [6])

$$\phi_{n,l}^{\text{I}(+,-)} = \frac{(-1)^{(n,n+1)}}{(n-1)!} \int_0^\infty u^{n-1}[A(u)\mathrm{e}^{uz} + B(u)\mathrm{e}^{-uz}](\cos, \sin)[u(x-(-1)^l b)]\mathrm{d}u \tag{5}$$

$$\phi_{n,l}^{\text{II}(+,-)} = \frac{(\cos, \sin)(n\theta_l)}{r_l^n} + \frac{(-1)^{(n,n+1)}}{(n-1)!} \int_0^\infty u^{n-1}C(u)\mathrm{e}^{uz}(\cos, \sin)[u(x-(-1)^l b)]\mathrm{d}u \tag{6}$$

where

$$A(u) = K(1+\sigma)(u+K)\mathrm{e}^{u(f-2h)}/(u-K)h(u), \quad B(u) = K(1+\sigma)\mathrm{e}^{uf}/h(u)$$
$$C(u) = (u+K)\mathrm{e}^{uf}[(u+K\sigma)\mathrm{e}^{-2uh} - u + K]/(u-K)h(u). \tag{7}$$

The dispersion equation is (cf. Linton and McIver [6])

$$H(u) \equiv (u-K)h(u) = 0 \tag{8}$$

where

$$h(u) = (u+K)\mathrm{e}^{-2uh} - u + K\sigma. \tag{9}$$

As x approaches $\pm\infty$, the far-field situations of the multipoles $\phi_{n,l}^{II(+,-)}$ with the cylinder located in region II are given by

$$\phi_{n,l}^{\mathrm{II}+} \sim \pi \mathrm{i}\frac{(-1)^n}{(n-1)!}[K^{n-1}C_K \mathrm{e}^{\pm \mathrm{i}K(x-(-1)^l b)+Kz} + k^{n-1}C_k \mathrm{e}^{\pm \mathrm{i}k(x-(-1)^l b)+kz}]$$

$$\phi_{n,l}^{\mathrm{II}-} \sim \mp\pi \frac{(-1)^n}{(n-1)!}[K^{n-1}C_K \mathrm{e}^{\pm \mathrm{i}K(x-(-1)^l b)+Kz} + k^{n-1}C_k \mathrm{e}^{\pm \mathrm{i}k(x-(-1)^l b)+kz}]$$

where $C_{(K,k)}$ represent the residues of $C(u)$ at $u = K$ and k, respectively. They have the following values:

$$C_K = \frac{2K\mathrm{e}^{Kf}(1+\sigma)\mathrm{e}^{-2Kh}}{2\mathrm{e}^{-2Kh} - 1 + \sigma}$$

$$C_k = \frac{(k+K)\mathrm{e}^{kf}[(k+K\sigma)\mathrm{e}^{-2kh} - k + K]}{(k-K)[\mathrm{e}^{-2kh}\{1 - 2h(k+K)\} - 1]}$$

Then we use the following Taylor expansions in Eq. (6):

$$\mathrm{e}^{uz}(\cos,\sin)[u(x-(-1)^l b)] = \mathrm{e}^{uf}\sum_{m=0}^{\infty}\frac{(-1)^{(m,m+1)}(ur_l)^m}{m!}(\cos,\sin)(m\theta_l) \quad (10)$$

Thus, Eq. (6) can be written as

$$\phi_{n,l}^{\mathrm{II}(+,-)} = \frac{(\cos,\sin)(n\theta_l)}{r_l^n} + \sum_{m=0}^{\infty} N_{nm}(r_l)^m(\cos,\sin)(m\theta_l) \quad (11)$$

$$\text{where} \quad N_{nm} = \frac{(-1)^{n+m}}{(n-1)!m!}\,⨍_0^{\infty} u^{n+m-1}C(u)\mathrm{e}^{uf}\,\mathrm{d}u \quad (12)$$

We have the expansions (cf. Li and Liu [15], (20))

$$\frac{\mathrm{e}^{\pm \mathrm{i}n\theta_j}}{r_j^n} = \sum_{m=0}^{\infty}\frac{(-1)^m(n+m-1)!}{(n-1)!m!(r_{jl})^{n+m}}(r_l)^m\mathrm{e}^{\mp \mathrm{i}(m\theta_l-(n+m)\theta_{jl})} \quad (13)$$

Here, $(r,\theta)_{(12,21)}$ have the values as in Li and Liu [15] for the study of Fig. 1b. Thus, for $j \neq l$, Eq. (6) can be expressed as

$$\phi_{n,j}^{\mathrm{II}(+,-)} = \sum_{m=0}^{\infty} D_{nm,jl}^{(+,-)}(r_l)^m\cos(m\theta_l) + \sum_{m=1}^{\infty} E_{nm,jl}^{(+,-)}(r_l)^m\sin(m\theta_l), \quad (14)$$

$$\text{where} \quad D_{nm,jl}^{(+,-)} = L_{nm,jl}^{(+,-)} + M_{nm,jl}^{(+,-)},\quad E_{nm,jl}^{(+,-)} = U_{nm,jl}^{(+,-)} + V_{nm,jl}^{(+,-)}, \quad (15)$$

$$L_{nm,jl}^{(+,-)} = (-,+)U_{nm,jl}^{(+,-)} = \frac{(-1)^m (n+m-1)!}{(n-1)!m!(2(-1)^{j+1}(b))^{n+m}} (\cos,\sin)\left[(n+m)\frac{\pi}{2}\right], \tag{16}$$

$$M_{nm,jl}^{(+,-)} = (+,-)V_{nm,jl}^{(-,+)} = (+,-)\frac{(-1)^{n+m}}{(n-1)!m!} \fint_0^\infty u^{n+m-1} C(u)(\cos,\sin)[2u(-1)^{j+1}(b)]\,du. \tag{17}$$

3.2 *Solution of Wave Diffraction*

$\phi_{\mathrm{D}}^{\mathrm{II}}(x,z)$ is given by (cf. Linton and McIver [6])

$$\phi_{\mathrm{D}}^{\mathrm{II}(K,k)} = \sum_{j=1}^{2}\sum_{n=1}^{\infty}(\chi_{n,j}^{(K,k)+}\phi_{n,j}^{\mathrm{II}+} + \chi_{n,j}^{(K,k)-}\phi_{n,j}^{\mathrm{II}-}) \tag{18}$$

with unknowns $\chi_{n,j}^{(K,k)\pm}$.

3.2.1 Incident Wave of Wave Number *K*

Using Taylor expansions of the exponential functions, the velocity potential of the incident wave of wave number K (Eq. (3)) is expressed as

$$\phi_{\mathrm{inc}}^{\mathrm{II}K} = 2\mathrm{e}^{Kf}\sum_{n=0}^{1}\cos\left[\frac{n\pi}{2} - K(-1)^l b\right]\sum_{m=0}^{\infty}\frac{(-1)^m K^m}{m!}(r_l)^m\cos\left[\frac{n\pi}{2} - m\theta_l\right] \tag{19}$$

Proceeding similarly as Li and Liu [15] with Eqs. (18), (19) and (4), we have

$$\begin{aligned}
&-\chi_{n,j}^{K(+,-)}(a)^{-2m} + \sum_{n=1}^{\infty}\chi_{n,j}^{K(+,-)}N_{nm} \\
&\quad + \sum_{j=1,\ j\neq l}^{2}\sum_{n=1}^{\infty}\left(\chi_{n,j}^{K+}(D,E)_{nm,jl}^{+} + \chi_{n,j}^{K-}(D,E)_{nm,jl}^{-}\right) \\
&= (-,+)2\mathrm{e}^{Kf}(\cos,\sin)[K(-1)^l b]\frac{(-1)^m K^m}{m!}, \quad m = 1,2,\ldots
\end{aligned} \tag{20}$$

Using the truncation method, the infinite system of linear equations in Eq. (20) is solved simultaneously for the unknown coefficients. Following the method of Li and Liu [15], we compute the sway (F_x^K) and heave (F_z^K) forces on the cylinder. Thus, using Eq. (20), we obtain the following:

$$F_x^K = \frac{\rho g H}{2} \int_0^{2\pi} \phi^{\mathrm{II}K}(a, \theta_1)(-a \sin\theta_1)\, \mathrm{d}\theta_1 = -\pi \rho g \chi_{1,1}^{K-} H \tag{21}$$

$$F_z^K = \frac{\rho g H}{2} \int_0^{2\pi} \phi^{\mathrm{II}K}(a, \theta_1)(a \cos\theta_1)\, \mathrm{d}\theta_1 = \pi \rho g \chi_{1,1}^{K+} H \tag{22}$$

where

$$\phi^{qK}(a, \theta_1) = \phi_{\mathrm{inc}}^{qK}(a, \theta_1) + \phi_{\mathrm{D}}^{qK}(a, \theta_1), \quad q = \mathrm{I}, \mathrm{II} \tag{23}$$

The non-dimensionalized form of F_x^K and F_z^K is (cf. Li and Liu [15])

$$M_{F_{(x,z)}}^K = \frac{|F_{(x,z)}^K|}{\pi \rho g H a} \tag{24}$$

3.2.2 Incident Wave of Wave Number *k*

The velocity potential of the incident wave of wave number k (Eq. (3)) can be similarly expressed as Eq. (19) with k instead of K. Then taking the same steps and utilizing Eq. (4), we obtain the respective system, Eq. (20) with the new set of unknowns $\chi_{n,j}^{k\pm}$. The right-hand expressions of the equations have k instead of K. $\chi_{n,j}^{k\pm}$ are calculated by solving this system of equations as in Sect. 3.2.1. The sway and heave forces ($F_{(x,z)}^k$) have the same expressions as in Eqs. (21) and (22) with K replaced by k. These are computed using the values of the new unknowns $\chi_{1,1}^{k\pm}$. Thus, the non-dimensionalized sway and heave forces are given by

$$M_{F_{(x,z)}}^k = \frac{|F_{(x,z)}^k|}{\pi \rho g H a} \tag{25}$$

4 Numerical Results

Multiple numerical examples were considered, with a density ratio ρ of 0.5. By employing the truncation method in Eq. (20) to truncate n and m and setting the truncated number as t', we determined the unknowns to compute the non-dimensionalized sway and heave forces. Their convergence analysis is presented in Table 1 specifying the runtimes (utilizing MATLAB R2022b software on an 11th Gen Intel(R) Core(TM) i5-1135G7 CPU with 8.00 GB of RAM, operating on Windows 11 Home Single Language version 21H2). Thus, for the numerical examples, the suitable truncation of the infinite system of equations was found at $m = n = t' = 6$. The use of

Table 1 Convergence of sway ($M^K_{F_x}$) and heave ($M^K_{F_z}$) forces for $f/a = -1.5$, $b/a = 10$, $h/a = 5$, and $\rho = 0.5$ for incident wave of wave number K

Truncated number	Run-time (s)	Sway force ($M^K_{F_x}$)				Heave force ($M^K_{F_z}$)			
		$Ka = 0.2$	$Ka = 0.4$	$Ka = 0.6$	$Ka = 0.8$	$Ka = 0.2$	$Ka = 0.4$	$Ka = 0.6$	$Ka = 0.8$
$t' = 1$	0.66	0.22274	0.43412	0.05998	0.45745	0.08295	0.31543	0.42467	0.06665
$t' = 2$	1.24	0.21496	0.48487	0.07006	0.44392	0.08307	0.33661	0.41892	0.06559
$t' = 3$	2.17	0.21378	0.49670	0.07815	0.43963	0.08333	0.34489	0.41978	0.06518
$t' = 4$	3.54	0.21366	0.49820	0.08051	0.43878	0.08338	0.34654	0.42043	0.06499
$t' = 5$	5.09	0.21365	0.49832	0.08099	0.43865	0.08339	0.34676	0.42062	0.06494
$t' = 6$	7.20	0.21365	0.49833	0.08106	0.43862	0.08339	0.34678	0.42066	0.06494
$t' = 7$	9.25	0.21365	0.49833	0.08106	0.43862	0.08339	0.34678	0.42066	0.06494

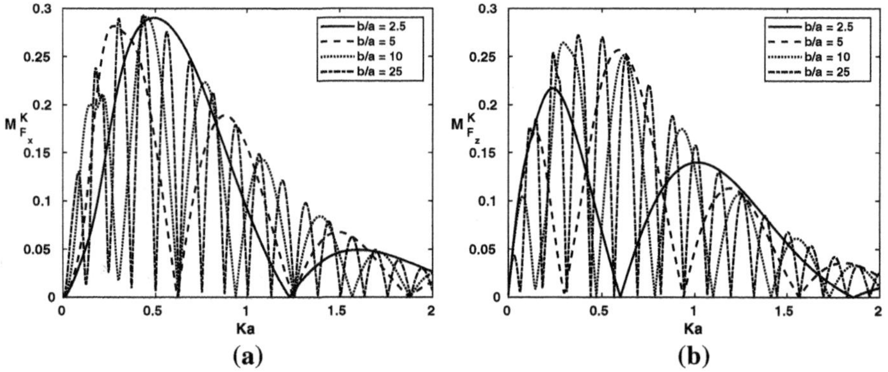

Fig. 2 Non-dimensional sway and heave forces on the cylinder for the incident wave of wave number K at $h/a = 5$, $f/a = -2.5$, and different spacing b/a

the multipole expansion method leads to quick convergence. The effects of various parameters on the aforementioned forces for both the free surface and interface incident wave modes are analyzed in Figs. 2, 3, 4, and 5. Additionally, a comparative study is presented to validate the method and the obtained results, ensuring agreement with previously published findings.

4.1 *Effect of Spacing Between Cylinder and Wall*

The variations of non-dimensionalized wave forces on the cylinder, $M^K_{F_{(x,z)}}$ and $M^k_{F_{(x,z)}}$ for incident waves of wave numbers K and k, respectively, with non-dimensionalized wave number Ka at $h/a = 5$ and $f/a = -2.5$ are plotted in Figs. 2 and 3, respectively, with distinguishable curves for the four different spacing b/a ($= 2.5,\ 5,\ 10,\ 25$). The wave exciting forces oscillate with increasing wave

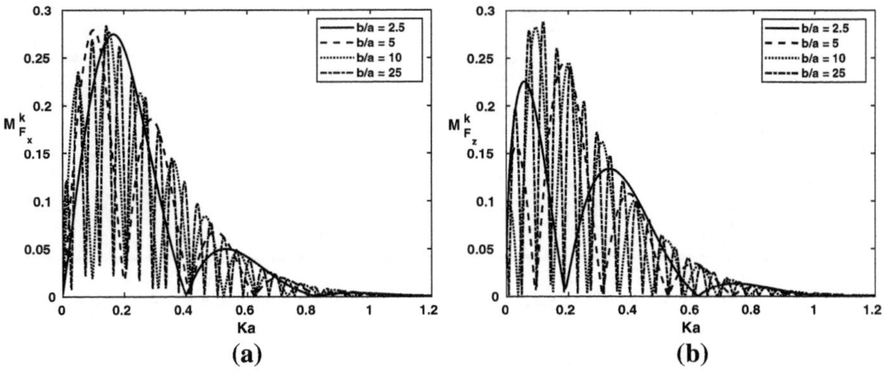

Fig. 3 Non-dimensional sway and heave forces on the cylinder for the incident wave of wave number k at $h/a = 5$, $f/a = -2.5$, and different spacing b/a

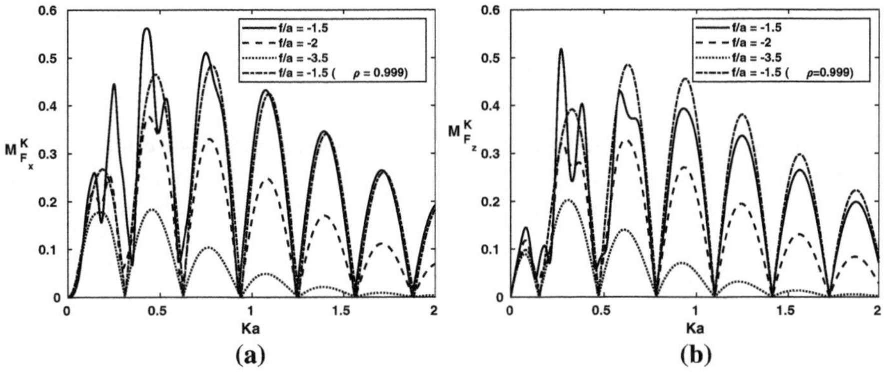

Fig. 4 Non-dimensional sway and heave forces on the cylinder for the incident wave of wave number K at $h/a = 5$, $b/a = 10$, and different submergence depths f/a

number Ka as an effect of both the incident wave and the reflected wave from the vertical wall. They even become null at certain wave numbers in all the figures.

Notably, when the spacing between the vertical wall and the cylinder is large ($b/a = 25$), even slight changes in the wave number of the incident wave alter the phase difference rapidly between the two waves. Additionally, the oscillations become more frequent as the spacing b/a increases. Comparing the forces for the incident waves with wave numbers K (Fig. 2) and k (Fig. 3), it is evident that the forces for wave number k converge to zero at a smaller wave number (higher frequency) compared to those for wave number K. Furthermore, it is observed that the peaks of the heave forces (Figs. 2b and 3b) are lower than the peaks of the sway forces (Figs. 2a and 3a) for incident waves of both wave numbers.

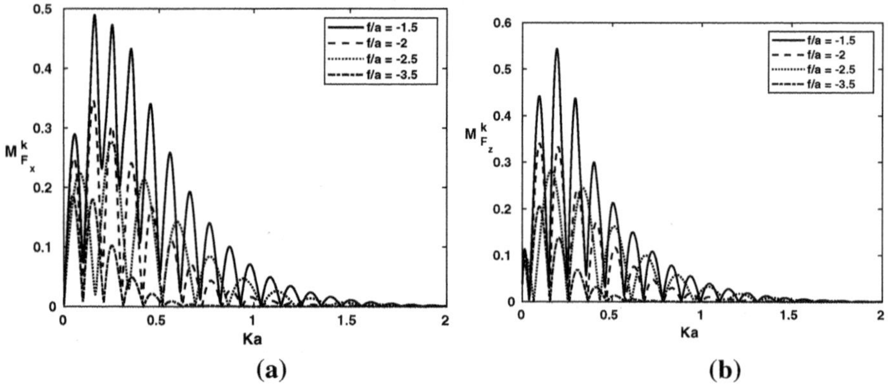

Fig. 5 Non-dimensional sway and heave forces on the cylinder for the incident wave of wave number k at $h/a = 5$, $b/a = 10$ and different submergence depths f/a.

4.2 Effect of Submergence Depth

Figures 4 and 5 illustrate the variations of non-dimensionalized wave forces on the cylinder, denoted as $M^K_{F_{(x,z)}}$ and $M^k_{F_{(x,z)}}$ for incident waves of wave numbers K and k, respectively, with non-dimensionalized wave number Ka at $h/a = 5$ and $b/a = 10$ for different submergence depth of the cylinder, f/a $(= -1.5, \; -2, \; -3.5)$ graphed with separate curves. Notably, $f/a = -3.5$ represents a larger depth of submergence for the cylinder compared to the other cases.

The peaks of both sway and heave forces tend to decrease with the increase in the submergence depth of the cylinder below the interface. Observing all the figures, we notice that the forces exhibit an oscillating behavior that differs from the results obtained in an unbounded fluid domain. When the cylinder is submerged near the interface ($f/a = -1.5$) and subjected to very small wave numbers, the sway and heave forces corresponding to the incident wave of wave number K oscillate but do not vanish (Fig. 4). These multiple oscillations arise due to the presence of the interface near the cylinder in this fluid model. Apart from that, for each distinct submergence depth in the current problem, the corresponding forces tend to vanish at the same wave numbers, particularly for the incident wave of wave number K. Moreover, Fig. 5 demonstrates that the forces corresponding to the incident wave of wave number k converge to zero.

By limiting the density ratio ρ to approximately 1 ($\rho = 0.999$) to represent the single-layer fluid model, the dimensionless sway and heave forces for the incident wave along the free surface and the submergence depth of the cylinder $f/a = -1.5$ are obtained. These forces are depicted as the fourth curves in Figs. 4a, b. Due to the relatively large uniform fluid depth $h = 5$ (as indicated in Li and Liu [15]) for small amplitude waves, it resembles an infinite-depth problem to some extent. Consequently, the obtained curves and the curves for the equivalent submergence depth

$f/h = 0.3$ in Fig. 11a, b of Li and Liu [15] exhibit a similar nature, albeit slightly decreased. The resemblance between the obtained curves in Fig. 4a, b and the published results provides validation for the newly calculated findings.

5 Conclusion

This study presented the analysis of non-dimensionalized sway (horizontal) and heave (vertical) forces on a cylinder (pipe bridge) placed below the interface and positioned in front of an infinite-length vertical wall, representing a cliff, in a density-stratified two-layer fluid. By applying the image principle and employing linear potential theory, we transformed the wave diffraction problem into an equivalent scenario of two symmetrical cylinders submerged below the interface in an open fluid domain. For both wave numbers' incident waves, the multipole expansions method ultimately gave the systems of linear equations with the unknowns. The unknowns were computed numerically by the suitable truncation of these equations whose convergence was analyzed in Table 1. The non-dimensional sway and heave forces were calculated for various numerical examples.

The method and solutions were validated by considering the density ratio ρ to represent a uniform-density fluid. The forces were meticulously examined and analyzed by varying the separation between the vertical wall and the submerged cylinder, as well as the submergence depth of the cylinder. The key findings of this study, which hold significant implications for engineering enhancements, can be summarized as follows:

- The wave exciting forces on the cylinder in the two-layer fluid bounded by a vertical wall exhibit distinct behavior compared to those in an unbounded two-layer fluid, which has been extensively studied in existing literature. These forces oscillate periodically in the presence of the vertical wall, vanishing at specific wave numbers. The periodicity remains consistent for different submergence depths of the cylinder, given a particular spacing between the cylinder and the vertical wall.
- For the submergence depth of $f/a = -1.5$, with $h/a = 5$ and $b/a = 10$, the forces experience multiple oscillations. This phenomenon arises due to the superposition of waves at the interface and on the surface of the cylinder, particularly when it is in close proximity to the interface. This results in the formation of standing waves, causing the forces to decrease between the periodic oscillations.
- When the present two-layer fluid model is limited to a uniform-density fluid with $\rho = 0.999$, the force curves for $h/a = 5$, $f/a = -1.5$ and $b/a = 10$ exhibit a nature similar to the results obtained for a single-layer fluid under equivalent parameters. However, it should be noted that the obtained results for the heave forces in this study show a slight decrease compared to the published results for a uniform-density finite-depth fluid.
- In general, the heave forces tend to be lower than the sway forces in all situations of the current problem. This characteristic can be attributed to the small amplitude

waves assumed in the linear potential theory. Consequently, the vertical displacement caused by any disturbance tends to be slightly smaller than other changes.

- The frequency of oscillation for both forces increases with the widening gap between the cylinder and the vertical wall. Moreover, the extremes of the forces are more pronounced when the cylinder is positioned closer to the interface. For the incident wave along the interface, both the sway and heave forces tend to completely disappear at large wave numbers (Ka).

Until now, the interaction of a submerged body with a vertical wall in a density-stratified fluid has remained unexplored. Further investigations can focus on considering the cylinder in the upper layer of the bounded two-layer fluid model. Additionally, studying the effect of a vertical wall on a half-immersed or interface-piercing cylinder would yield interesting insights. The scope of this research can be extended to address a three-layer fluid problem, which is relevant to offshore structures in the ocean.

Acknowledgements The authors express their gratitude to the editor and the reviewer for their valuable input and diligent review of the manuscript, which greatly contributed to its improvement in its current form.

References

1. Dean WR (1948) On the reflexion of surface waves by a submerged circular cylinder. Math Proc Camb Philos 44(4):483–491
2. Ursell F (1950) Surface waves on deep water in the presence of a submerged circular cylinder-I. Proc Camb Philos Soc 46:141–152
3. Friis A, Grue J, Palm E (1991) Application of Fourier transform to the second order 2D wave diffraction problem. In: Tulin's Festschrqt MP, Miloh T (eds) Mathematical approaches in hydrodynamics, pp 209–227
4. Stokes GG (1847) On the theory of oscillatory waves. Trans Camb Philos Soc 8:441–455
5. Lamb H (1932) Hydrodynamics, art 231. Cambridge University Press, Cambridge
6. Linton CM, McIver M (1995) The interaction of waves with horizontal cylinders in two-layer fluids. J Fluid Mech 304:213–229
7. Linton CM, McIver P (2001) Handbook of mathematical techniques for wave/structure interactions. Chapman and Hall/CRC, New York
8. Feng B, Zheng YH, You YG, Wu BJ, Sheng SW (2008) Interaction of linear waves with an array of infinitely long horizontal circular cylinders in a two-layer fluid of infinite depth. Ocean Eng 35(10):963–982
9. Das D, Mandal BN (2007) Wave scattering by a horizontal circular cylinder in a two-layer fluid with an ice-cover. Int J Eng Sci 45(10):842–872
10. Das D, Mandal BN (2010) Wave radiation by a sphere submerged in a two-layer ocean with an ice-cover. Appl Ocean Res 32(3):358–366
11. Sahu M, Das D (2021) Hydrodynamic forces on a submerged circular cylinder in two-layer fluid with an ice-cover. J Univ Shanghai Sci Technol 23:282–294
12. Zheng S, Zhang Y (2015) Wave diffraction from a truncated cylinder in front of a vertical wall. Ocean Eng 104:329–343
13. Cong P, Bai W, Teng B (2019) Analytical modeling of water wave interaction with a bottom-mounted surface-piercing porous cylinder in front of a vertical wall. J Fluids Struct 88:292–314

14. Cong P, Chen L, Gou Y (2020) Hydrodynamic interaction among multiple columns in front of a vertical wall. Ocean Eng 197:106877
15. Li A, Liu Y (2020) Water wave diffraction and radiation by a submerged horizontal circular cylinder in front a vertical wall. In: Proceedings of the ASME 2020 39th international conference on ocean, offshore and arctic engineering. Ocean engineering, vol 6B
16. Li A, Sun X, Liu Y (2021) Water wave diffraction and radiation by a submerged sphere in front of a vertical wall. Appl Ocean Res 114:102818
17. Sturova IV, Motygin OV. Radiation problem for an interface-piercing cylinder in a two-layer fluid
18. Sharma M, Kaligatla RB, Sahoo T (2020) Wave interaction with a submerged floating tunnel in the presence of a bottom mounted submerged porous breakwater. Appl Ocean Res 96:102069
19. Tabssum S, Kaligatla RB, Sahoo T (2020) Surface gravity wave interaction with a partial porous breakwater in the presence of bottom undulation. J Eng Mech 146(9):04020088
20. Chakrabarti A, Daripa P, Hamsapriye (2005) Trapped modes in a channel containing three layers of fluids and a submerged cylinder. Z Angew Math Phys 56:1084–1097
21. Das D (2016) Wave scattering by a horizontal circular cylinder in a three-layer fluid. J Ocean Eng Sci 1(2):135–148
22. Das D, Sahu M (2021) Wave radiation by a sphere in three-layer fluid. Appl Ocean Res 107:102492
23. Majumder M, Das D (2022) Cut-off frequencies for oblique waves scattering in three-layer fluid. Int J Appl Eng Res 17(1):36–40
24. Ghosh M, Sahu M, Das D (2022) Wave scattering by a submerged sphere in three-layer fluid. J Mar Sci Appl 21(1):37–50
25. Ghosh M, Das D (2023) Wave radiation by a horizontal circular cylinder in a three-layer fluid. Geophys Astrophys Fluid Dyn

A Study of Magnetohydrodynamic Nanofluid Flow Over Stretching Sheet Using Homotopy Analysis Method

Parthkumar P. Sartanpara and Ramakanta Meher

Abstract In this article, the study of the magnetohydrodynamic nanofluid flow over a stretching sheet is carried out with nanoparticles such as Cu and Ag. By utilizing similarity transformation, the system of partial differential equations (PDEs) is transformed into a system of ordinary differential equations (ODEs). The robust homotopy analysis method has been applied to analyze the velocity profile and the temperature profile at different amounts of nanoparticle volume fraction, magnetic parameters, and Eckert number.

Keywords Magnetohydrodynamic flow · Nanofluid · Heat transfer · Stretching sheet · Homotopy analysis method

Nomenculture

u	Velocity component along x-axis
v	Velocity component along y-axis
T	Temperature of nanofluid
T_w	Temperature at surface of sheet
T_∞	Ambient temperature
u_w	Stretching velocity in x-direction
B	Magnetic field
μ_{nf}	Viscosity of nanofluid
μ_f	Viscosity of base fluid
α_{nf}	Thermal diffusivity
ρ_{nf}	Density of nanofluid
ρ_f	Density of base fluid
ρ_s	Density of solid fraction
ϕ	Nanoparticles volume fraction
$(\rho C_p)_{nf}$	Heat capacity of the nanofluid

P. P. Sartanpara (✉) · R. Meher
Department of Mathematics, Sardar Vallabhbhai National Institute of Technology, Surat, Gujarat 395007, India
e-mail: parthsartanpara@gmail.com

R. K. Ray et al. (eds.), *Advances in Theoretical and Applied Mechanics*, Lecture Notes in Mechanical Engineering, https://doi.org/10.1007/978-981-97-0418-7_11

k_{nf}	Thermal conductivity of nanofluid
k_f	Thermal conductivity of base fluid
k_s	Thermal conductivity of solid fraction
Pr	Prandtl number
ν_f	Kinematic viscosity of fluid
M	Magnetic parameter
Re_x	Reynolds number
σ	Electrical conductivity
Ec	Eckert number
ζ	Dimensionless similarity variable
ψ	Stream function
g	Dimensionless stream function
θ	Dimensionless temperature

Subscript

∞	Condition at free stream
w	Condition at the surface
nf	Nanofluid
f	Base fluid
s	Solid nanoparticles

1 Introduction

In many science and engineering processes, namely extrusion of polymer, wire drawing, continuous casting, food and paper manufacturing, stretching of plastic film, production of glass fiber, etc., there are metallurgical processes, drawing continuous filaments through quiescent fluids, blowing glass, manufacturing sheets of plastic and rubber, growing of crystal and fiber spinning with constant cooling. Sub-sequential stretch is used to produce these sheets to reach the desired thickness. The pace of stretching and cooling during the process depends on the intended final result. The stretching sheet model is useful for understanding the fluid flow and heat transfer in these processes. Additionally, stretching sheet model is very helpful to study bio-mechanics and aerospace problems.

Sakiadis [1] is credited with developing the boundary layer equations for axisymmetric flows in 2D and being the first to investigate boundary layer flow over a stretched plate that moves at a constant velocity. The effects of slip on magnetohydrodynamic (MHD) "boundary layer flow" across an exponentially stretched sheet were studied by Mukhopadhyay [2]. Ali investigated the "thermal boundary layer" of a "power law fluid" over a stretched surface [3]. Elbashbeshy [4], who looked at the flow and "heat transfer" characteristics over stretched surfaces, offered a fresh angle to this inquiry. Later, the stretching sheet problem with various effects became the center of attention of the researchers.

The stretching sheet problem can be expressed in mathematical expression using a system of PDEs. Using similarity transformations, the system of nonlinear ODEs can be presented. The solution of a nonlinear ODE system is very complex and hard to find. To overcome this complexity, researchers have used many analytical and numerical methods. As numerical methods needed more computational work, analytical methods became the center of attention. Many analytical methods such as "homotopy perturbation method" [5], "Adomian decomposition method" [6], "variational iteration method" [7], and "homotopy analysis method"(HAM) [8, 9] have been used by researchers. As HAM has the advantage of providing faster convergence, HAM is widely used to solve nonlinear problems arising in engineering and science [10–13]. Researcher also modified the HAM to get faster solution [14–22] and very accurate in real world problems.

The stretching sheet problem has been studied in finite domain by Nourazar et al. [23] with using the analytical approach called homotopy perturbation method (HPM). HAM has upper hand over HPM as it has convergence parameter h, which controls the convergence of series solution and provide faster solution. Thus, HAM would be very applicable to study the stretching sheet fluid flow problem in infinite domain. This paper aims to handle a "two-dimensional" imposed "boundary layer MHD" problem caused by an "incompressible nanofluid" flow with MHD in the presence of a magnetic field across a horizontal flat plate using the homotopy analysis method. The current issue considers the "viscous dissipation" effect over a horizontally extended flat plate and a "incompressible nanofluid" in the presence of a "magnetic field". Findings for the "boundary layer flow" of "nanofluid" using HAM, including MHD with changing magnetic field and "viscous dissipation" effect, have not yet been reported.

2 Homotopy Analysis Method

Consider nonlinear Equation (1) to understand method [8, 9]

$$N[f(\eta)] = 0, \tag{1}$$

where η stands for independent variables and N is a nonlinear operator with unknown function $f(\eta)$. All boundary and starting conditions, which can be handled the same way, are neglected here for the sake of clarity and simplification. The deformation equation of zero order with homotopy concept is

$$(1 - p)L[\phi(\eta; p) - f_0(\eta)] = hpH(\eta)N[\phi(\eta; p)], \tag{2}$$

where $h \neq 0$ is an "auxiliary parameter", $H(\eta) \neq 0$ is an "auxiliary function", and $p \in [0, 1]$ is "embedding parameter". L is an "auxiliary linear operator", $f_0(\eta)$ is an "initial guess" of $f(\eta)$, and $\phi(\eta; p)$ is a "unknown function". It must be noted that the choice of the initial guess, the auxiliary linear operator, h, and $H(\eta)$ are without restriction.

Now,

$$\text{at } p = 0, \quad \phi(\eta; 0) = f_0(\eta) \text{ and at } p = 1, \quad \phi(\eta; 1) = f(\eta). \tag{3}$$

So, as p changes its value from 0 to 1, $\phi(\eta; p)$ starts changing from $f_0(\eta)$ to solution $f(\eta)$. Writing $\phi(\eta; p)$ in series form using "Taylor series" expansion in the form of p, one gets

$$\phi(\eta; p) = f_0(\eta) + \sum_{m=1}^{\infty} f_m(\eta) p^m, \tag{4}$$

where

$$f_m(\eta) = \frac{1}{m!} \frac{\partial^m \phi(\eta; p)}{\partial p^m}\bigg|_{p=0} \tag{5}$$

If the L operator, h and $H(\eta)$ are being chosen properly that the above series (4) converges at $p = 1$, then it can be written

$$f(\eta) = f_0(\eta) + \sum_{m=1}^{\infty} f_m(\eta), \tag{6}$$

Equation (6) should be one of the solutions of Eq. (1). The controlling equation can be derived from the "zero-order deformation equation", as illustrated in Eq. (6). Define vector,

$$\bar{f}(\eta) = \{f_0(\eta), f_1(\eta), f_2(\eta), \ldots, f_m(\eta)\} \tag{7}$$

Differentiating Equation (2) m-times w.r.t. p, dividing with $m!$ and putting $p = 0$, the following "mth-order deformation equation" be obtained:

$$L[f_m(\eta) - \chi_m f_{m-1}(\eta)] = hH(\eta) R_m(\bar{f}_{m-1}) \tag{8}$$

where $R_m(\bar{f}_{m-1})$ stands for

$$R_m(\bar{f}_{m-1}) = \frac{1}{(m-1)!} \frac{\partial^{m-1} N[\phi(\eta; p)]}{\partial p^{m-1}}\bigg|_{p=0} \tag{9}$$

and

$$\chi_m = \begin{cases} 0, & m \leq 1 \\ 1, & m \in (1, \infty). \end{cases} \tag{10}$$

It should be noticed that the linear "mth-order deformation equation" controls $f_m(\eta)$ for $m \leq 1$.

3 Problem Formulation

To study the nanofluid flow over stretching sheet, let us consider a "two-dimensional steady boundary layer flow" an "incompressible", "viscous", and "electrically conducting nanofluid" conducted by "stretching sheet". With the assumption, surface temperature T_w and ambient temperature T_∞ are as shown in Fig. 1. A variable magnetic field $B(x)$ is being applied to the surface sheet. The fundamental equation of mass can be written as [24]

$$\Delta \cdot V = 0, \tag{11}$$

the fundamental equation of momentum can be written as [24]

$$\rho_{nf}\left((V \cdot \Delta)V + \frac{\partial V}{\partial t}\right) = \Delta \cdot S + F, \tag{12}$$

and the fundamental equation of energy can be written as [24]

$$(\rho Cp)_{nf}\left((V \cdot \Delta)T + \frac{\partial T}{\partial t}\right) = \Delta(k\nabla T) \tag{13}$$

Here V is fluid velocity vector, S is Cauchy stress tensor, F is the interaction of the applied magnetic field with velocity of fluid produced Lorentz force, and k is thermal conductivity.

Using Eqs. (11)–(13), following equations govern the boundary layer approximation for velocity and temperature [23]:

$$\frac{\partial u}{\partial x} + \frac{\partial v}{\partial y} = 0, \tag{14}$$

$$u\frac{\partial u}{\partial x} + v\frac{\partial u}{\partial y} = \frac{\mu_{nf}}{\rho_{nf}}\frac{\partial^2 u}{\partial y^2} - \frac{\sigma B(x)^2 u}{\rho_{nf}}, \tag{15}$$

$$u\frac{\partial T}{\partial x} + v\frac{\partial T}{\partial y} = \alpha_{nf}\frac{\partial^2 T}{\partial y^2} + \frac{\mu_{nf}}{(\rho C_p)_{nf}}\left(\frac{\partial u}{\partial y}\right)^2. \tag{16}$$

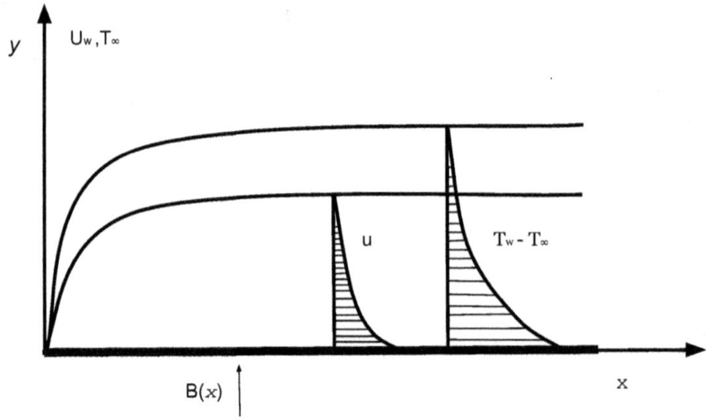

Fig. 1 Graphical representation of physical model

Here, the continuity equation is presented in Eq. (14), u is components of velocity in the direction of x, and v is the components of velocity in the direction of y, as shown in Fig. 1. "Two-dimensional momentum equation" in the presence of a variable "magnetic field" can be presented by Eq. (15), where ρ_{nf}, μ_{nf}, and σ are the the "density of nanofluid", "dynamic viscosity", and "electrical conductivity", respectively. Magnetic field $B(x)$ is acting in perpendicular direction to stretching sheet as shown in Fig. 1. Equation (16) stands for energy equation of two dimensions including the viscous dissipation term, with T temperature, $(\rho Cp)_{nf}$ is the "nanofluid heat capacity" and α_{nf} is "thermal diffusivity".
The conditions at boundary are derived as

$$\begin{gathered} u = u_w = ax^m, \quad v = 0, \quad T = T_w, \quad \text{at} \ \ y = 0, \\ u \to 0, \quad T \to T_\infty, \quad \text{as} \ \ y \to \infty, \end{gathered} \tag{17}$$

Here, velocity component on the surface of stretching sheet in x-direction is presented by u_w, a, and m that are constants, T_w is temperature of plate, and T_∞ is ambient temperature. The "dynamic viscosity" μ_{nf}, "nanofluid density" ρ_{nf}, "thermal conductivity" k_{nf}, and the "heat capacity" $(\rho Cp)_{nf}$ are defined as following:

$$\begin{gathered} \mu_{nf} = \frac{\mu_f}{(1-\phi)^{5/2}}, \quad \rho_{nf} = \phi\rho_s + \rho_f(1-\phi), \quad \frac{k_{nf}}{k_f} = \frac{2k_f + k_s + 2\phi(k_s - k_f)}{2k_f + k_s - 2\phi(k_s - k_f)} \\ (\rho C_p)_{nf} = (\rho C_p)_f(1-\phi) + (\rho C_p)_s\phi, \end{gathered} \tag{18}$$

where ϕ is "volume fraction of nanoparticles", μ_f is "dynamic viscosity" of fluid, ρ_s is "density of nanoparticles" and ρ_f is "density of fluid", $(\rho Cp)_f$ and $(\rho Cp)_s$ are "heat capacity" of fluid and nanoparticles respectively, and k_s and k_f are the "thermal conductivities" of "nanoparticles" and fluid, respectively.

Here, magnetic field $B(x)$ with B_0 and m as constants varies according to following function:

$$B(x) = B_0\sqrt{x^{m-1}}, \tag{19}$$

Defining following the dimensionless similarity variable and to get transformation the governing equation into system of ODE:

$$\eta = \frac{y}{x}Re_x^{\frac{1}{2}}, \quad Re_x = \frac{\rho_f u_w(x)}{\mu_f}x. \tag{20}$$

The dimensionless velocity and temperature using steam function can be defined as

$$f(\eta) = \frac{\psi(x, y)Re_x^{\frac{1}{2}}}{u_w(x)}, \quad \theta(\eta) = \frac{T - T_\infty}{T_w - T_\infty}, \tag{21}$$

where stream function $\psi(x, y)$ can be derived as

$$u = \frac{\partial\psi}{\partial y}, \qquad v = -\frac{\partial\psi}{\partial x} \tag{22}$$

Using similarity transformation, we can write Eqs. (14)–(16) into following ODE:

$$f''' + Af''f - B(f')^2 - Cf' = 0, \tag{23}$$

$$\theta'' + D\theta' f + E(f'')^2 = 0, \tag{24}$$

with conditions

$$f'(0) = 1, \quad f(0) = 0, \quad f'(\infty) \to 0, \quad \theta(0) = 1, \quad \theta(\infty) \to 0 \tag{25}$$

where the coefficients are expressed as the following:

$$A = \left(\phi\frac{\rho_s}{\rho_f} + (1-\phi)\right)(1-\phi)^{\frac{5}{2}}\left(\frac{m+1}{2}\right)^2, \quad B = \left(\phi\frac{\rho_s}{\rho_f} + (1-\phi)\right)(1-\phi)^{\frac{5}{2}}m,$$
$$C = \left(M(1-\phi)^{\frac{5}{2}}\right), \quad D = Pr\left(\phi\frac{(\rho C_p)_s}{(\rho C_p)_f} + (1-\phi)\right), \quad E = \frac{EcPr}{(1-\phi)^{\frac{5}{2}}} \tag{26}$$

Here, M is dimensionless magnetic parameter, Re_x is "Reynolds number", Ec is "Eckert number", and Pr is "Prandtl number", and they can be expressed as

$$M = \frac{\sigma \cdot B_0^2}{\rho_f \cdot a}, \quad Pr = \frac{((\rho C_p)_f)}{K_f}v_f, \quad Ec = \frac{u_w^2}{C_p\Delta T}, \quad Re_x = \frac{\rho_f u_w}{\mu_f}x. \tag{27}$$

The physical quantities such as skin friction coefficient Cf and Nusselt number Nu can be presented as

$$f''(0) = Cf\sqrt{\frac{Re}{2}}, \quad -\frac{Nu}{\sqrt{Re}} = \theta'(0) \tag{28}$$

4 Semi-analytical Simulation

For having analytical simulation using HAM [8, 9], the initial approximation must be satisfying the boundary conditions shown in Eq. (25). Thus, initial approximation can be written as

$$f_0(\eta) = 1 - e^{-\eta}, \quad \theta_0(\eta) = e^{-\eta}. \tag{29}$$

The linear operator can be presented as

$$L_f = f''' - f', \quad L_\theta = \theta'' - \theta. \tag{30}$$

Our linear operator L_f and L_θ follows the conditions stated below

$$L_f(C_1 + C_2e^{-\eta} + C_3e^{\eta}) = 0, \quad L_\theta(C_4e^{-\eta} + C_5e^{\eta}) = 0 \tag{31}$$

where C_i are constants $(i = 1, 2, 3, 4, 5)$. The solution of velocity profile can be written in series form as

$$f(\eta) = 1 - e^{-\eta} - \frac{1}{6}h_f\left(\begin{array}{c} A(3\eta e^{-\eta} + e^{-\eta} + e^{-2\eta} - 2) + B(2e^{-\eta} - e^{-2\eta} - 1) \\ +3C(\eta e^{-\eta} + e^{-\eta} - 1) - 3\eta e^{-\eta} - 3e^{-2\eta} + 3 \end{array}\right) + \cdots, \tag{32}$$

and temperature profile can be written in following series solution

$$\theta(\eta) = e^{-\eta} + \frac{1}{6}h_\theta\left(D(2e^{-\eta} + 3\eta - 2) + E(2e^{-\eta} - 2) - 3\eta\right) + \cdots \tag{33}$$

Table 1 Properties

Properties	Unit	Fluid phase	Nanoparticles	
		H_2O	Cu	Ag
ρ	kg m^{-3}	997.1	8933	10500
k	W m^{-1} K^{-1}	0.613	400	429
C_p	J kg^{-1} K^{-1}	4179	385	235

Table 2 Comparison of the solution of HAM and HPM [23] solution at $\mathrm{Pr} = 6.2, m = 0, \mathrm{Ec} = 0.1, M = 0.2$ for Al_2O_3-water for volume fraction $\phi = 0.2$

ζ	$f'(\zeta)$		$\theta(\zeta)$	
	HAM	HPM[23]	HAM	HPM [23]
0	1	1	1	1
0.2	0.9008946630	0.911545	0.8453061281	0.830011
0.4	0.8135044124	0.827806	0.7476814893	0.747524
0.6	0.7345868455	0.748988	0.6810377986	0.664384
0.8	0.6624134204	0.675187	0.6307979312	0.628484
1.0	0.5960586448	0.606404	0.5889067772	0.593911

Table 3 Skin friction coefficient and Nusselt number for different term approximations

Term approxi.	$f''(0)$	$-\theta'(0)$
1	−1.10317	0.67309
2	−1.09457	0.66328
3	−1.09556	0.66853
4	−1.09542	0.67237
5	−1.09544	0.67349
6	−1.09544	0.67336
7	−1.09544	0.67305
8	−1.09544	0.67292
9	−1.09544	0.67292
10	−1.09544	0.67292

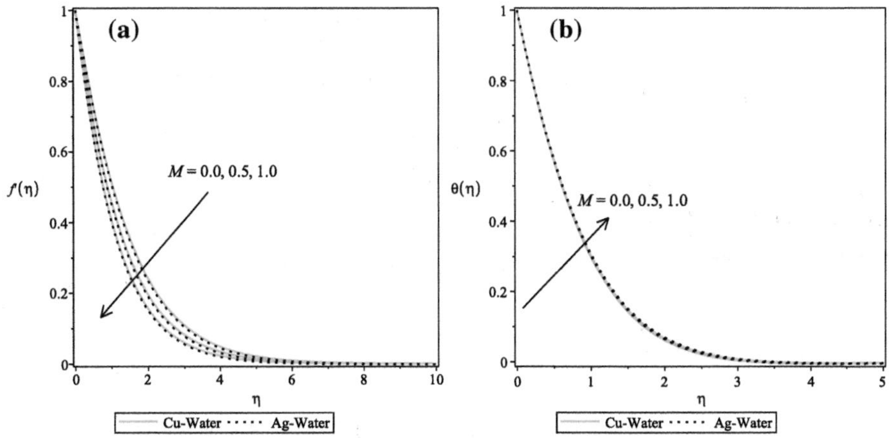

Fig. 2 Effect of unsteady magnetic field at $Pr = 6.2, \phi = 0.2$ $Ec = 0.1$, and $M = 0.5$ on profiles of **a** velocity, **b** temperature

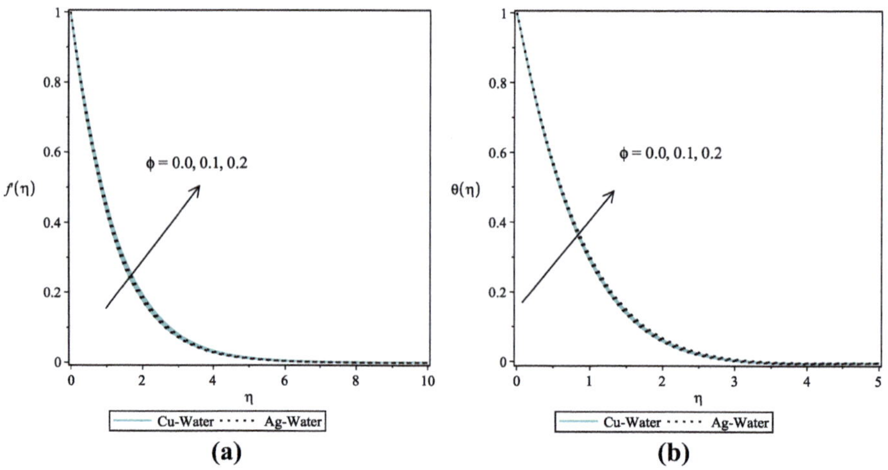

Fig. 3 Effect of volume fraction at $Pr = 6.2$, $m = 0$, $Ec = 0.1$, and $M = 0.5$ on profiles of **a** velocity, **b** temperature

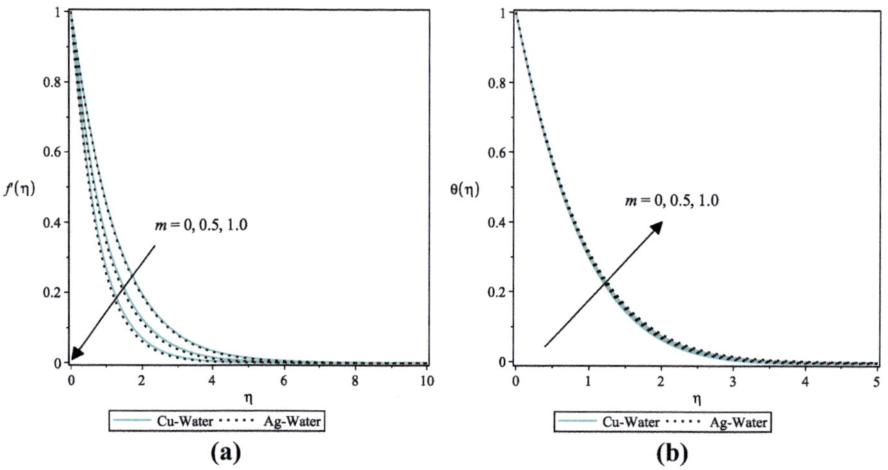

Fig. 4 Effect of unsteady magnetic field at $Pr = 6.2$, $\phi = 0.2$ $Ec = 0.1$, and $M = 0.5$ on profiles of **a**, velocity **b** temperature

5 Results and Discussion

In this section, we will discuss the graphical solution obtained using HAM. The physical properties of nanoparticles and base fluids are given in Table 1. For the Al_2O_3-water nanofluid instance, Table 2 examines the velocity profile and temperature profile for $\phi = 0.2$ with $\mathrm{Pr} = 6.2$, $M = 0.2$, $\mathrm{Ec} = 0.1$, and $m = 0$ and compares the HAM solution to HPM solution. While we have chosen an unbounded region of

Fig. 5 Eckert number effect on $\theta(\eta)$ at $M = 0.5$, $Pr = 6.2$, $\phi = 0.1$, and $m = 0$

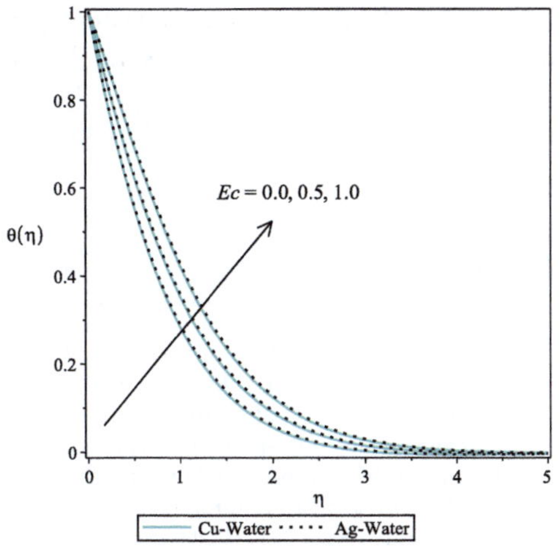

$\eta \in [0, \infty)$, the method of HPM [23] was adopted in a bounded region of $\eta \in [0, 5]$. Table 2 explains HAM's effectiveness because the produced findings are extremely convenient with the data that were previously provided. It is observable that solution obtained using HAM is very accurate and very effective approach to such kind of problems arising in fluid flow problem. Table 3 presents the skin friction coefficient and Nusselt number for different term approximations which clearly indicates that the series solution provides convergent solution as terms increase. From Fig. 2a, it is observable that the similar behavior of velocity profile is seen, and while increasing magnetic parameter for Ag-water and Cu-water, the velocity profile decreases but temperature profile increases as the magnetic parameter rises for Ag-water and Cu-water, as shown in Fig. 2b. It is also noticeable from Fig. 2b that temperature profile of Cu-water is lightly lower than Ag-water. Figure 3a, b indicated that with increasing volume fraction of nanoparticles behavior of velocity profile and temperature profile increases, respectively. Figure 4 presents the unsteady magnetic field effect on velocity profile (Fig. 4a) and temperature profile (Fig. 4b). From Fig. 5, the Eckert number also affects the temperature profile with proportionally, as temperature increasing with the using larger Eckert number.

6 Conclusions

Here, HAM has been employed to analyze the effect of MHD nanofluid flow over a stretching sheet under heat transfer effect. The results have good agreement with available published results. The results of dimensionless velocity profile, temperature profile for different nanoparticles are presented with graphical representation and numerical representation. The obtained result can conclude following analysis. As the magnetic field be stronger, the velocity profile's nature is decreasing but the nature of temperature profile is increasing. Temperature profile's nature can be enhanced with increasing the Eckert number. The effect of volume fraction is proportional to the velocity profile and temperature profile.

References

1. Sakiadis BC (1961) Boundary-layer behavior on continuous solid surfaces: I. Boundary layer equations for two-dimensional and axisymmetric flow. AIChE J 7(1):26-8
2. Mukhopadhyay S (2013) Slip effects on MHD boundary layer flow over an exponentially stretching sheet with suction/blowing and thermal radiation. Ain Shams Eng J 4(3):485–91
3. Ali ME (1995) On thermal boundary layer on a power-law stretched surface with suction or injection. Int J Heat Fluid Flow 16(4):280–90
4. Elbashbeshy EM (2001) Heat transfer over an exponentially stretching continuous surface with suction. Arch Mech 53(6):643–51
5. He JH (2003) Homotopy perturbation method: a new nonlinear analytical technique. Appl Math Comput 135(1):73–9
6. Wazwaz AM (1999) A reliable modification of Adomian decomposition method. Appl Math Comput 102(1):77–86
7. Wazwaz AM (2014) The variational iteration method for solving linear and nonlinear ODEs and scientific models with variable coefficients. Central Euro J Eng 4(1):64–71
8. Liao S (1998) Homotopy analysis method: a new analytic method for nonlinear problems. Appl Math Mech 19(10):957–62
9. Liao S (2003) Beyond perturbation: introduction to the homotopy analysis method. Chapman and Hall/CRC
10. Sartanpara P, Meher R (2022) Computational study of Klein-Gordan equation using homotopy analysis method. In: AIP conference proceedings 2022 Oct 7, vol 2451, no 1. AIP Publishing LLC, p 020030
11. Verma L, Meher R (2022) Effect of heat transfer on Jeffery-Hamel Cu/Ag-water nanofluid flow with uncertain volume fraction using the double parametric fuzzy homotopy analysis method. Euro Phys J Plus 137(3):372
12. Kumar A, Meher R (2022) Solving non-linear partial differential equations using homotopy analysis method (HAM). Nonlinear dynamics and applications 2022. Springer, Cham, pp 955–964
13. Prajapati VJ, Meher R (2023) An analytical investigation of time-fractional Sharma-Tasso-Olever equation arising in physical sciences. In: 2023 International conference on fractional differentiation and its applications (ICFDA) 2023 Mar 14. IEEE, pp 1–6
14. Iyiola OS, Soh ME, Enyi CD (2013) Generalised homotopy analysis method (q-HAM) for solving foam drainage equation of time fractional type. Math Eng Sci Aerosp (MESA) 4(4)
15. Sartanpara PP, Meher R (2023) A robust computational approach for Zakharov-Kuznetsov equations of ion-acoustic waves in a magnetized plasma via the Shehu transform. J Ocean Eng Sci 8(1):79–90
16. Prajapati VJ, Meher R (2022) Solution of time-fractional Rosenau-Hyman model using a robust homotopy approach via formable transform. Iran J Sci Technol Trans A Sci 46(5):1431–44
17. Sartanpara PP, Meher R (2023) A robust fuzzy-fractional approach for the atmospheric internal wave model. J Ocean Eng Sci 8(3):308–22
18. Sartanpara PP, Meher R (2023) Analytical study of time fractional Fisher equation using homotopy approach with a generalized transform. In: 2023 international conference on fractional differentiation and its applications (ICFDA) 2023 Mar 14. IEEE, pp 1–6
19. Sartanpara PP, Meher R, Meher SK (2022) The generalized time-fractional Fornberg-Whitham equation: an analytic approach. Partial Differ Eqn Appl Math 5:100350

20. Golmankhaneh AK, Porghoveh NA, Baleanu D (2013) Mean square solutions of second-order random differential equations by using homotopy analysis method. Rom Rep Phys 65(2):350–362
21. Sartanpara PP, Meher R (2023) Solution of generalised fuzzy fractional Kaup-Kupershmidt equation using a robust multi parametric approach and a novel transform. Math Comput Simul 1(205):939–69
22. Kumar D, Singh J, Tanwar K, Baleanu D (2019) A new fractional exothermic reactions model having constant heat source in porous media with power, exponential and Mittag-Leffler laws. Int J Heat Mass Transf 138:1222–1227
23. Nourazar SS, Habibi Matin M, Simiari M (2011) The HPM applied to MHD nanofluid flow over a horizontal stretching plate. J Appl Math 2011
24. Bouslimi J, Omri M, Mohamed RA, Mahmoud KH, Abo-Dahab SM, Soliman MS (2021) MHD Williamson nanofluid flow over a stretching sheet through a porous medium under effects of joule heating, nonlinear thermal radiation, heat generation/absorption, and chemical reaction. Adv Math Phys 21:2021

Solid Mechanics

Mechanical Characterization of Iron Flakes Filled Polyurethane Composites

Abhinav Singh, Vipin Chandra, and Rajesh Kitey

Abstract Tensile and Fracture Toughness Tests are conducted on Iron-Flakes filled polyurethane composite of micrometer thickness (200–300 μm) thin film sheets having different weight fractions of iron-flakes filler percentages (10, 20, 30, 40, 50 and 60%) in the PU matrix. The PU-matrix is prepared by adding polyurethane resin and PU-hardener in (100:70) ratio by weight. Xylene as a solvent is added (20–40%) by solution weight to reduce the PU-matrix viscosity. Iron-Flakes fillers are reinforced at different weight fractions in the PU-matrices composite system. The tensile tests are conducted Using ASTM D412 on a UTM machine. Here we considered the compliance correction for the direct tensile test of the thin film sheet. The density of newly made composite materials is calculated by using the pycnometer method followed by ASTM D2734-16. It is observed that Young's modulus, density, tensile strength, and fracture toughness of the PU/Iron-flakes composite increase with the increase in the filler weight fraction percent in the matrix. In contrast, the elongation percentage of the composites decreases. It is suggested that in the presence of high filler content in the matrix, the mechanical parameters and fracture toughness strength have improved the property of PU/Iron-flakes composites.

Keywords Polyurethane · Fracture toughness · Tensile test · Young's modulus · Density · Tensile strength

1 Introduction

Polyurethane (PU) is a widely used polymeric material as a type of coating for various everyday applications such as fibers, biomaterials, constructions, and automobiles. PU coatings on automobiles improve color retention, high gloss, corrosion resistance, and scratch resistance. Polyurethane paints are painted on the external surface of the aircraft to stop air leakage. Such paints and coatings also powerfully resist the coarse force, which becomes a stern situation for aircraft slashed with debris or equipment.

A. Singh · V. Chandra (✉) · R. Kitey
Department of Aerospace Engineering, Indian Institute of Technology Kanpur, Kanpur, India
e-mail: chandrav@iitk.ac.in

R. K. Ray et al. (eds.), *Advances in Theoretical and Applied Mechanics*, Lecture Notes in Mechanical Engineering, https://doi.org/10.1007/978-981-97-0418-7_12

Composites of filled polymers are used in various civil and defense applications. The usage of filled polymers in the aviation and automobile sector has grown manifolds due to their high specific strength and specific stiffness.

Putson et al. [1, 2] used copper particles as filler to reinforce the Polyurethane. The Composites' ability to resist the applied forces leading to its failure increases as the filler content increases, and the elongation at the break of the specimens showed a gradual reduction as the copper filler content was increased in the composites irrespective of the particle size. Maganty et al. [3, 4] experimented using the nano-particle size of silica added with polyurethane resin by liquid suspension, which improves the mechanical properties of polyurethane. Nano particle-containing coatings improved nano hardness, elastic modulus, and tensile strength by 20 and 30%, corresponding to 6.74 and 9.75 wt% silica nanoparticles. At 20 wt% suspension or lower, nano-particle filler in the matrix was homogeneous, and brittleness in the composite material was less. Youssef et al. [5, 6], in their research work, have reported the consequences of the addition of the Organic filler Cu (II) phthalocyanines CuPc in Polyurethane composites. The three compositions of PU composite films containing up to 1, 15, and 30 wt% of CuPc have different thermal properties and tensile behavior compared with the PU film as the reference material. The test specimens were prepared of 2 mm thickness. The tensile strength of neat PU and the combination of CuPc/PU composite films increased from 31.1 to 58.3 MPa for different weight fractions. The elongation of the specimens at break increased from 910 to 1230% due to the increment in the CuPc content from 1 to 30%. The values of Young's and storage moduli increased as the CuPc filler content increased, indicating that the samples had good rigidity, high mechanical parameters, and excellent thermal stability. Oprea et al. [7, 8] have examined the mechanical properties of black carbon and cork filler reinforced with a PU composite system. The cork is between 1 and 15% and carbon black 2 and 3% by weight in composites. Adding cork and carbon black fillers to PU composites increases Young's modulus while decreasing the elongation at break. Kheng et al. [9] conducted the fracture toughness test of layer-by-layer prepared thin films with elastic polyurethane and poly (acrylic acid) as the stiffening agent. A single-edge-notch tension (SENT) specimen is used to study mode I crack propagation as a function of applied loading. The experimental result shows that the PU/PAA samples have a maximum stress of 16 MPa, whereas the pure PU has 4 MPa. The mode-I fracture toughness for the PU/PAA films was 65 N/mm, while for the pure PU, this value was 61 N/mm.

Based on the literature review, it is observed that the fracture strength and tensile strength of pure PU have improved by reinforcing the metal filler in the pure PU composite system. The principal objective of this work is to develop the micrometer thickness thin film sheet of the PU/iron-flakes composite system and study the effect of the filler on the mechanical properties of the neat PU composite system.

2 Materials

2.1 Manufacturing Thin Film Sheets and Samples

Polyurethane thin film sheet is prepared using polyurethane resin (AURACAP PCB PU) of density 0.98 gm/cm^3 and PU-hardener of density 1.06 gm/cm^3 in the ratio of 100:70 by mass. Xylene is added as a solvent in the range of (20–40%) by total weight mass to make a less viscous and homogeneous solution. A stirrer is used to mix the solution. Iron-flake filler is reinforced into the matrix at different weight fractions. When the desired uniformity is achieved, add a few drops of an accelerator to the solution for fast curing. The mixture is poured using a hand spray gun, maintaining air pressure of 3–5 bar, maintaining an appropriate distance, and slowly moving a hand from one end to another end over the silicon-sprayed transparent sheet pasted below the stencil of dimension (100 mm × 15 mm × 1 mm). Repeat this process according to the required thickness. After spraying the mixture, leave the samples for 24 h at room temperature. The cured sheets are placed over the flat surface for two weeks before machining the tensile and fracture tests. For each weight fraction (0, 10, 20, 30, 40, 50, and 60%) of the filler, at least five samples are made, and two different (rectangular and dog-bone) shapes are considered in this study for testing the material. The samples of PU/Fe-flakes composites are shown in Fig. 1a, b, and the setup of Tensile Testing in Fig. 1c.

3 Experimental Methods and Measurements

3.1 Tensile Test

The tensile tests are conducted on flat-type specimens following the ASTM D412 or D882-12. Samples of dimensions 100 mm × 15 mm × (thickness $\leq$ 300 μm) are machined and loaded with 250 N. Pneumatic grips hold the samples, maintaining the 4–5 bar pressure. Tensile experiments are carried out in displacement-controlled mode, maintaining the crosshead speed of 2 mm/min by using UTM Tinius Olsen at an ambient temperature of 25 °C. Young's modulus of composites is measured by the initial slopes of the stress versus strain curve. The maximum tensile strength (σ_{max}) tests are conducted on dogbone-type specimens following the ASTM D638. The dogbone sample has a shoulder at each end of the specimens and a gauge section (33 mm) in between these shoulders. The crosshead speed of 2 mm/min is maintained using UTM Tinius Olsen and loaded with 250 N load cell at an ambient temperature of 25 °C. The maximum force at the point of break is used to calculate the maximum tensile strength.

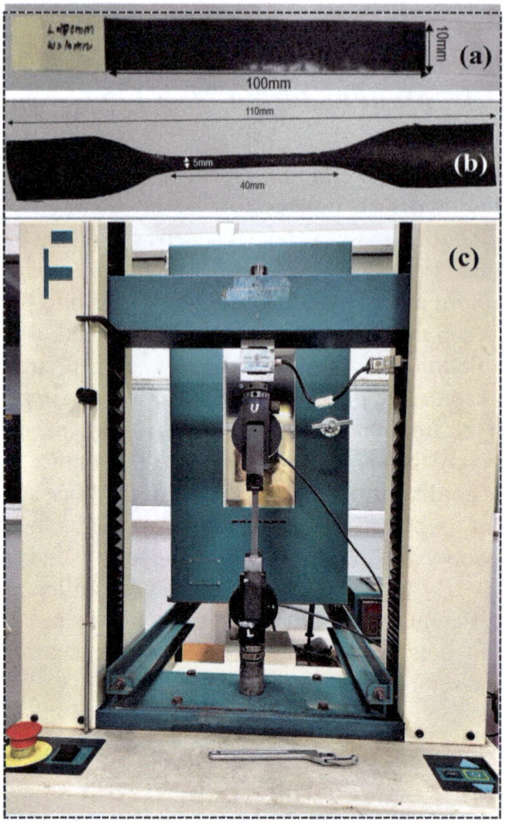

Fig. 1 PU/iron-flakes composite specimen: **(a)** flat and **(b)** dogbone; **(c)** tensile testing setup

3.2 *Fracture Test*

Single Edge Notch Tension (SENT) test specimens of dimensions 100 mm × 15 mm × (thickness ≤ 300 μm) are machined by following the ASTM D412 standards as shown in Fig. 2a. To cause a natural crack, initially, an edge notch of 4.5 mm is marked by a scalpel at the equidistant of the sample. Initiation of a straight natural crack within the prescribed range of a/w ratio (0.45–0.55) is being carefully monitored, where a and w represent the initial crack length and width of the specimen, respectively. The test specimens are loaded symmetrically using a pneumatic grip under displacement control mode by employing UTM Tinius Olsen. The high pixel resolution camera is employed for recording the time of initializing the crack, as shown in Fig. 2b. Stress Intensity Factor (SIF), K_I, is determined by using linear elastic fracture mechanics,

$$K_I = f(a/w)\sigma\sqrt{\pi a} \tag{1}$$

where,

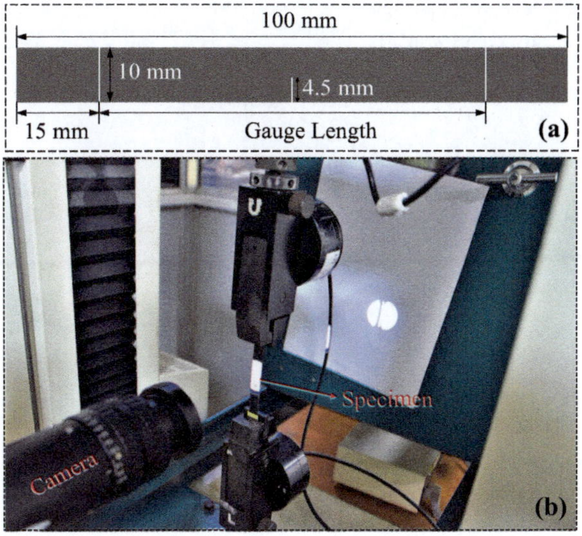

Fig. 2 Fracture toughness testing **(a)** sample specimen dimensions, and **(b)** test setup

$$f\left(\frac{a}{w}\right) = 1.12 - 0.23\left(\frac{a}{w}\right) + 10.55\left(\frac{a}{w}\right)^2 - 21.72\left(\frac{a}{w}\right)^3 + 30.41\left(\frac{a}{w}\right)^4$$

In Eq. (1), $f\left(\frac{a}{w}\right)$ is the geometric parameter. To find the critical stress intensity factor K_I the load P is replaced by the load corresponding to the fracture initiation.

4 Results and Sections

4.1 Density Analysis

Density is an essential parameter in deciding the properties of the composite. The theoretical density measurement in the composite material is standardized with the mixture rule for finding the relative proportion of the reinforced filler and matrix system. In contrast, experimental density is calculated by using the pycnometer method. The void presence and non-uniform distribution of the filler in the composite system could be the reason for the difference between the magnitude of theoretical and experimental densities, as shown in Fig. 3.

ASTM standard D2734-16 can be applied to find the void percentage in different weight fractions of the composites. As our composites are made by spraying on the mold surface to make micron-thickness film in that case the void's presence is unavoidable, but the difference is not very large, hence it meets the desired properties of the composites.

Table 1 summarizes the all specimen's density calculations and their comparison.

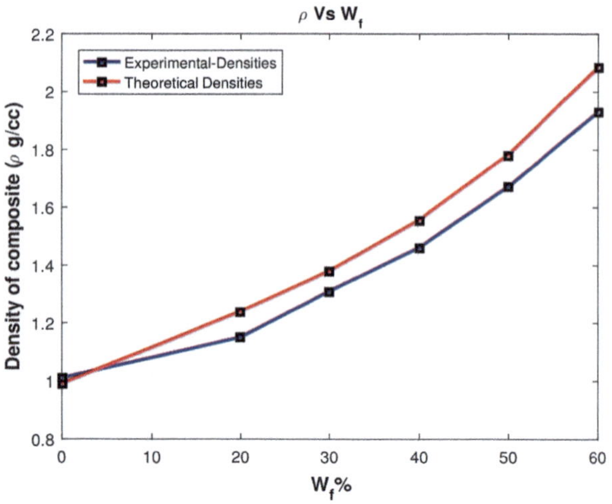

Fig. 3 Graph between density versus $w_f(\%)$

Table 1 Densities of iron-flakes filled composites

Material type	Weight fraction (%)	Experimental density $\rho(\text{gm/cm}^3)$	Theoretical density $\rho(\text{gm/cm}^3)$
Neat polyurethane	0	1.01	0.98
PU + iron − flakes	20	1.15	1.22
	30	1.31	1.37
	40	1.46	1.55
	50	1.67	1.78
	60	1.93	2.08

4.2 Compliance Correction and Young's Modulus Analysis

The machine crosshead displacement adds not only to the elongation of the specimen but also the slippage of the grip and the displacement of machine components because of the system compliance, which exaggerates the elongation of the specimen, which results in a lesser E_u (elastic modulus) than the actual value. To correct elastic modulus and other parameters, we implemented a modified correction method by Langley et al. [10] and Sergueeva et al. [11] built on the standard ASTM D3379. So, the corrected tensile strain could be demonstrated as given below,

$$\in = \frac{\Delta L^*}{L} \tag{2}$$

$$\in = \frac{\Delta L}{L} \times \left(1 - \frac{C_s}{C_a}\right) \tag{3}$$

where ΔL^* is the true specimen elongation, i.e., $\frac{FL}{AE}$, ΔL the machine crosshead displacement, i.e., $\frac{FL}{AE_u}$, F is the load, E the corrected tensile modulus, C_s the system compliance, and C_a the apparent compliance. The initial slope is determined as the apparent compliance (C_a), in a traditional displacement-load curve extracted from the tensile testing, it is represented in the given equation form as shown below,

$$C_a = \frac{\Delta L}{F} \tag{4}$$

The value of C_s can be obtained by the intercept at zero-gauge length, i.e.,

$$C_a = \frac{L}{AE} + C_s \tag{5}$$

The obtained modulus can then be corrected by C_s, as

$$E = \frac{L}{A(C_a - C_s)} \tag{6}$$

Table 2 is the summary of all the tested specimens of different weight fractions of filler content with the PU composite system before and after compliance correction of E. The Young's modulus increases with the increase of filler percentage in neat PU as shown in Fig. 4.

4.3 Maximum Tensile Strength Analysis

The tensile characteristic is based on the strength properties of composites. The maximum tensile strength of neat PU and PU/Fe-flakes composite system is conducted on UTM Tinius Olsen, and it is analyzed by plotting the graph between stress versus strain. Consistently increasing slope of the curves indicates that the composite's stiffness and strength increase with increasing filler in neat PU. The testing is done after curing samples for two weeks at room temperature 25 °C. Each composition of samples has 3–5 specimens testing.

The strength of the neat PU is low as compared to the Filler reinforced PU composite system. The maximum tensile strength increases with the increase in the composition of weight fraction percent in the PU matrix. Still 50% weight of the filler in PU matrix the mixture is homogeneous and after increasing beyond that the mixture is not mixing properly and the samples breakdown in between while taken out from the stencils. The max. tensile strength of all tested samples is shown in Fig. 5.

Table 2 Young's modulus comparison table

Filler	Weight fraction of filler (%)	Number of specimens tested	*E* before C.C.	*E* after C.C.	Error%
Neat PU	0	Specimen1	2.84	2.96	4
		Specimen2	3.37	3.49	3
		Specimen3	3.96	4.15	5
		Average	3.40	3.53	4
PU + Fe-Flakes	20	Specimen1	7.09	7.40	4
		Specimen2	7.75	8.01	3
		Specimen3	8.20	8.90	7
		Average	7.68	8.08	5
	30	Specimen1	11.51	12.46	8
		Specimen2	12.01	13.12	9
		Specimen3	11.98	12.66	6
		Average	11.83	12.74	7
	40	Specimen1	13.85	14.84	7
		Specimen2	12.34	13.24	7
		Specimen3	11.83	13.01	9
		Average	12.67	13.70	8
	50	Specimen1	17.98	20.42	13
		Specimen2	16.24	18.26	12
		Specimen3	16.27	17.98	10
		Average	16.80	18.88	12
	60	Specimen1	34.46	40.60	18
		Specimen2	32.83	42.63	30
		Specimen3	30.12	38.65	27
		Average	32.47	40.62	25

4.4 Fracture Analysis of Composites

The composites fracture behavior is analyzed by plotting load versus crosshead displacement graph obtained from fracture test data as shown in Fig. 6a.

Approximately linear curve until fracture initiation is observed for all cases. This is followed by a spontaneous drop in the load for the neat PU and less than 20% weight fraction of PU/Fe-flakes specimens. On the contrary, continually decreasing crosshead displacement consequent to fracture initiation is observed for a higher weight fraction percent of filler in PU matrices. Figure 6b, values obtained from the LEFM method; the critical fracture toughness values increase with the weight fraction percent of filler in PU matrices. Figure 7, is one of the parts of the fracture test performed at various levels of loading.

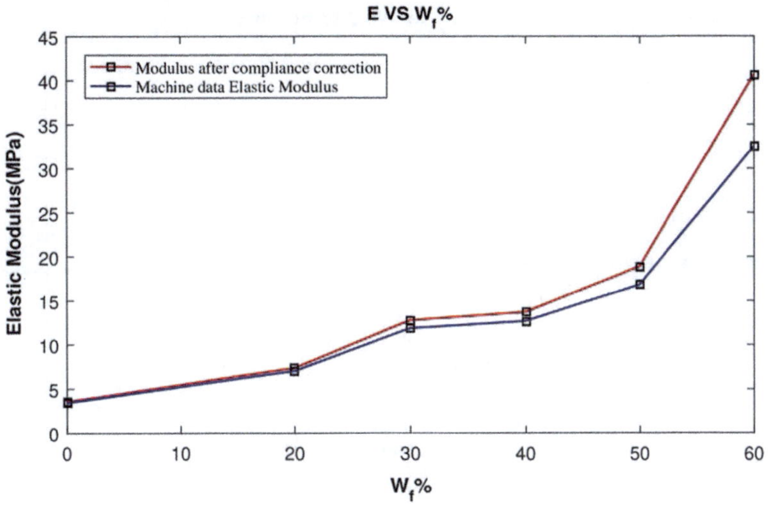

Fig. 4 Young's modulus versus weight fraction

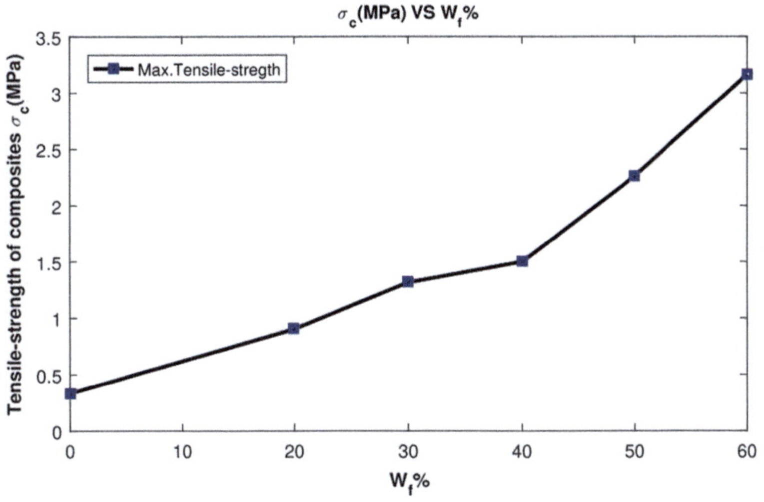

Fig. 5 Tensile strength versus weight fraction

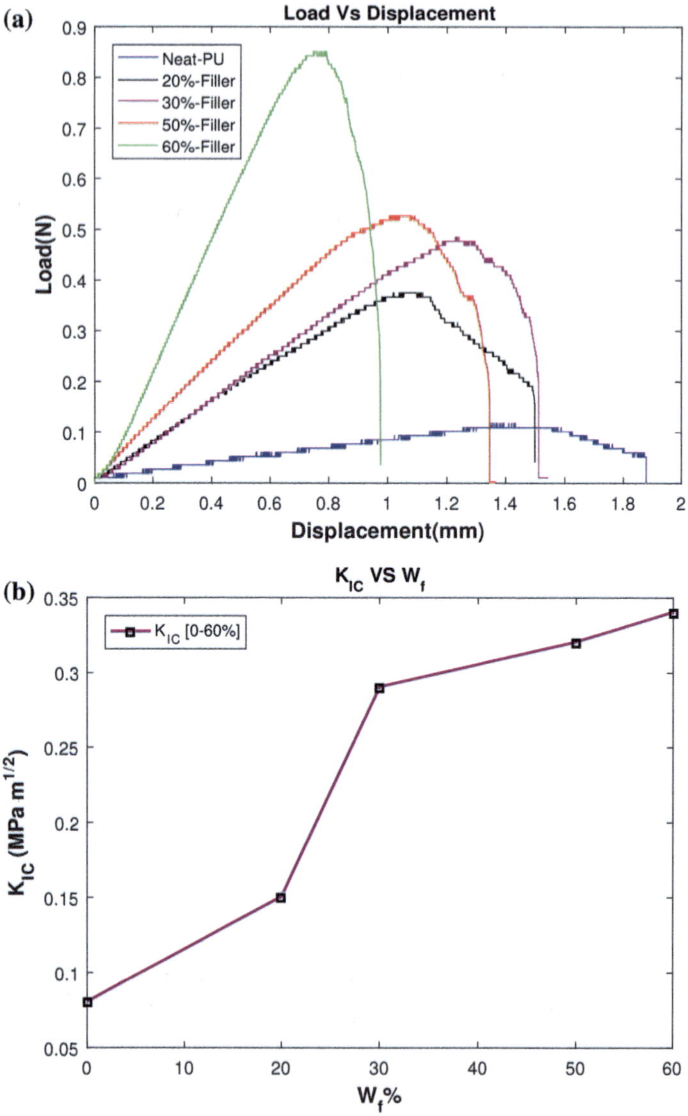

Fig. 6 SENT testing results of PU and PU/Fe-flakes **(a)** load versus crosshead displacement, and **(b)** fracture toughness versus weight fraction

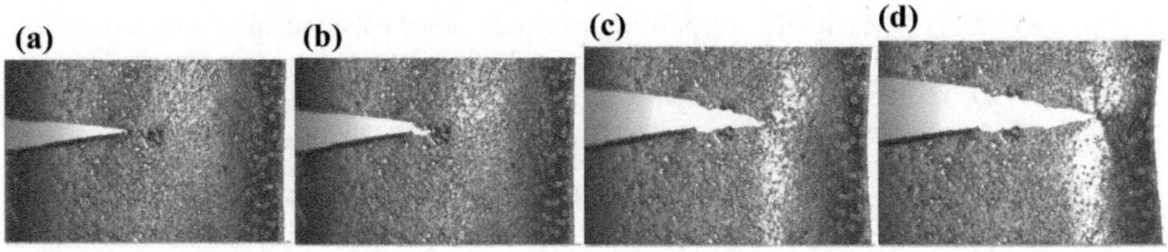

Fig. 7 Crack propagation at various levels of loading

5 Conclusion

The PU matrix is prepared by adding PU resin and PU hardener in (100:70) proportion by weight. Iron-flakes fillers are reinforced with PU-matrix in the five different weight fractions (20, 30, 40, 50, 60%) using xylene solvent for the homogeneous mixture. Two different shapes of samples are prepared for testing. The Tensile tests are conducted by following the ASTM D638 and D412. The elongation at break is corrected for Young's modulus by using compliance correction method. The composites exhibit monotonically increasing Young's modulus, maximum tensile strength and other mechanical properties with the increase in filler weight percentage into PU matrix, up to 50% the matrix and filler mixture is homogeneous and showed promising results after curing the samples. The post-crack initiation area under the load versus displacement curve indicates the composite's resistance to crack propagation. The stress Intensity Factor increases with the increase in filler weight percent in PU-matrix. Also, the force required to initiate the crack increases. It is suggested that the tensile strength is dominant in the presence of the filler up to 50% of the property of PU/iron-flake composites, whereas the elongation at break decreases. The fracture property of PU/iron-flake composites improved by 4–5 times higher than the neat PU composite.

References

1. Putson C, Jaaoh D, Meauma N, Muensit N (2012) Effect of micro-and nano-particle fillers at low percolation threshold on the dielectric and mechanical properties of polyurethane/copper composites. J Inorg Organomet Polym Mater 22(6):1300–1307
2. Putson C, Lebrun L, Guyomar D, Muensit N, Cottinet P-J, Seveyrat L, Guiffard B (2011) J Appl Phys 109:024104
3. Maganty S, Roma MPC, Meschter SJ, Starkey D, Gomez M, Edwards DG, Ekin A, Elsken K, Cho J (2016) Enhanced mechanical properties of polyurethane composite coatings through nano-silica addition. Progr Organic Coatings 90:243–251
4. Petrovic ZS, Javni I, Waddon A, Banhegyi G (2000) Structure and properties of polyurethane-silica nanocomposites. J Appl Polym Sci 76(2):133–151
5. Youssef TE, Al-Turaif H, Wazzan AAA (2015) Investigations on the structural and mechanical properties of polyurethane resins based on cu (ii) phthalocyanines. Int J Polymer Sci 2015
6. Hartmann B, Duffy JV, Lee GF, Balizer E (1998) Thermal and dynamic mechanical properties of polyurethane. J Appl Polym Sci 35:1829–1852

7. Oprea S (2008) Effects of fillers on polyurethane resin-based polyurethane elastomeric bearing materials for passive isolation. J Compos Mater 42(25):2673–2685
8. Rozman HD, Yeo YS, Tay GS, Abubakar A (2003) The mechanical and physical properties of polyurethane composites based on rice husk and polyethylene glycol. Polym Testing 22:617–623
9. Kheng E, Iyer HR, Podsiadlo P, Kaushik AK, Kotov NA, Arruda EM, Waas AM (2010) Fracture toughness of exponential layer-by-layer polyurethane/poly (acrylic acid) nanocomposite films. Eng Fract Mech 77(16):3227–3245
10. Li C-T, Langley NR (1985) Improvement in fiber testing of high-modulus single filament materials. J Am Ceram Soc 68:C-202–C-204
11. Sergueeva AV, Zhou J, Meacham BE, Branagan DJ (2009) Gage length and sample size effect on measured properties during tensile testing. Mater Sci Eng A 526:79–83

Optimal Partial Safety Factor Calibration of Structure Considering Series System Failure

P. P. Roy, P. Kumar, and S. Bhattacharjya

Abstract Partial safety factors (PSFs) given in Indian Code of Practice for plain and reinforced concrete (IS 456: 2000) are the same for live load as well as for dead load for limit states of collapse as well as for limit states for serviceability. However, it is obvious that uncertainty in live load is more compared to dead load. Hence, proper calibration of code is a must for safe and economic design of structure. In this work, considering uncertainty in live load is more, partial safety factors using the first order reliability method are calculated. In the first module, PSFs are calculated for a single limit state function i.e., flexure only for a simply supported beam under UDL. The results show significant variations in PSFs with variations in loadings. In the second module, a set of combined PSFs are obtained using multivariate normal distribution concept for flexure as well as the shear limit state under static loading as well as under dynamic loading for the same simply supported beam. Here also, it is observed that the PSFs vary with variations in loads and changes in geometric properties. To overcome this problem, an optimization has been performed for different live load to dead load ratios using Baker's approach. Finally, a unique set of PSFs are worked out which is valid for all load ratios under static and dynamic loads. A comparison is also made with IS code specified PSFs, which shows that the proposed set of PSFs yields more economical section than the existing PSFs of IS Code.

Keywords Partial safety factor · Reliability analysis · Code calibration

P. P. Roy (✉) · P. Kumar · S. Bhattacharjya
Department of Civil Engineering, Indian Institute of Engineering, Science and Technology, Shibpur, Howrah 711103, India
e-mail: 2020cepr024.partha@students.iiests.ac.in

S. Bhattacharjya
e-mail: soumya@civil.iiests.ac.in

R. K. Ray et al. (eds.), *Advances in Theoretical and Applied Mechanics*, Lecture Notes in Mechanical Engineering, https://doi.org/10.1007/978-981-97-0418-7_13

1 Introduction

Structural engineers always aim to design safe structures. The structures are designed for the loads which are anticipated to act on that structure. But the loads are not at all deterministic, rather probabilistic in nature. To account for the variation in the load, the PSFs are considered. Partial safety factors cannot be taken as an arbitrary value. Also, if the maximum load is considered for designing a structure, it becomes uneconomical as this load will act on the structure once or twice. So, it is wiser to calibrate PSFs for a specified level of risk so that the obtained designs become safe and economic.

In India, the reinforced concrete (RC) structures are designed using code IS 456 [1]. The PSFs given in the code are the same for Live Load (LL) as well as for Dead Load (DL) for all load cases and all limit states. But it is obvious that uncertainty in LL is more compared to DL. Hence, the PSFs of LL should be more than that of DL. However, this fact is not reflected in the present version of the IS code. So, proper calibration of the code is required to reveal the effect of uncertainty more realistically. While referring to other internationally accepted code of practice for reinforced concrete it is evident that other than the Indian Code, we can use higher PSF for LL compared to DL (refer Table 1). It can be noted that the perspectives are different for different foreign standards. Moreover, the level of uncertainty in RC construction is different in different countries due to the source of materials, workmen skills, etc. Thus, for the Indian scenario, PSFs will be different. As a result, a separate calibration of PSFs is inevitably required, rather than adopting the PSFs just from a foreign code.

The PSFs can be calculated by reliability methods. There are a lot of methods by which one can compute appropriate PSFs using the reliability concept. Two most common methods, namely Monte Carlo Simulation (MCS) and first order reliability method (FORM) are often used. The former is computationally extensive [2]. Hence, in this study, a FORM-based PSF evaluation strategy is followed. Many efforts have been noted to explain the methods for calculating a more reliable PSF for different structures (e.g. Reinforced beam, prestressed beam, retaining wall etc.). Ranganathan [3] explained FORM-based code calibration for the simple structure. But only one limit state of failure and one type of loading is considered at a time. Bhattacharya

Table 1 PSFs of DL and LL for limit state of collapse

Codes	Limit state of collapse	
	DL	LL
IS 456: 2000 (Indian)	1.50	1.50
ACI 318: 2019 (American)	1.20	1.60
EC2: 2004 (European)	1.35	1.50
CSA A23.3: 2004 (Canadian)	1.25	1.50
CoP-HK: 2004 (Hong Kong)	1.40	1.60
AS/NZS 1170:2002 (ANZ)	1.20	1.50

et al. [4] reviewed the reliability of various existing structures and available analytical methods for determining target reliabilities illustrated with the US Navy's Mobile Offshore Base concept. Kim and Suh [5] evaluated the target reliability indices and PSFs for caisson sliding of a vertical breakwater using FORM. Agarwal and Bhattacharya [6] obtained PSFs for a rectangular prestressed concrete beam in an ultimate flexural limit state using the FORM and Gradient Projection Method. Kotes et al. [7] supported Eurocodes for calculating PSFs for design of bridges. Melchers and Beck [8] presented FORM-based methods for evaluating PSFs for safety analysis of existing structures. In this context, the book of Vereecken et al. [9], in addressing PSFs for temporary scaffold structures, can also be also considered. Cheng and Yao [10] studied the PSF of load for RC columns through a deterministic optimization design. Jacinto et al. [11] develop a set of analytical expressions for the probabilistic models to calibrate PSF using FORM. In this regard, Lim et al. [12] characterised disaster resilience by reliability, redundancy, and recoverability and undertaken safety analysis using these newly defined measures, which is worth mentioning.

The above literature address PSF evaluation using a single limit state of failure. But, in practice, there may be multiple limit states of failure of collapse, which in series characterize the actual failure of the system. In this regard, Harthy and Frangopol [13] used second moment reliability methods for components and systems integrated with optimization tools in the design of prestressed concrete beams. Hu and Du [14] developed a new system reliability method to accurately estimate product reliability using multivariate normal distribution. Sýkora et al. [15] used two reliability levels that are needed: (1) the minimum level below which the structure is unreliable and should be upgraded, and (2) the target level indicating an optimum upgrade strategy. It is recommended that these levels be established using economic optimisation and the marginal life-saving costs principle, as both the approaches take into account the costs of safety measures and the failure consequences. It is noted that, an optimization strategy is required to evaluate a set of PSF, which will be applicable for a wide range of loads and structures.

Keeping the above in board, an attempt is made in this paper to evaluate PSFs, which will consider multiple limit states of failure, simultaneously. The proposed PSFs should be invariant overloads, and applicable for a wide range of loading conditions. A FORM-based method is adopted as it is simple and widely used. Multi-variate normal distribution concept is used to evaluate PSFs for a series system. Baker's approach is adopted to yield a set of optimal PSFs, which is invariant over various load ranges. To elucidate the method, a simply supported beam under static load (uniformly distributed load and point load), and dynamic loads (harmonic, impulse, and moving) is considered. The limit states of flexure and shear are simultaneously taken into account. With the proposed set of PSFs in this study, a design comparison is made with that by the IS:456-2000.

2 The FORM-Based Optimal PSF Evaluation

FORM calculates the reliability by mapping the failure surface onto the standard normal space and then by approximating it with a tangent hyperplane at the design point. Provided that the reliability problem is well- behaved and straightforward, the solutions obtained by FORM are reasonably close to that obtained by the simulation method. Consider the fundamental case of the two variable problem i.e.

$$G = R - S \tag{1}$$

Here, G is the limit state equation, R is the resistance (capacity) effect, and S is the load effect. Using the Haofer-Lind method [16], reliability index can be defined as:

$$\beta = \frac{(\mu_R - \mu_S)}{\sqrt{(\sigma_R^2 + \sigma_S^2)}} \tag{2}$$

In the above, μ and σ denote mean and standard deviation, respectively. Equation (2) is valid for normal distribution of load and resistance. The probability of failure P_{fi} for the ith limit state is given by,

$$P_{fi} = \Phi(-\beta_i) \tag{3}$$

where, Φ is the cumulative distribution function of the standard normal variable.

The task of a system designer is to accurately predict system reliability encompassing all the component reliabilities. The system may be in series or in parallel arrangement. The series system is a weak system because failure of a single component leads to failure of a complete system. Thus, the series system is also called as the weakest link model. The proposed study deals with a simply supported beam only. Hence, a series system is considered. The probability of failure in the series system is given as [17],

$$P_{fs} = 1 - \prod_{(i=1)}^{n} \left(1 - P_{fi}\right) \tag{4}$$

where P_{fs} is the probability of system failure and P_{fi} is the probability of failure of individual component failure. Using, the multivariate normal distribution concept of [14] the mean ($\boldsymbol{\mu}$) and covariance matrix ($\boldsymbol{\Sigma}$) of system failure is expressed as

$$\boldsymbol{\mu} = (-\beta_1, -\beta_2, \ldots, -\beta_m); \quad \boldsymbol{\Sigma} = \begin{bmatrix} 1.0 & \rho_{12} & .. & \rho_{1m} \\ \rho_{21} & \rho_{22} & .. & \rho_{2m} \\ .. & .. & .. & .. \\ \rho_{m1} & \rho_{m2} & .. & \rho_{mm} \end{bmatrix} \tag{5}$$

where, β_i refers to the reliability index of the i-th limit state function and ρij denotes the correlation coefficient between the i-th and j-th limit state functions. The combined reliability is then given by [14]

$$R_i = \phi_k(-\mu_i, \Sigma) \quad \text{with} \quad \beta_i = \phi^{-1}(R_i) \tag{6}$$

where, ϕ_k is the probability density function of the normally distributed variate. Denoting ε as a small step size, a coefficient, α_i, is expressed as,

$$\alpha_i = \frac{\partial \beta}{\partial \varepsilon} = \frac{\beta_{ij} - \beta_i}{\varepsilon} \tag{7}$$

The PSFs can be obtained as [14],

$$\Upsilon_{i,j} = \left[\Phi_{Lj}^{-1}\{\phi(-\beta_i \alpha_{i,Lj})\}\right]/\mu_{Lj}, \quad \forall j \in p, \tag{8}$$

As mentioned, PSFs vary with change in load values. But for convenience of industrial community, the final PSFs should be constant at least over a group of design problems with varying loading types and values. Thus, to propose a unique set of PSFs, an optimization scheme is adopted as per Baker's approach [11]. Here, a logarithmic function is minimized to arrive at the best possible set of PSFs, which is expressed as,

$$S = \sum_i \{\log_{10}(P_{ft}) - \log_{10}(p_{fi})\}_i^2 \times W_i \tag{9}$$

where, P_{ft} is probability of failure corresponding to determined reliability indexes, p_{fi} is probability of failure corresponding to target reliability index and W_i is relative weight.

3 Numerical Study

In this section, a set of PSFs is computed for a RC simply supported beam under static load and dynamic load. Limit states of failure for flexure and shear are considered with various loadings—(i) uniformly distributed load, (ii) concentrated load, (iii) an impulse load, (iv) time-varying sinusoidal load, and (v) moving load due to traffic. The combined PSFs are evaluated as per Eq. (8), and the optimization is performed by minimizing Eq. (9) for all the above load cases. For each case, the optimal PSF for DL and LL is recommended. Finally, taking all the considered load on a single board, a unique set of PSFs is proposed.

To start with, only a limit state of flexure under an uniformly distributed load is considered. The beam (Fig. 1) has cross-sectional width b, depth d, DL W_D, LL W_L, span l and grade of concrete f_{ck}. The limit state function is given by [1]:

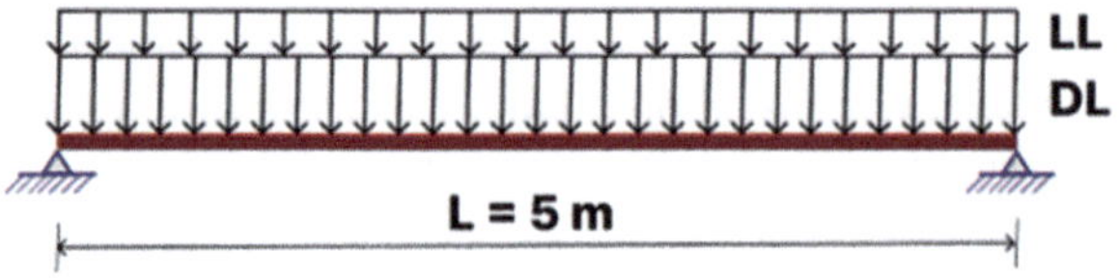

Fig. 1 Simply supported beam subjected to uniformly distributed DL and LL

$$0.1338 f_{ck} b d^2 - W_D l^2/8 - W_L l^2/8 = 0 \tag{10}$$

The deterministic length of the beam is considered as 5.0 m. The target reliability index (β) is 4.0 The uncertain variables are listed in Table 2. The obtained PSFs are presented for a wide range of DL and LL in Fig. 2a–d. It is observed that for a particular live load if dead load on a structure increases the PSF for live load decreases but the same for dead load increases. But, in all the cases the PSFs of DL and LL vary significantly with load levels.

Table 2 Uncertain variables considered

Variable	Mean/nominal	Nominal value (X)	Mean (μ)	Standard deviation (σ)
f_{ck}	1.34	20.0 (MPa)	26.8 (MPa)	2.68 (MPa)
W_D	1.05	15.0 (kN)	15.75 (kN)	2.12 (kN)
W_L	0.70	25.0 (kN)	17.50 (kN)	2.18 (kN)

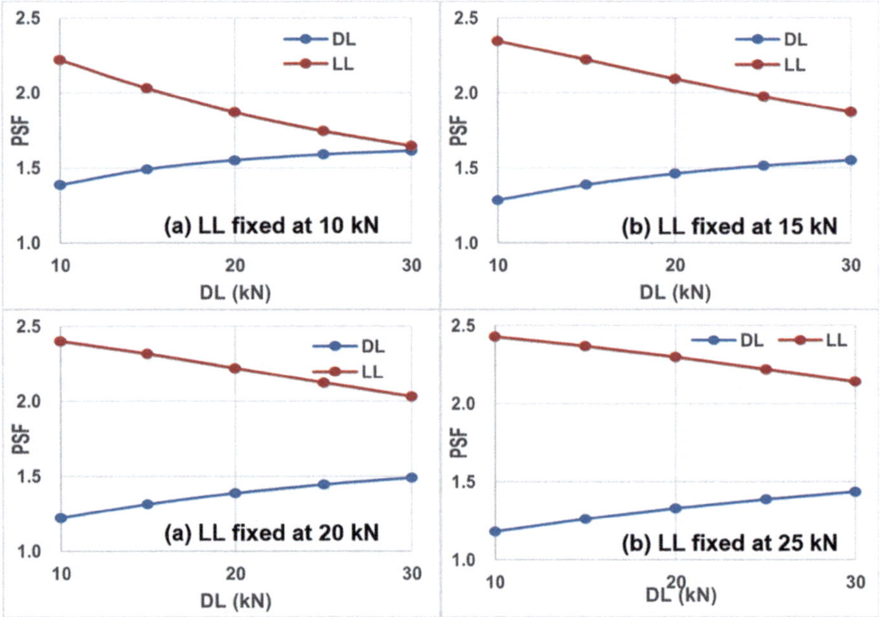

Fig. 2 **a–d** PSF of DL and LL for fixed valued of LL

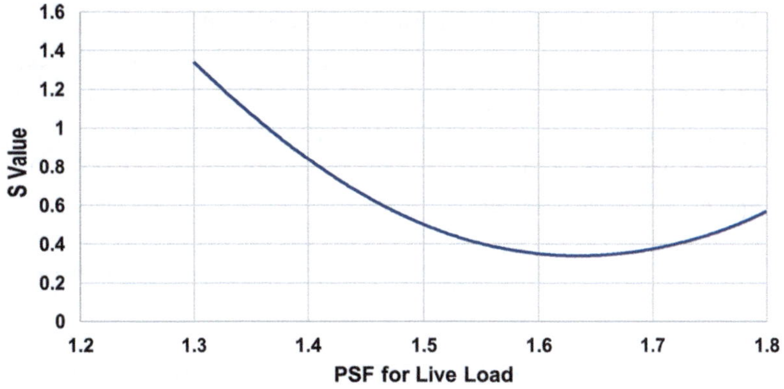

Fig. 3 *S* value for dead load PSF = 1.2

An S- curve is presented in Fig. 3, which yields PSF of LL as 1.3, when PSF of DL is 1.2 (PSF of DL is taken 1.2 in most of the foreign design standards).

Figures 2 and 3 are obtained for all the load cases (ii)–(v). For load case (ii), the point load for LL is taken as $W_L \times l$ acting at midspan to preserve the equivalency. For the load case (iii), the impulse load is of the same magnitude for that of the point load LL, exerted for a time of 0.1 secs at midspan. In load case, (iv) a sinusoidal harmonic load is considered to act at the midpoint of the beam with a frequency of 10 cps for 100 cycles. In the load case (v), a moving load of magnitude $W_L \times l$ is considered to move from left to right with a step increment of 0.1 m along the length. The bending moment and shear forces for load cases (iiii)–(v) are obtained by the response surface method following [2]. The uniform design method is adopted as design of experiment [2]. The obtained PSFs are summarized in Table 3.

It can be noted from Table 3 that the PSF for all other load cases except point loading is found to be 1.3 and for point load it is 1.4. The corresponding PSF for dead load is 1.2. Using these partial safety factors the required depth of the RC beam is calculated and presented in Table 4. The depth requirement by the IS:456-2000 is also presented in the same table. It can be clearly observed that the proposed PSF yields more economic design. The depth requirement by the proposed PSF is 26% less than the present IS:456-2000-based design approach.

Table 3 Optimal PSFs for various load

S. No.	Types of loadings	PSF for LL
I	Uniformly distributed load	1.3
II	Point load	1.4
III	Impulse load	1.3
IV	Harmonic load	1.3
V	Moving load	1.3

Table 4 Comparison of calibrated PSF with PSF suggested in IS code

	Proposed PSFs	PSF of IS code
PSF for DL + LL (point load)	1.2 DL + 1.4 LL	1.5 DL + 1.5 LL
Design depth of beam for point load	310 mm	418 mm
PSF for DL + LL (all other cases)	1.2 DL + 1.3 LL	1.5 DL + 1.5 LL
Design depth of beam for all other cases	307 mm	418 mm

4 Conclusion

A PSF evaluation strategy using FORM, multi-variate normal distribution, and Baker's S-curve-based optimization approach is presented. The method is tested for a simply supported RC beam subjected to various forms of loading—uniformly distributed load, point load, impulse load, harmonic load, and moving traffic load. Limit states of flexure and shear are simultaneously considered. The present study predicts PSF for DL as 1.2, and PSF for LL as 1.3 for all the static and dynamic load cases, except for the point load case. The PSF for LL is 1.4 for the point load case. Based on these findings, a design comparison is made with that by the IS:456-2000. The results show that using the PSF sets proposed in this study, one can have more economical design, maintaining the desired target reliability requirement. The depth requirement of the beam also becomes almost 26% less than the conventional design approach.

However, here FORM is used to assess reliability but for higher uncertainty levels. For more complex structures with multi-dimensional limit state functions subjected stochastic dynamic load, significant error may be induced by the FORM. In such cases, second order reliability analysis or Monte Carlo simulation can be performed. The study can be also extended by including continuous beams or frame structures. In the present case, only flexure and shear limit states are used, which leaves the scope of including other limit states. In fact, for arriving at a definite set of PSF, an exhaustive and rigorous numerical and practical experimentation is needed, encompassing various forms of structures and loadings. The present small study is only the stimulator of such a need for safe economic design perspective of India.

References

1. BIS, IS 456:2000 (2000) Indian standard plain and reinforced concrete—code of practice, 4th rev. Bureau of Indian Standards, New Delhi
2. Bhandari A, Datta G, Bhattacharjya S (2018) Efficient wind fragility analysis of RC high rise building through meta modelling. Wind Struct 27(3):199–211
3. Ranganathan R.: Structural reliability: analysis and design, 2nd impression
4. Bhattacharya B, Basu R, Ma K (2001) Developing target reliability for novel structures: the case of the mobile offshore base. Mar Struct 14

5. Kim SW, Suh KD (2011) Evaluation of target reliability indices and partial safety factors for sliding of caisson breakwaters. J Coast Res
6. Agarwal G, Bhattacharya B (2010) Partial safety factor design of rectangular partial prestressed concrete beams in ultimate flexural limit state. J Struct Eng 37(4)
7. Koteš P, Prokop J, Strieška M, Vičan J (2017) Calibration of partial safety factors according to Eurocodes. MATEC Web Conf 117
8. Melchers RE, Beck AT (2018) Structural reliability analysis and prediction, 3rd edn. Wiley
9. Vereecken E, Botte W, Droogné D, Caspeele R (2019) Reliability-based calibration of partial factors for the design of temporary scaffold structures. Struct Infrastr Eng 16
10. Cheng Z, Yao J (2021) Study on partial factor of load for reinforced concrete columns. Math Probl Eng
11. Jacinto L, Santos LO, Neves LAC (2021) Calibration of partial safety factors using FORM, 29 RPEE, Series III, No 14
12. Lim S, Kim T, Song J (2022) System-reliability-based disaster resilience analysis: framework and applications to structural systems. Struct Safety 96
13. Al-Harthy AS, Frangopol DM (1997) Integrating system reliability and optimization in prestressed concrete design. Comput Struct
14. Hu Z, Du X (2019) A partial safety factor method for system reliability prediction with outsourced components. ASCE-ASME J Risk Uncertainty Eng Syst Part B Mech Eng 5
15. Sýkora M, Diamantidis D, Holický M, Jung K (2016) Target reliability for existing structures considering economic and societal aspects. Struct Infrastr Eng 13
16. Hasofer AM, Lind NC (1974) Exact and invariant second-moment code format. J Eng Mech Div 100
17. Thoft-Christensen P, Baker MJ (1982) Structural reliability theory and its applications. Springer-Verlag, Berlin Heidelberg, New York
18. Chakkodi, Ranganathan R (1988) Partial safety factors for RCC design. Int J Struct 8
19. Haldar A, Mahadevan S (2000) Reliability assessment using stochastic finite element analysis. Wiley, New York

A Design Guideline of Laminated Composite Plate Roof with Cut-Out

Asheequl Irshad, Syed Furquan Ghous, Meghashree Patra, and Sanjoy Das Neogi

Abstract The application of laminated composite materials has grown in popularity in a variety of weight-sensitive engineering sectors due to their exceptional stiffness-to-weight ratio, which distinguishes them from other materials. In order to accommodate various utilities, opening or cut-out in the plate has become an essential requirement in various industrial structure. This paper intends to conduct a detailed study of the stress behavior of composite plates with various locations of opening in them for uniformly distributed load intensity. Finite element parametric investigation is undertaken out to determine the stress generation, strain and deflection for cross and angle ply orientations considering simply supported boundary conditions to determine the most suitable position of cut-out.

Keywords Laminated composite plate · Cut-out · Uniform load

1 Introduction

Composite material has gained immense popularity among the weight-sensitive branches of engineering. Composite has certain distinct advantages compared to other traditional materials used in structure, such as steel. This includes low specific weight, high specific strength and good weathering resistance. The most notable property of the composite is that it can be tailored as per the requirement of the user. Among the others, civil engineers too have chosen this material to construct the roofing entities where huge column-free spaces and such weathering resistance is required. Opening/cut-out in the roofs are inevitable when different utilities are

A. Irshad (✉)
University College London, London, UK
e-mail: asheequl.irshad.23@ucl.ac.uk

S. F. Ghous
Bauhaus-Universitat Weimar, Weimar, Germany

M. Patra · S. Das Neogi
Techno International New Town, Kolkata, India

R. K. Ray et al. (eds.), *Advances in Theoretical and Applied Mechanics*, Lecture Notes in Mechanical Engineering, https://doi.org/10.1007/978-981-97-0418-7_14

to be permitted in different levels of elevation. Additionally, in many industries, the use of these openings frequently necessitates the inclusion of cutouts to allow for the passage of light, service lines, and, in some cases, even to change the resonant frequency. The edges of these cutouts must be reinforced to manage stress concentration to ensure structural integrity. Furthermore, the precise positioning of the cutout or opening is an important issue to keep in mind during the design process. The position of cut-out causing the lowest stress and deflection generation can be stated as the most optimized location. This helps the practicing civil engineers in construction keeping the safety and economical aspect in mind. A comprehensive review of the existing literature was done to gain a thorough understanding of the current state of knowledge and the information available in open publications that can contribute to the successful achievement of current objectives. This review likely involved an in-depth examination of relevant academic papers, journals, books, and other sources to gather insights, findings, and methodologies that can be valuable in this study. Pandya and Kant [1] conducted a finite element analysis on laminated composite plates, using a higher-order displacement model. They employed a C0 continuous displacement finite element formulation and introduced a higher-order approach to analyze the flexure of thick indeterminate laminated composite plates subjected to transverse stress. This methodology likely allowed for more accurate and detailed investigations of the behavior of these complex structures, providing valuable insights into their mechanical properties and performance under various loading conditions. Akavci [2–4] contributed to the field by presenting a theory of first-order shear deformation for symmetrically laminated composite plates placed on elastic foundations. This theory introduced two novel hyperbolic shear displacement models specifically designed for orthotropic laminated composite plates. These models, named HPSDT1 and HPSDT2, were utilized to investigate the bending behavior of orthotropic laminated composite plates. The research also included separate analyses of elastic foundation buckling and free vibration characteristics of symmetric and anti-symmetric laminated composite plates. By developing and applying these advanced displacement models, the study aimed to enhance the understanding of the structural behavior and response of laminated composite plates in real-world applications and loading scenarios. In their work, Ying et al. [5] presented accurate solutions for the bending and free vibration behavior of dynamically graded beams placed on a Winkler-Pasternak elastic foundation. They utilized the two-dimensional theory of elasticity to develop these precise solutions. The study likely aimed to investigate the mechanical response and dynamic characteristics of such beams with spatially varying properties resting on a specific type of elastic foundation. By employing the theory of elasticity and providing precise solutions, the research aimed to offer valuable insights into the behavior of dynamically graded beams and their interaction with the underlying foundation, which can be relevant for various engineering and structural applications. In their study, Reddy et al. [6] conducted a bending analysis of laminated composite plates using the finite element method. The main objective of this research was to investigate the influence of transverse shear deformation on the deflection and stresses of laminated composite plates under the action of an evenly distributed load. Transverse shear deformation is a significant consideration in the analysis of thin plates, particularly in the case of laminated composites, where different layers have varying material properties and orientations. The finite element method allowed the researchers to simulate and analyze the complex behavior of these plates, taking into account the effects of transverse shear, and provided insights into the mechanical response and structural performance of laminated composite plates under distributed loading conditions. The findings of this study can be valuable for designing and optimizing composite structures in engineering applications where deflection and stress distribution are critical factors to consider. In their research, Ghughal et al. [7] utilized the trigonometric shear deformation theory (TSDT) to consider the combined effects of longitudinal normal strain and shear deformation in their analysis. The TSDT is an advanced approach that accounts for both bending and shear deformations, making it suitable for studying the behavior of thin and thick plates under various loading conditions. On the other hand, Khateeb and Zenkour [8] conducted a bending analysis of complex plates resting on elastic foundations. The study likely involved investigating the mechanical response and deformation characteristics of these plates when subjected to external loads while considering the interaction with the underlying elastic foundation. Understanding how complex plates behave under such conditions can be crucial in various engineering applications where these structures are commonly used. Both studies contribute to the understanding of the behavior of plates and composite structures, with Ghughal et al. focusing on the incorporation of shear deformation effects through the TSDT and Khateeb and Zenkour examining the interaction with elastic foundations. Their findings can have implications in optimizing the design and performance of plates in real-world engineering applications. In their work, Sayyad [9] investigated the cylindrical bending behavior of orthotropic plates using the nth-order plate theory. This advanced theory allows for more accurate and comprehensive analysis of plate deformation compared to simpler plate theories. To demonstrate the precision and efficiency of the nth-order plate theory, the researcher performed calculations of the displacement and stresses for a simply supported plate that extends infinitely in the y-direction. The plate was subjected to a sinusoidally distributed load. By using this specific example, Sayyad likely aimed to showcase how the nth-order plate theory can accurately predict the behavior of orthotropic plates under complex loading conditions. The findings of this study are valuable in engineering applications where orthotropic plates are commonly utilized, such as in aerospace, civil, and mechanical engineering. Understanding the precise bending behavior and stress distribution in these plates can help in optimizing their design and performance for different practical scenarios. In their study, Ghumare et al. [10] explored the cylindrical bending behavior of laminated composite plates using an advanced shear deformation theory. This theory considered the effects of shear deformation in the displacement field by employing parabolic functions along the thickness coordinate. The use of advanced shear deformation theories is essential for accurately predicting the behavior of laminated composite plates, especially when dealing with complex loading conditions. In contrast, the classical theory of plates (CPT) assumes that plane sections that are initially normal to the mid-surface of the plate will remain plane and parallel to it even after deformation. The CPT is a simplified theory that neglects certain deformation effects, such as transverse shear deformation, which may not be sufficient to accurately describe the behavior of thin laminated composite plates. By using the advanced shear deformation theory, Ghumare et al. aimed to provide a more comprehensive and accurate analysis of the cylindrical bending response of laminated composite plates. This research likely contributes to the understanding and design optimization of composite structures in various engineering applications. The findings of this study can be relevant in industries such as aerospace, automotive, and marine engineering, where laminated composite materials are commonly used to achieve lightweight and high-strength structures. In their work, Setoodeh [11] introduced a closed form method for analyzing the bending and free vibration characteristics of simply supported rectangular laminated composite plates. The proposed method allowed for a more straightforward and efficient approach to calculate the deflection and vibration modes of these plates compared to numerical methods. To conduct their investigations, the study focused on symmetric and anti-symmetric laminates, which are configurations of laminated composite plates with distinct stacking sequences of different materials. The static behavior (bending) and dynamic behavior (free vibration) of both types of laminates were analyzed using an improved first-order shear deformation theory. This theory is more refined than classical plate theories as it considers the effects of transverse shear deformation, which can be crucial in accurately predicting the behavior of thin laminated composite plates. By utilizing the closed form method and the improved shear deformation theory, Setoodeh aimed to provide valuable insights into the structural response of laminated composite plates in different loading conditions. Such studies contribute to the design and optimization of composite structures in engineering applications where laminates are utilized for their high-strength and lightweight properties, such as in aerospace, automotive, and civil engineering.

Above survey of literature reveals the fact that laminated composite plate have been studied for different boundary condition and also for various loading conditions to study the structural responses, but the civil engineering form of composite plate as a roofing entity with cut-out remained scanty. Providing a cut-out at different locations will result in a different response exhibited by the structure. Since the requirement of the cut-out remains inevitable, the analysis of the plate with cutout in different locations is very important to properly assess the most suitable position.

Table 1 Nondimensional fundamental frequencies of plate with a cutout

		$a/h = 10$			$a/h = 1000$			
a′/a	Reddy [12]	Present			Reddy [12]	Present		
		8 × 8	10 × 10	12 × 12		8 × 8	10 × 10	12 × 12
0.0	19.07	19.45	19.34	18.375	19.75	20.11	19.41	19.56
0.2	18.67	18.38	18.47	18.24	19.20	19.77	19.67	19.27
0.4	20.24	20.67	19.8	19.53	20.80	21.35	21.23	20.93
0.5	22.80	22.24	22.6	22.67	23.51	24.27	23.11	23.61

$a/b = 1$, $\upsilon = 0.3$

This paper aims to conduct a numerical study in the stress, strain and deformation analysis of a laminated composite plate with various locations of cut-outs. For this purpose, a problem was first solved for the validation of the present formulation which is presented in Table 1 wherein the results are compared with a published literature [12].

2 Numerical Example

For this purpose, a problem was first solved for the validation of the present formulation which is shown in Table 1.

A square Graphite/epoxy laminated composite plate with different placements of openings have been analyzed using Finite Element Analysis in Ansys Workbench 2022. The typical dimensions of the square plates are (1000 mm × 1000 mm) with a thickness of 10 mm. Square-shaped cut-outs of dimensions (100 mm × 100 mm) are provided. The lamination is considered as five-plied [0°/90°/0°/90°/0°] and five-plied [45°/− 45°/45°/− 45°/45°]. The slab in the numerical model has an equal imposed load across its surface. To apply this load conveniently, pressure was utilized for the loading. The applied load was specifically determined to be 10 kN/m^2. Additionally, all four edges of the plate are simply supported, meaning they are free to rotate but not to displace vertically, providing stable support for the slab.

The material properties of the graphite/epoxy lamina are given below:

$$E_1 = 120\,\text{GPa};\ E_2 = 7.9\,\text{GPa}$$
$$V_{12} = 0.30;\ G_{12} = G_{23} = G_{13} = 5.5\,\text{GPa}$$

Due to the different placements of square cut-outs in the plate, these are referred and classified into different types of panels:

Panel Type I—Plate with square cut-out at center;
Panel Type II—Plate with square cut-out at the middle of span, along the edge;
Panel Type III—Plate with square cut-out at the corner.

The geometry and ply-orientations are illustrated in Fig. 1.

Under the action of the applied load, maximum and minimum values of stress, strain and deformation for different types of panels as mentioned above has been recorded. The mesh sizing is considered as 10 mm which was optimized according to accurate results and low execution time.

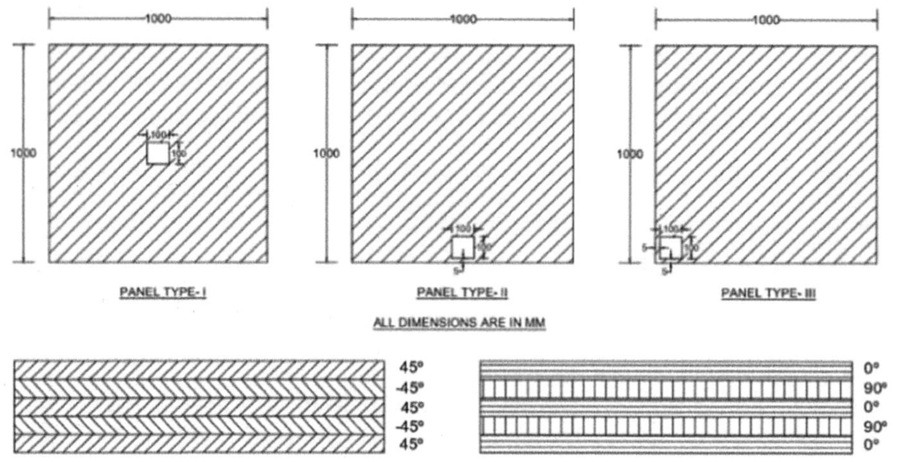

Fig. 1 Plate with different positions of cut-out and ply-orientations

3 Results and Discussions

Under the action of the applied load, the maximum and minimum values of stress, strain and deformation for the different ply orientation of the different panel types as mentioned above have been recorded. With respect to the values obtained the panels are marked on a scale of 1–6 where 6 is awarded to the panel with the least values of stress, strain, deformation and vice-versa. Finally, the marks are added up and ranked to see which position of opening is most appropriate. Figures 2, 3 and 4 show the Major Principal Stress contours, Major Principal Strain contours and Deflection contours respectively for both cross and angle ply orientations of all the 3 panel types. Tables 2, 3 and 4 shows the comparative values of the stated parameters from the detailed numerical analysis.

It is evident from the above Figures and Tables that maximum stress and strain is concentrated around the edges of the openings or the center of the plate and gradually decreases as they move towards the corner. The panel types were marked accordingly and added up. In Table 5 a rank has been assigned to all the Panel types to give a selection guideline in different available options.

4 Conclusions

1. The finite element software employed in this study is well-suited for analyzing composite plate roof panels with cuts, as it produces results that closely match those obtained from the benchmark problems.

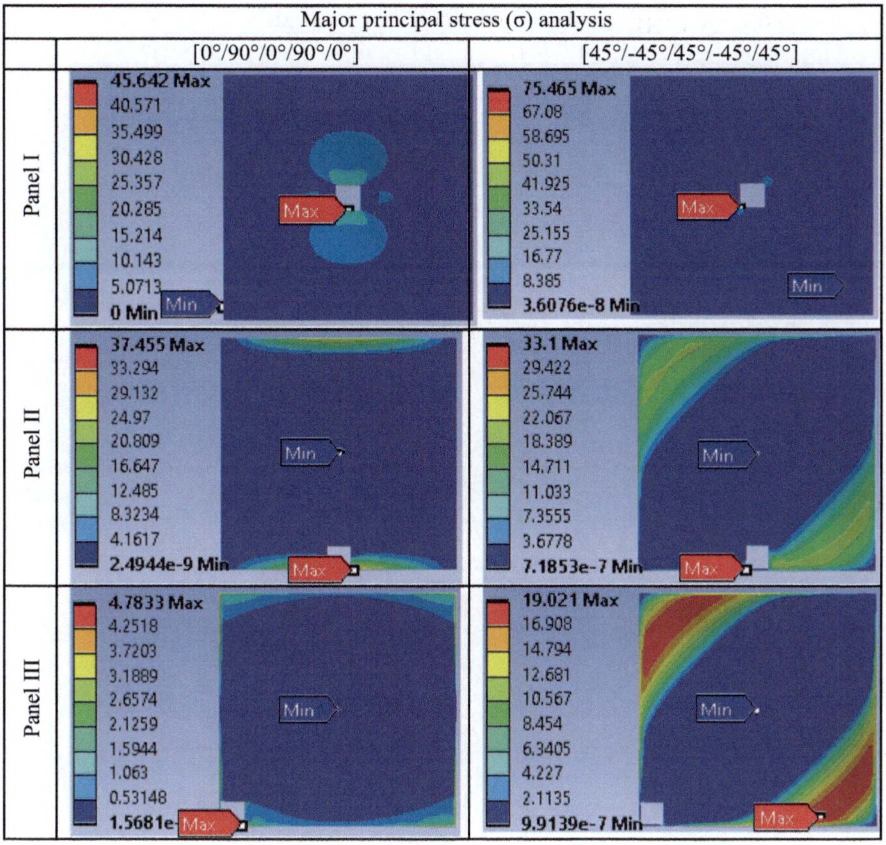

Fig. 2 Major principal stress contours

2. Generally, Angle-plied (AP) plates results in greater stress generation than Cross-plied (CP) plates. For CP plates, the higher stress region is witnessed surrounding the cut-outs and along the spans, whereas, for AP plates, the stress is significantly caused at the corners.
3. For both ply-orientations, strain contour is identical but the values are higher is case of AP plates.
4. Deflection is significantly caused at the center of the plate, but AP plates results in lower deflection than CP plates
5. It evident that the corner opening (Panel III) is the most favorable for both ply-orientation cases.
6. Similarly, the Panel I secured the lowest rank making centre cut-out the most unfavorable position for providing a cut-out.
7. The angle ply orientation of Panel II shows intermediate range of stress, strain and deformation and can be stated as the best alternative of Panel I. Table 4 will help an engineer regarding the eccentricity of a cut-out that can be allowed.

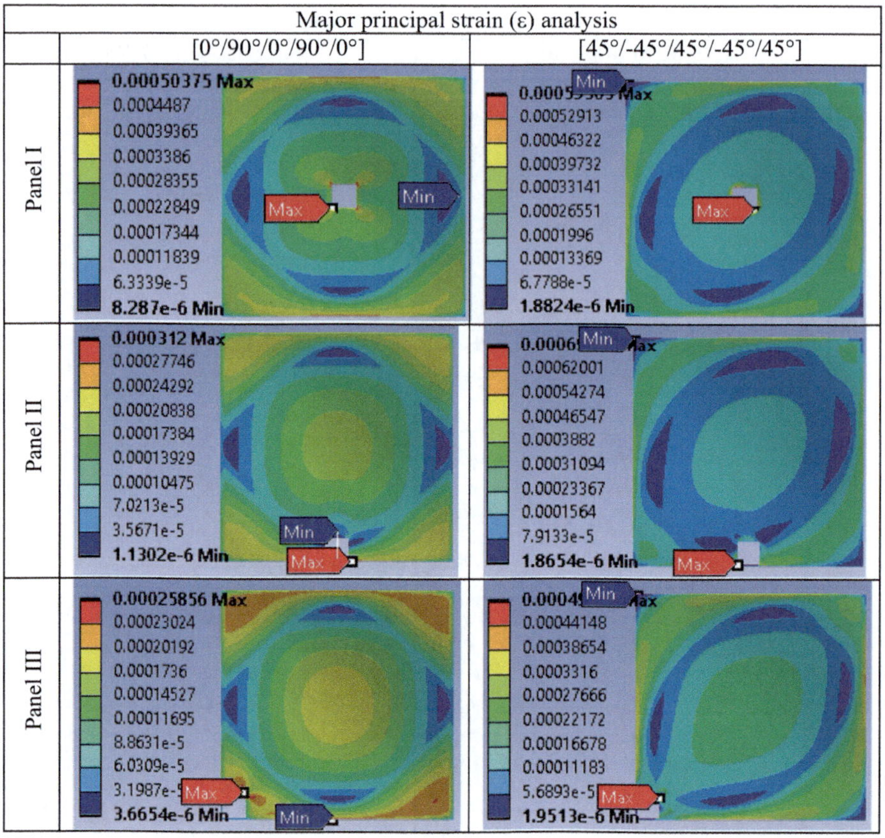

Fig. 3 Major principal stain contours

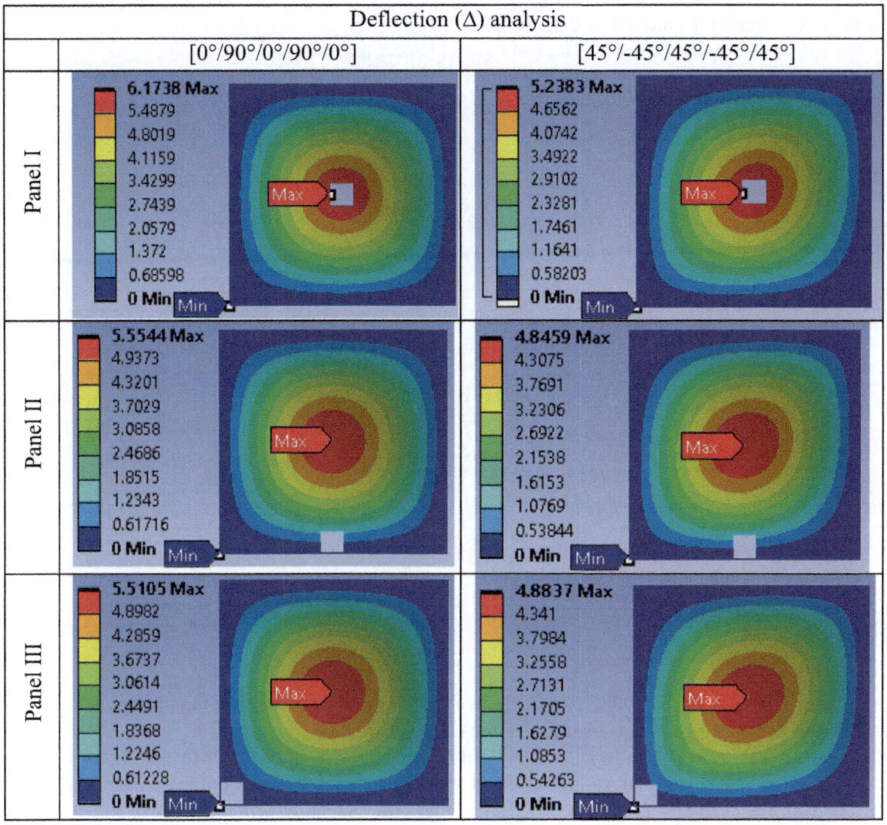

Fig. 4 Deflection contours

Table 2 Comparative analysis of maximum principal stresses

Type	CP		Marks	AP		Marks
	Max. (MPa)	Min. (MPa)		Max. (MPa)	Min. (MPa)	
I	45.642	0	2	92.668	0	1
II	37.455	0	3	33.1	0	4
III	25.33	0	6	29.373	0	5

Table 3 Comparative analysis of maximum principal strains

Type	CP		Marks	AP		Marks
	Max.	Min.		Max.	Min.	
I	6.51E − 04	5.42E − 07	3	1.07E − 03	5.95E − 07	1
II	6.24E − 04	4.01E − 07	4	6.97E − 04	2.22E − 07	2
III	4.23E − 04	1.65E − 06	6	4.96E − 04	6.17E − 07	5

Table 4 Comparative analysis of deflections

Type	CP		Marks	AP		Marks
	Max. (mm)	Min. (mm)		Max. (mm)	Min. (mm)	
I	6.1738	0	1	5.2383	0	4
II	5.5544	0	2	4.8459	0	6
III	5.5105	0	3	4.8837	0	5

Table 5 Rank table

Panel type	Ply-orientation	Total marks secured	Rank
I	[0°/90°/0°/90°/0°]	6	4th
	[45°/– 45°/45°/– 45°/45°]	6	4th
II	[0°/90°/0°/90°/0°]	9	3rd
	[45°/ – 45°/45°/ – 45°/45°]	12	2nd
III	[0°/90°/0°/90°/0°]	15	1st
	[45°/ – 45°/45°/ – 45°/45°]	15	1st

References

1. Pandya BN, Kant T (1988) Finite element analysis of laminated composite plates using a higher-order displacement model. Compos Sci Technol 32:137–155
2. Akavci SS (2007) Buckling and free vibration analysis of symmetric and antisymmetric laminated composite plates on an elastic foundation. J Reinf Plast Compos 26:1907–1919
3. Akavci SS, Yerli HR, Dogan A (2007) The first order shear deformation theory for symmetrically laminated composite plates on elastic foundation. Arab J Sci Eng 32(2B):341–348
4. Akavci SS (2010) Two new hyperbolic shear displacement models for orthotropic laminated composite plates. Mech Compos Mater 46(2):215–226
5. Ying J, Lu CF (2008) Two-dimensional elasticity solutions for functionally graded beams resting on elastic foundations. Compos Struct 84:209–219
6. Reddy SB, Reddy RA (2012) Bending analysis of laminated composite plates using finite element method. Int J Eng Sci Technol 4(2):177–190
7. Ghugal YM, Sayyad AS (2013) Stress analysis of thick laminated plates using trigonometric shear deformation theory. Int J Appl Mech 5(1):1350003, 1–23
8. Khateeb AS, Zenkour MA (2014) A refined four-unknown plate theory for advanced plates resting on elastic foundations in hydrothermal environment. Compos Struct 111:240–248
9. Sayyad AS, Ghumare SM, Sasane ST (2014) Cylindrical bending of orthotropic plate strip based on nth-order plate theory. J Mater Eng Struct 1:47–57
10. Ghumare SM, Naik NS (2015) Cylindrical bending of laminated composite plates using refined shear deformation theory. J Appl Exper Mech 1:10–17
11. Setoodeh AR, Azizi A (2015) Bending and free vibration analyses of rectangular laminated composite plates resting on elastic foundation using a refined shear deformation theory. Iran J Mater Forming 2(2):1–13
12. Reddy JN (1982) Large amplitude flexural vibration of layered composite plates with cut outs. J Sound Vib 83(1):1–10

Reduced-Order Dynamics of MFC-Actuated Plate

Paulomi Mukherjee, Aghna Mukherjee, A. Arockiarajan, and Shaikh Faruque Ali

Abstract Macrofibre composite (MFC) actuated smart materials are highly flexible. Many degrees of freedom are required to analyse continuous flexible structures and monitor their dynamics which makes the process computationally expensive. Hence, the motivation of this study is to reduce the computational cost of real-time finite element solutions of such structures without compromising accuracy. Reduced-order modelling is an excellent alternative to mimic a complicated system into an efficient reduced-order system. In this study, ordinary differential equations are generated for MFC-actuated plate. The equations of motion are solved using the Newmark integration scheme. Two different reduction schemes have been used, i.e. system equivalent reduction and expansion process (SEREP) and proper orthogonal decomposition (POD). Both methods are shown to be equally efficient in substantially reducing the degrees of freedom of the entire structure and preserving all dynamic characteristics in the reduced system, along with high computational efficiency.

1 Introduction

Piezoelectric materials can produce charge under applied mechanical loads [1]. Piezoelectric materials (PZT) are attached to the materials which enables it for sensing and actuating. There are other materials like shape memory alloys (SMAs) [2], electroactive polymers (EAPs) [3], soft and hard magnetic materials [4] which can be similarly used for designing active shell structures. Smart material-integrated struc-

P. Mukherjee · A. Arockiarajan · S. Faruque Ali (✉)
Indian Institute of Technology Madras, Chennai 600036, India
e-mail: sfali@iitm.ac.in

P. Mukherjee
e-mail: am20d030@smail.iitm.ac.in

A. Arockiarajan
e-mail: aarajan@iitm.ac.in

A. Mukherjee
ETH Zurich, Zurich, Switzerland

R. K. Ray et al. (eds.), *Advances in Theoretical and Applied Mechanics*, Lecture Notes in Mechanical Engineering, https://doi.org/10.1007/978-981-97-0418-7_15

tures with actuation or sensing capabilities can be classified as smart structures. As the smart structures are primarily thin-walled, hence they can be modelled based on the classical lamination plate theory (CLPT) [5], first-order shear deformation plate theory (FOSDT) [6], etc. These structures' multi-physics aspects have often been modelled using linear constitutive laws [2]. In other examples in the literature, varying quadratic electrical potential was used with the FOSDT hypothesis for modelling MFC-actuated plates [7]. Mukherjee et al. [8] developed a model to capture the mechanics of a bistable shell with magnetic particles actuated by an external magnetic field using CLPT within a Rayleigh–Ritz framework. Apart from analytical models based on shell theories, several researchers have focused on using FE methods for analysing these smart structures. Carrera et al. [9] developed a geometrically nonlinear FE model for static and dynamic analysis of MFC-actuated shells considering a nonlinear electric potential.

Smart structures are modelled as continuous systems; hence, the complete FE model contains many degrees of freedom (dofs). Hence, it becomes computationally inefficient for real-time applications. To reduce the extensive computation cost along with maintaining accuracy, researchers have proposed reduced-order modelling (ROM). Several methods have been reported for different dynamical systems. Sastry et al. [10] presented a detailed formulation of different ROMs, namely the Guyan reduction, the dynamic condensation, the improved reduced system (IRS), the dynamic IRS (DIRS) and the SEREP. Linear dynamic analysis and application of SEREP in control are studied by Patel et al. [11]. The proper orthogonal decomposition (POD) [12], which is called principle component analysis (PCA), is used for nonlinear systems. For POD, singular value decomposition (SVD) [13] is used, which is cumbersome for large order matrices. Many other reduction techniques like component mode synthesis (CMS) [14] and independent component analysis (IMC) [15] have been used in the literature. All these reduction methods are problem-specific due to the advantage and shortcomings of the respective techniques.

In this study, FEM formulation for composite plates with integrated MFC actuators using FOSDT has been developed. Subsequently, two different reduced-order modelling schemes, SEREP and POD, are applied to the model. Static and dynamic analyses are performed to understand the efficiency of reduction schemes in capturing the full-order characteristics while substantially reducing the computational cost.

1.1 Mathematical Formulation

In this study, a plate with MFC patches on either side of the laminate is studied using FOSDT as shown in Fig. 1. A brief description of the formulation is discussed in the following section.

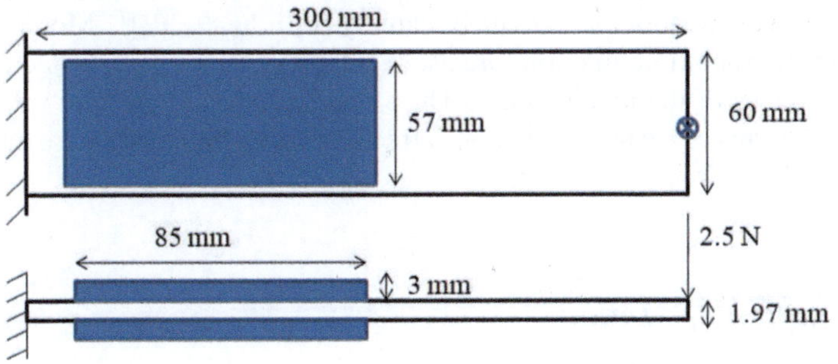

Fig. 1 Cantilever plate with attached MFC patches in bimorph configuration

1.2 Constitutive Relationship

A linear piezoelectric constitutive relation is used as large deformation is not considered in this case. Here, d_{33} type MFC has been used. The linear constitutive equation of the MFCs for sensing and actuation purposes are given by the following equations [16]:

$$\sigma_{ij} = C_{ijkl}\varepsilon_{kl} - e_{kij}E_k \tag{1}$$

$$D_i = e_{ijk}\varepsilon_{jk} + \chi_{ji}E_j, \tag{2}$$

where σ_{ij} and ϵ_{ij} denote the stress tensor and the strain tensor components, respectively, D_i and E_j denote the displacement due to applied electric field and the electric field components, respectively. The coefficients C_{ijkl}, e_{kij} and χ_{ji} denote the elastic compliance constants, the piezoelectric constants and the dielectric constants, respectively.

1.3 Displacement Field

According to FOSDT, the displacement fields are defined as:

$$u(x, y, z, t) = u_o(x, y, t) + z\theta_x(x, y, t) \tag{3}$$

$$v(x, y, z, t) = v_o(x, y, t) + z\theta_y(x, y, t) \tag{4}$$

$$w(x, y, z, t) = w_o(x, y, t), \tag{5}$$

where u_o, v_o, w_o denote the displacements of a point on the plane $z = 0$, θ_x and θ_y are the rotations of transverse normal about y and x-axis, respectively. In this work, 4-noded quadratic elements have been used to discretize the domain. At every node 5 degrees of freedom are selected, namely $[u, v, w, \theta_x, \theta_y]^T$. Moreover, linear Lagrangian interpolation functions are used to define each of the selected degrees of freedom in terms of the nodal values. The stress and strain fields are subsequently calculated in terms of the nodal degrees of freedom using the constitutive relationship defined in the previous subsection.

1.4 Equation of Motion

The equations of motion for the system are derived using Hamilton's principle given as follows [17]:

$$\delta \int_{t_1}^{t_2} (T - U + W_{nc})\mathrm{d}t = 0, \tag{6}$$

where T denotes the kinetic energy of the system and U is the potential energy and W_{nc} represents work done by non-conservative forces. The kinetic energy T of the system is calculated as:

$$T = \frac{1}{2} \int_{v} \rho(\dot{u}^2 + \dot{v}^2 + \dot{w}^2)\mathrm{d}v. \tag{7}$$

The potential energy is defined as:

$$U = \frac{1}{2} \int_{v} (\{\varepsilon\}^T \{\sigma\} - [E]^T \{D\})\mathrm{d}v. \tag{8}$$

The displacement fields in Eqs. 3–5 are substituted in Eqs. 7 and 8. Subsequently, using Eq. 6, the equations of motion of the composite plate actuated by the MFC are obtained as:

$$M_{uu}\ddot{\Delta} + K_{uu}\Delta + K_{u\phi}\phi = F \tag{9}$$

$$K_{\phi u}\Delta + K_{\phi\phi}\phi = Q_a, \tag{10}$$

where Eqs. 9 and 10 are the actuation equation and sensing equation, respectively. In this study, actuator equation is used. M_{uu} is the global mass matrix, K_{uu} is the global stiffness matrix corresponding to mechanical degrees of freedom, $K_{u\phi}$ is the global stiffness matrix due to electro-mechanical coupling and $K_{\phi\phi}$ the global stiffness matrix due to electrical dofs alone. Also, Δ is the vector consisting of transverse displacement and rotation of the structural part of the system. Here, F is the externally applied load on the system and Q_a is the vector of electrical charge. Ignoring the damping effect of structure and substituting the value of ϕ from Eq. 10

into Eq. 9, the governing second-order differential equation for the undamped system is given by

$$M\ddot{X} + KX = F, \tag{11}$$

where $K \in R^{N\times N}$, $M \in R^{N\times N}$ and $X \in R^{N\times 1}$ as mentioned in Eq. 11 are element stiffness, mass matrix and displacement vector, respectively.

2 Model Order Reduction Technique

Consider a model of a linear system with N degrees of freedom. In order to develop a reduced model, N degrees of freedom system is replaced with an equivalent n degrees of freedom system such that $n < N$. The transformation can be performed in several ways, among them two of the methods are discussed in the following section.

2.1 *System Equivalent Reduction and Expansion Technique (SEREP)*

SEREP is a reduction process where several normal modes are reserved. The process is completed in two steps, i.e. reduction in the number of modes and reduction in degrees of freedom [11]. Considering a N degrees of freedom finite element model for a linear vibrating system. From the full-order mass and stiffness matrix, modal matrix $\phi \in R^{N\times N}$ can be obtained. Using the modal matrix, the system can be represented in modal coordinate system, $q \in R^{N\times 1}$, as

$$X = \phi_{N\times N} q_{N\times 1}. \tag{12}$$

Let $\Phi \in R^{N\times N}$ denote the modal matrix and p modes are retained among the N modes. The degrees of freedom, X, can be partitioned into master and slave degrees of freedom. Therefore, dofs can be written as

$$X = \begin{Bmatrix} X_m \\ X_s \end{Bmatrix}. \tag{13}$$

Then, $\Phi \in R^{N\times p}$ is the complete modal matrix with p modes, $\Phi_m \in R^{m\times p}$ is the master modal matrix and $\Phi_s \in R^{(N-m)\times p}$ is the slave modal matrix. Introducing a $(p \times 1)$ vector of generalized coordinates $Z(t)$, therefore the displacement vector is given by

$$X(t) = \begin{Bmatrix} X_m(t) \\ X_s(t) \end{Bmatrix} = \begin{bmatrix} \Phi_m \\ \Phi_s \end{bmatrix} \{Z(t)\}. \tag{14}$$

Solving the first row,

$$\{Z(t)\} = [\Phi_m]^{+}\{X_m(t)\}, \tag{15}$$

where $[\Phi_m]^{+}$ is the pseudoinverse of Φ_m, substituting Eq. 15 in Eq. 14 we get:

$$X(t) = \begin{Bmatrix} X_m(t) \\ X_s(t) \end{Bmatrix} = \Psi X_m = \begin{bmatrix} \Phi_m \\ \Phi_s \end{bmatrix} \left[\Phi_m^T \Phi_m\right]^{-1} \Phi_m^T \{X_m(t)\}. \tag{16}$$

The transformation matrix is thus given by:

$$\Psi = \begin{bmatrix} \Phi_m \\ \Phi_s \end{bmatrix} \left[\Phi_m^T\ \Phi_m\right]^{-1} \Phi_m^T. \tag{17}$$

The reduced mass and stiffness matrices are given by:

$$M_r = \Psi^T M \Psi,\ K_r = \Psi^T K \Psi.$$

In the present study, damping of the system is not considered.

2.2 Proper Orthogonal Decomposition (POD)

In this procedure, a transient response of all degrees of freedom is obtained using the Newmark-beta method [17], by providing initial boundary conditions. Then the full-order system can be expressed in the form of a matrix as shown

$$O = \begin{pmatrix} Z_1(t_1) & \dots & Z_m(t_1) \\ \vdots & \ddots & \vdots \\ Z_1(t_n) & \cdots & Z_m(t_n) \end{pmatrix}. \tag{18}$$

Here m is the number of degrees of freedom and n denotes the number of time steps considered. Then singular value decomposition (SVD) is applied on O, which divides the matrix into three different submatrices.

$$[W, \delta, V] = SVD(O), \tag{19}$$

where W and V are orthogonal matrices. δ is a diagonal matrix, with, $\delta_1 > \delta_2 > \delta_3 \cdots > \delta_m > 0$. Now selecting the p important proper orthogonal modes, from W matrix, the transformation matrix can be written in the form below

$$\Psi = W(:, 1 : p). \tag{20}$$

The reduced mass and stiffness matrices are

$$M_r = \Psi^T M \Psi, K_r = \Psi^T K \Psi.$$

SEREP is a suitable reduction technique for linear systems, and POD is useful in the nonlinear system as well. The accuracy of these methods has been highlighted in the subsequent sections.

3 Results and Discussion

In this section, two plates are analysed, one is a simply-supported composite plate and the other is a cantilever plate attached with MFC on either side. Static and dynamic analyses are performed to understand the efficiency of the reduced-order model in capturing system characteristics with reduced computational cost.

3.1 *Analysis of Simply-Supported Composite Plate*

An example of a simply-supported composite plate as shown in Fig. 2 has been solved numerically using FOSDT and Neumark Beta time integration method. The properties of the composite plate are mentioned in Table 1. The plate dimension is 250×250×10 mm. After the convergence study, the plate is discretized into 110 nodes and each node has 5 dofs. The following studies are done on the same geometry.

Study 1: The static analysis of the simply-supported plate under point load at the midpoint is performed numerically in MATLAB and ABAQUS®. The comparison of the centre line normalized displacement is shown in Fig. 2b. Normalization is done

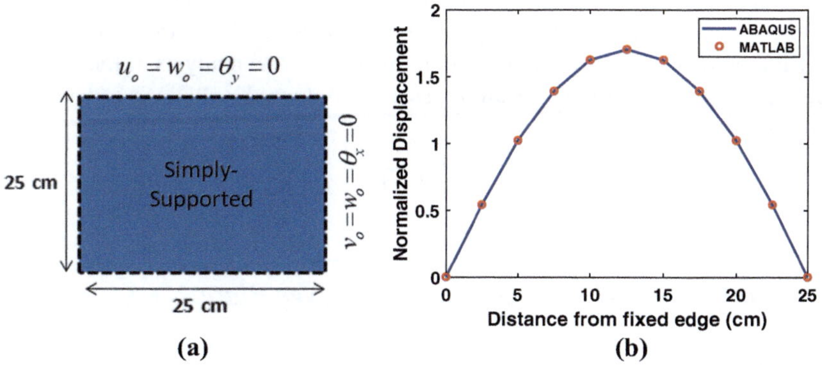

Fig. 2 (**a**) Schematic of simply-supported plate. (**b**) Midpoint static displacement comparison between MATLAB and ABAQUS® under point load applied on the mid-point

Table 1 Properties of plates and MFC

MFC-d_{33}	Aluminium plate	Composite plate
$E_1 = E_2 = 15.2$ GPa	$E_1 = E_2 = 70$ GPa	$E_1 = 15.2$ GPa
$E_3 = 29.4$	$\nu_{32} = \nu_{31} = 0.32$	$E_2 = 2.1$ GPa
$\nu_{32} = \nu_{31} = 0.312$	$G_{31} = 26.515$ GPa	$\nu_{12} = 0.25$
$G_{32} = G_{31} = 6.06$	**Mild steel plate**	$G_{12} = 0.5E_2$
$d_{33} = 467 \times 10^{-12}$	$E_1 = E_2 = 210$ GPa	$\rho = 800\,\mathrm{Ns^2/m^4}$
$d_{32} = -210 \times 10^{-12}$	$\nu_{32} = \nu_{31} = 0.29$	
$\alpha = 0.75 \times 10^{-6}/°\mathrm{C}$		

using Eq. 21, where w is the actual transverse displacement, a and h are plate width and thickness, respectively, and q is the applied load. Results from both methods are matching closely which validates the present FEM formulation.

$$\overline{w} = w\frac{100E_2h^3}{qa^4}. \tag{21}$$

Study 2: Frequency analysis of the same plate is conducted on full-order model as well as on reduced-order models using SEREP and POD. The first five natural frequencies are compared as shown in Table 2. Reduced-order model frequencies are matching accurately with the full-order model which signifies the efficiency of the reduced-order model.

Study 3: The time history of the above-mentioned simply-supported plate is evaluated under step loading of 10 N applied at the mid-point of the plate. The transverse displacements are normalized using Eq. 21. The normalized time history is compared between full-order and reduced-order models as shown in Fig. 3a. The analysis using the three methods is repeated for a triangular load. The load used for this study is given in Eq. 22. The mid-point normalized displacement curve between the full-order model and reduced models is shown in Fig. 3b. It is shown, the reduced-order model can accurately replicate the full-order model with a lesser number of degrees of freedom. The computation time is improved by 99.27% for POD and 98.3% for SEREP as shown in Table 3. SEREP requires eigenvalue analysis of the full system which is an inevitable time-consuming step as compared to POD as reflected in the result as well.

$$\overline{q} = 10\left(\frac{1-tt}{TT}\right). \tag{22}$$

3.2 Analysis on MFC-Actuated Cantilever Plate

Here, an MFC-actuated cantilever plate in a bimorph configuration is analysed numerically with the FEM formulation as discussed in Sect. 2 and with ABAQUS®. For modelling the MFC patch in ABAQUS®, an equivalent modelling approach is used.

Table 2 First five natural frequency comparisons of composite simply-supported plate

Mode	Full-order model (Hz)	SEREP (Hz)	POD (Hz)
1	1208.3	1208.3	1208.3
2	3393.2	3393.2	3393.2
3	3434.5	3434.5	3434.5
4	4928.8	4928.8	4928.8
5	7275.5	7275.5	7275.5

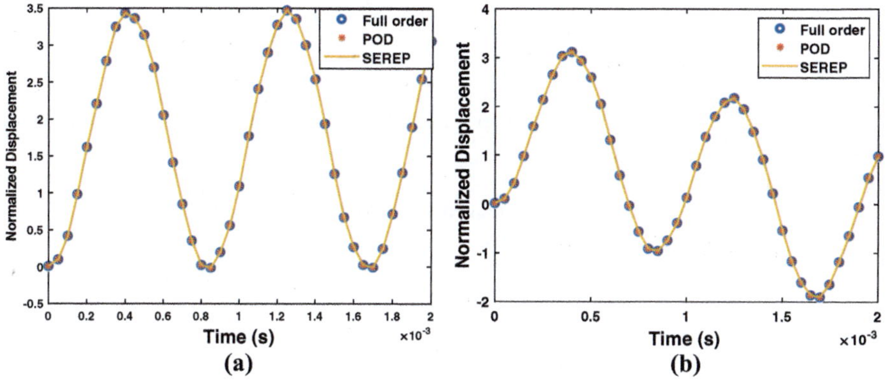

Fig. 3 Mid-point time history comparison of full-order and reduced-order model of simply-supported plate under (**a**) step loading applied at the centre. (**b**) Triangular loading applied at the centre

Table 3 Analysis time comparison between full-order and reduced-order model

	Full-order model	SEREP	POD
Dimension	550×550	15×15	5×5
Analysis time (s)	0.412	0.007	0.003

The actuation using an applied potential difference is mimicked using an equivalent coefficient of thermal expansion. The selected CTE is equal to the free strain per volt that the MFC produces on the application of a potential difference across it. The details of the equivalent modelling technique can be found in [18]. For validation, an aluminium cantilever plate of dimension $300 \times 60 \times 1.97$ mm, properties mentioned in Table 1, with two MFC-d_{33} patches attached in a bimorph configuration, as shown in Fig. 4a is analysed. A potential difference of 400 V is applied across the terminals of the top MFC while the bottom is short-circuited. Figure 4b shows the centerline deflection of the bimorph obtained using the Full-order FE model and ABAQUS®. The result shows good agreement with each other thus validating the FEM formulation.

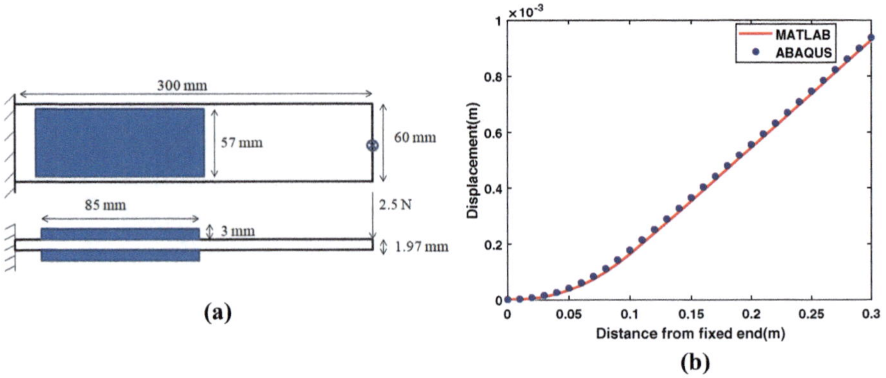

Fig. 4 **a** Schematic of MFC-actuated cantilever plate. **b** Free edge static displacement comparison between MATLAB and ABAQUS®

Table 4 Comparison of first five natural frequencies of MFC-actuated cantilever plate

Mode	Full-order model (Hz)	SEREP (Hz)	POD (Hz)
1	0.0113	0.0113	0.0113
2	0.0635	0.0635	0.0635
3	0.0857	0.0857	0.0857
4	0.1711	0.1711	0.1711
5	0.2479	0.2479	0.2479

$$\mathrm{MAC}_i = \frac{\left|{v_{ir}}^T v_{if}\right|^2}{({v_{ir}}^T v_{ir})({v_{if}}^T v_{if})} \tag{23}$$

Study 2: Frequency analysis of the MFC-actuated cantilever plate with mild steel as the base plate (properties are mentioned in Table 1) is performed and results are compared with reduced-order modelling methods as shown in Table 4. The first five natural frequency obtained using SEREP exactly matches the full-order model. Along with that mode shapes have been compared between the SEREP and the full-order model using modal assurance criteria (MAC) as shown in Fig. 5. The MAC definition is given in Eq. 23, where v_{ir} and v_{if} are the ith normal modes computed using the SEREP and the full-order system, respectively. Unit MAC value represents the exact match of the modes between full-order and reduced-order models. From Fig. 5, it is noticed that the MAC value is near to unity which denotes that the modes are retained in reduced older models as well.

$$q = 400 \sin\left(\frac{\pi}{50} t\right). \tag{24}$$

Study 3: The time series of the free edge midpoint of a mild steel cantilever plate attached with MFC patches in bimorph configuration is compared between full-

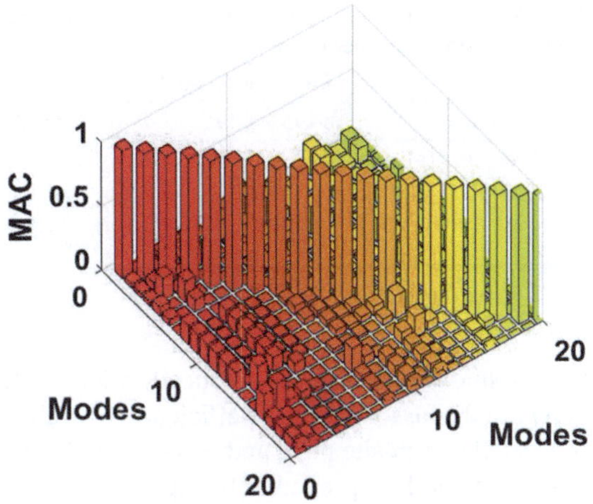

Fig. 5 MAC plot for first 20 modes of MFC-actuated cantilever plate

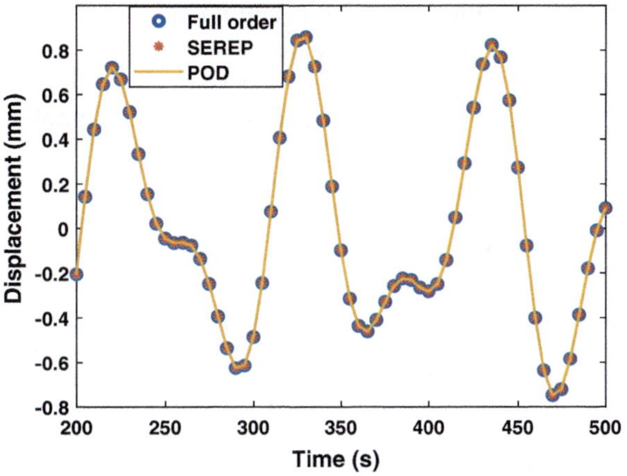

Fig. 6 Free edge midpoint time history comparison between full-order and reduced-order model of MFC actuated cantilever plate

order model and reduced-order models under sinusoidal voltage applied on the top MFC while the bottom one is short-circuited. The applied voltage variation is given in Eq. 24. From Fig. 6, it is observed that the reduced-order models can accurately predict the time history of the free edge midpoint. In a reduced-order model, matrix size can be reduced drastically and the computation efficiency can be increased by 96% and 94.31% for SEREP and POD, respectively, as shown in Table 5. The plots are overlapping on each other hence error norm between the full-order model and the reduced models is observed, and it has come to be 2.7643 and 3.03 for SEREP and POD, respectively, which describes the efficiency of the reduced-order modelling.

Table 5 Comparison of required analysis time of MFC-actuated cantilever plate between full-order model and reduced-order models

	Full-order model	SEREP	POD
Dimension	1495 × 1495	20 × 20	20 × 20
Analysis time (s)	29	1.1	1.65

4 Conclusion

In this study, two different reduced-order modelling techniques like SEREP and POD are discussed and their efficiency is evaluated by applying them on a simply-supported composite plate and on one MFC-actuated cantilever plate. The accuracy of both methods is proved to be high in both case studies. The reduction process is different in both methods although the computation cost has been reduced by more than 95% without compromising the accuracy of the response successfully. These methods can be efficiently used to replicate real-time applications and for solving highly complex time-consuming analyses. Here, only linear problems have been solved using ROMs which can be extended to nonlinear problems as well.

References

1. Cady WG (2018) Piezoelectricity: volume two: an introduction to the theory and applications of electromechanical phenomena in crystals. Courier Dover Publications
2. Mukherjee A, Ali SF, Arockiarajan A (2020) Modeling of integrated shape memory alloy and macro-fiber composite actuated trailing edge. Smart Mater Struct 29(8):085005
3. Viswamurthy S, Ganguli R (2007) Modeling and compensation of piezoceramic actuator hysteresis for helicopter vibration control. Sens Actuators A: Phys 135(2):801–810
4. López-Ortega A, Estrader M, Salazar-Alvarez G, Roca AG, Nogués J (2015) Applications of exchange coupled bi-magnetic hard/soft and soft/hard magnetic core/shell nanoparticles. Phys Rep 553:1–32
5. Raja S, Ikeda T, Dwarakanathan D (2011) Deflection and vibration control of laminated plates using extension and shear actuated fiber composites. Smart Mater Res 2011
6. Hashemi SH, Es'Haghi M, Karimi M (2010) Closed-form vibration analysis of thick annular functionally graded plates with integrated piezoelectric layers. Int J Mech Sci 52(3):410–428
7. Marinković D, Köppe H, Gabbert U (2007) Accurate modeling of the electric field within piezoelectric layers for active composite structures. J Intell Mater Syst Struct 18(5):503–513
8. Mukherjee A, Sudersan S, Ali S, Arockiarajan A (2021) Magnetic actuation of switchable bistable structures: a numerical study. Smart Mater Struct 30(7):075025
9. Carrera E (1997) An improved Reissner-Mindlin-type model for the electromechanical analysis of multilayered plates including piezo-layers. J Intell Mater Syst Struct 8(3):232–248
10. Sastry C, Mahapatra DR, Gopalakrishnan S, Ramamurthy T (2003) An iterative system equivalent reduction expansion process for extraction of high frequency response from reduced order finite element model. Comput Methods Appl Mech Eng 192(15):1821–1840
11. Patel H, Kumar R, Ali SF (2020) Serep integrated control of flexible structures. IFAC-PapersOnLine 53(1):51–56

12. Hsieh WW (2001) Nonlinear principal component analysis by neural networks. Tellus A 53(5):599–615
13. Han S, Feeny B (2003) Application of proper orthogonal decomposition to structural vibration analysis. Mech Syst Signal Process 17(5):989–1001
14. Farhat C, Géradin M (1994) On a component mode synthesis method and its application to incompatible substructures. Comput Struct 51(5):459–473
15. Comon P (1994) Independent component analysis, a new concept? Signal Process 36(3):287–314
16. Zhang S-Q, Wang Z-X, Qin X-S, Zhao G-Z, Schmidt R (2016) Geometrically nonlinear analysis of composite laminated structures with multiple macro-fiber composite (MFC) actuators. Compos Struct 150:62–72
17. Reddy JN (2003) Mechanics of laminated composite plates and shells: theory and analysis. CRC Press
18. Portela P, Camanho P, Weaver P, Bond I (2008) Analysis of morphing, multi stable structures actuated by piezoelectric patches. Comput Struct 86(3–5):347–356

Analysing Machine Learning Approaches for Lamb Wave-Based Damage Detection

Kamal Kishor Prajapati, Akshay Rai, and Mira Mitra

Abstract One of the most efficient strategies to automate the structural health monitoring (SHM) process is the data-driven approach. It requires a data acquisition system, preprocessing techniques for turning the data into useful features, and a machine learning (ML) algorithm for creating a correlation between the structure's final condition and the extracted features. In this study, a variety of ML classification algorithms such as, 'Support Vector Classifier (SVC)', 'Random Forest Classifier', 'Extra Tree Classifier', 'Extreme Gradient Boosting Classifier (XGB Classifier)' and 'Artificial Neural Network (ANN)' have been analysed to determine the best algorithm for identification of damage in a thin aluminium plate. SVC, Random Forest Classifier, and Extra Tree Classifiers have achieved an accuracy of 50%, whereas XGBoost classifier has achieved an accuracy of 58%. It is observed that ANN is able to successfully differentiate between the lamb wave responses of damaged and undamaged samples of thin aluminium plates.

Keywords Damage detection · Lamb wave · ANN · Random forest · XGBoost · SVM · Data-driven approaches · SHM · Machine learning

1 Introduction

A large number of aerospace, mechanical, and civil structures have a fixed working life. If structures are used after this time, the system's functionality or effectiveness may decline. In some scenarios, it can cause a sudden system collapse, which could have major life-threatening or detrimental economic repercussions. In order to avoid these conditions, sensors such as strain gauges, piezoelectric transducers, and optical fibres are permanently integrated into the system. These sensors enable the structural

K. K. Prajapati (✉) · A. Rai · M. Mitra
Department of Aerospace Engineering, Indian Institute of Technology Kharagpur, 721302 Kharagpur, West Bengal, India
e-mail: kkpkgp@gmail.com

M. Mitra
e-mail: mira@aero.iitkgp.ac.in

R. K. Ray et al. (eds.), *Advances in Theoretical and Applied Mechanics*, Lecture Notes in Mechanical Engineering, https://doi.org/10.1007/978-981-97-0418-7_16

health monitoring (SHM) system to detect, locate and measure damage in structures at an early stage [1–3].

The data-driven approach of SHM needs past data in order to identify and characterize the damage. It depends on the trends or patterns of the data to detect the end-life behaviour of the structure. A SHM system based on data-driven techniques requires a data acquisition system to gather sensor data, preprocessing methods to extract useful features, and machine learning (ML) algorithms to identify the presence of damage [4, 5].

Numerous data-driven ML techniques, including Artificial Neural Networks (ANNs) and Support Vector Machines (SVMs), have been suggested and put into practice to automate the damage identification process [6]. To identify the crack damage in aluminium plate, Lu et al. [7] have developed an ANN model based on inverse analysis. Nick et al. [8] have used three supervised learning approaches (SVM, feedforward neural networks (FNN) and naive Bayes classifiers), and two unsupervised techniques (self-organizing maps (SOM) and k-means) to address the problem associated with SHM. Nick et al. [9] have implemented and compared support vector machines (SVMs), feedforward neural networks (FNNs), naive Bayes classifiers (NBs), AdaBoost, and random forests to identify the type along with the severity of the damage. Qian et al. [10] have proposed an ANN model for assessing the magnitude of the damage using lamb wave analysis.

In our study, we have developed and evaluated different classification models like 'Support Vector Classifier (SVC)', Extreme Gradient Boosting Classifier (XGB Classifier), 'Random Forest Classifier', 'Extra Tree Classifier', along with Artificial Neural Network (ANN). A comparative study between these models has also been done. The common steps required to develop and evaluate the model's performance can be shown in Fig. 1.

The remainder of the articles is structured as follows. Section 2 deals with the methodology, including dataset description, data preprocessing, and training and testing dataset formation. Section 3 focuses on results and discussion, which include the evaluation of different classification algorithms along with a brief introduction. Comparisons between these models have been analysed as the conclusion in Sect. 4.

2 Methodology

This section of the article discusses the dataset's description, preprocessing, and construction details for both the training and testing datasets.

2.1 Dataset Description

The primary objective of this study is to examine several ML methods for lamb wave-based SHM. A training and testing dataset is the main prerequisite for this purpose.

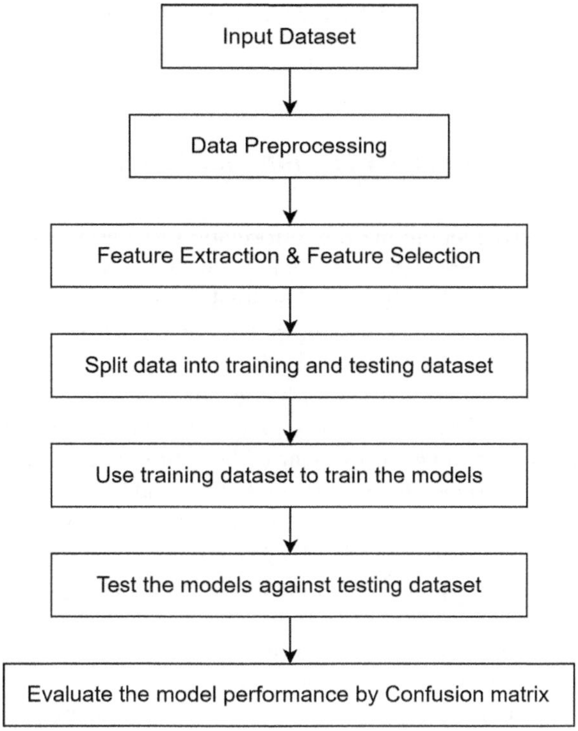

Fig. 1 Flow diagram

The dataset utilized to create various machine learning models is identical to that used in the author's earlier work [11]. The dataset is constructed using the lamb wave response of a 1.6 mm thin square aluminium plate of side 1200 mm under pristine and damaged condition by finite element simulation and experiment. The lamb wave response is actuated and measured using piezoelectric (PZT) transducers in a pitch-catch manner. The dataset is a collection of 264 samples with equal number of samples recorded in pristine and damaged condition. Out of these 264 samples, 216 samples are obtained using FE simulation, whereas 48 samples are obtained experimentally. Two different methods have been used to create damage in aluminium plate during the experiments. In first method, a nut-bolt has been bonded to the structure, and in second method, a notch has been created. In each case, 24 samples have been recorded.

2.2 Data Preprocessing

Each sample of the dataset contains a 1-D time domain response having 1925 data points. The response data and recorded time value are normalized against their maximum absolute value. Every data sample has a unique ID (referred to as sampleID) and condition. The binary values of 0 and 1 are allocated to the 'condition' label, representing the pristine condition and damaged condition of the aluminium plate, respectively.

2.3 Feature Extraction and Feature Selection

Directly feeding the raw information to classifiers may cause overfitting and poor accuracy. From these raw data points, it is necessary to identify some key parameters that can accurately reflect the final health of the structure. For example, the mean of the signal is a key parameter of a signal. These essential parameters are known as features, and the extraction of these features is known as feature extraction.

Feature extraction can be accomplished through manual or automated preprocessing of data points. The process of manually extracting features frequently calls for a significant investment of both time and effort, in addition to the knowledge and experience relevant to the field. To address this, an automated feature extraction tool, namely Time Series Feature Extraction on the basis of Scalable Hypothesis tests (Tsfresh) is used [12]. When the feature extraction sub-module of the Tsfresh package is applied to the data, 787 features are generated.

As high number of features often result in overfitting and increased computational costs during training of the classifier. Hence, to reduce the number of features feature selection technique is used.

Feature selection refers to the process of identifying the characteristics or features that have a substantial effect on the given output variable. This can also be done manually or automatically. Feature select module of Tsfresh package has been used for automatic feature selection. Using the feature select sub-module of Tsfresh, the total number of features is lowered to 43.

2.4 Construction of Training and Testing Dataset

After feature selection, the obtained features are used to train and test classification models. To avoid overtraining the models, only 72 of the 216 simulated samples are used during training. The nut-bolt dataset having 24 samples is used for testing the models. These samples were only introduced to the model during testing to assess the model's generalization capability. As a result, there are three times as much training data as testing data.

3 Results and Discussion

In our study, various classification algorithms are employed, and their performance is analysed. Some of the commonly used classification algorithms named 'Support

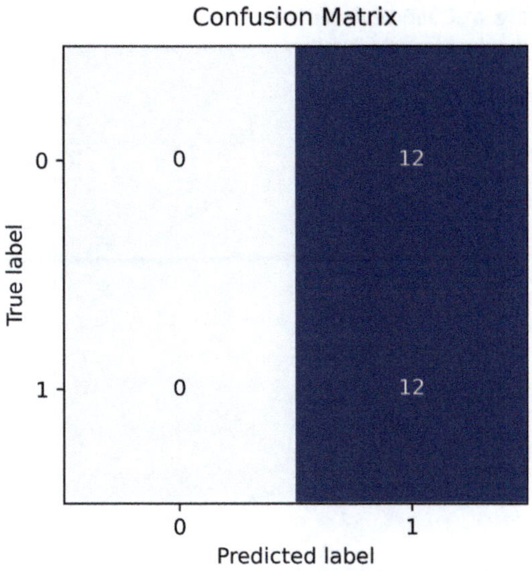

Fig. 2 Confusion matrix for SVC

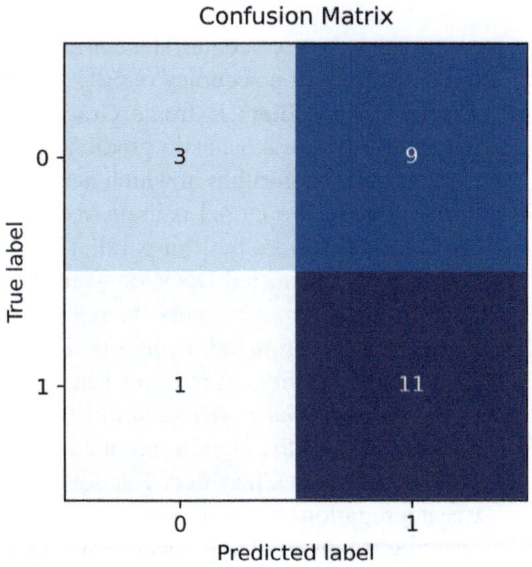

Fig. 3 Confusion matrix for XGBoost classifier

Vector Classifier (SVC)', Extreme Gradient Boosting Classifier (XGB Classifier), 'Random Forest Classifier', 'Extra Tree Classifier', along with Artificial Neural Network (ANN) have been applied to the data. The performance of these algorithms has been evaluated in this section.

1. **Support Vector Classfier**: Support Vector Classifier can be applied in both classification and regression applications. It is generally used in many classification problems. SVC fits the provided data and returns the best-fit hyperplane that

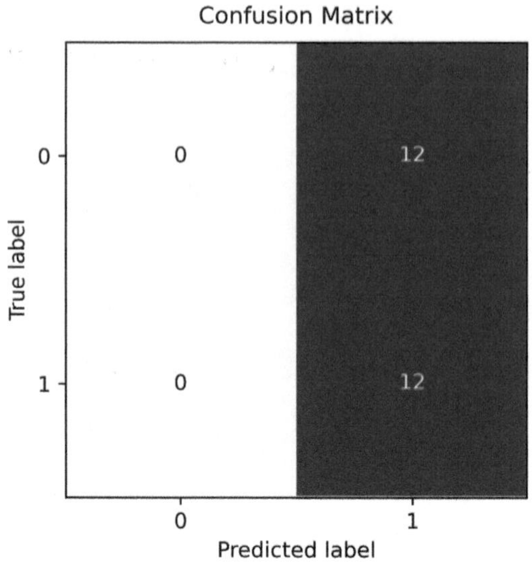

Fig. 4 Confusion matrix for Random Forest Classifier

categorizes the given data. The confusion matrix is depicted in Fig. 2. The SVC has a classification accuracy of 50% when determining the state of the plate.

2. **XGBoost Classifier**: Extreme Gradient Boosting (XGBoost) is a successful adaptation of the stochastic gradient boosting ML algorithm. It is a decision tree ensemble algorithm in which new trees correct the error in the existing tree in the model. The model is improved by adding trees until it can no longer be improved. Trees are built in parallel with XGBoost rather than sequentially as with Gradient Boosted Decision Tree. XGBoost differs from gradient boosting by employing a more aggressive regularization strategy to reduce overfitting. It is a perfectly optimized technique for producing better results. In comparison with other classifiers, it requires less amount of computational power and time. The confusion matrix depicted in Fig. 3 demonstrates that the model classifies the condition of the aluminium plate with 58% accuracy.
3. **Random Forest Classifier**: Random Forest Classifier is a variant of decision tree aggregation (bagging) based on ensemble learning. In bagging, each successive tree is constructed independently that used a bootstrap sample of the given dataset, with no reliance on previous trees. Random forests introduce randomness to the bagging technique. Each node in a traditional decision tree classifier is split using the best split among all variables. However, in Random Forest, the best predictor is from a randomly chosen subset of predictors at each node. As a result, a large number of independent decision trees are formed. Each decision tree makes predictions based on the new data. By aggregating (majority of votes for classification) the model predicts the final output. The performance of the classifier can be shown by confusion matrix as shown in Fig. 4. The classifier has an accuracy of 50%.

Fig. 5 Confusion matrix for Extra Tree Classifier

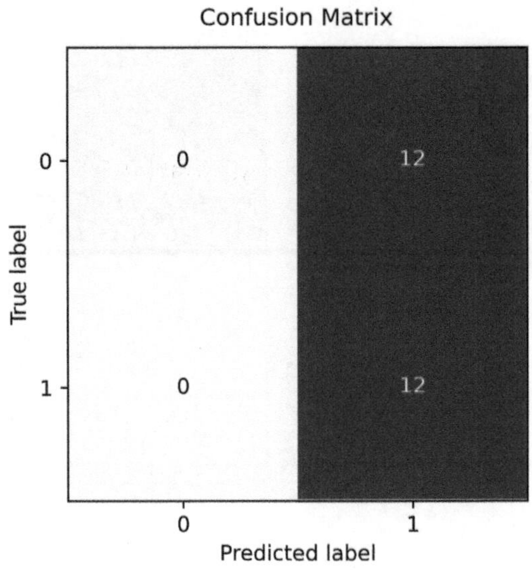

4. **Extra Tree Classifier**: Extra Trees classifier is an ensemble of decision trees like random forest. From a given set of training samples, it generates an ample quantity of unpruned decision trees. It makes the final prediction by taking the majority of votes from decision tree output. The Extra Tree Classifier fits each decision tree to the entire training dataset, as opposed to random forest, which develops decision trees by using a bootstrap sample of the training dataset. Unlike random forests, it selects the split points at random. The accuracy of the model is 50%. The confusion matrix of the model is shown in Fig. 5.
5. **Artificial Neural Network**: ANN is a set of three or more interconnected layers. The input layer, where we feed the training dataset, is the first layer. The layer, which gives us the final output or prediction, is the output layer. Depending on the level of complexity, one or more hidden layers may exist between the input and output layers. An ANN learns by adjusting the weight of the connection and improves the performance by updating the weights. The developed ANN has 43, 86, and 1 neuron in the input, hidden, and output layers respectively. The detailed architecture of ANN is explained in [13]. The ANN predicts the presence of damage with 100% accuracy. The confusion matrix is shown in Fig. 6.

The comparison between the performances of these classification models has been shown in Fig. 7. Five of the most commonly used classification algorithms have been selected for this purpose. It is observed that the ANN is the best algorithm to serve the required purpose with 100% accuracy. SVC, Random Forest Classifier, and Extra Tree Classifier have achieved an accuracy of 50%, whereas XGBoost classifier has achieved an accuracy of 58%.

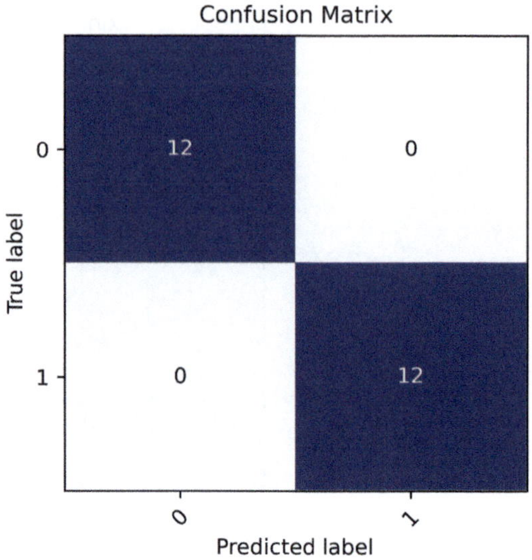

Fig. 6 Confusion matrix for ANN

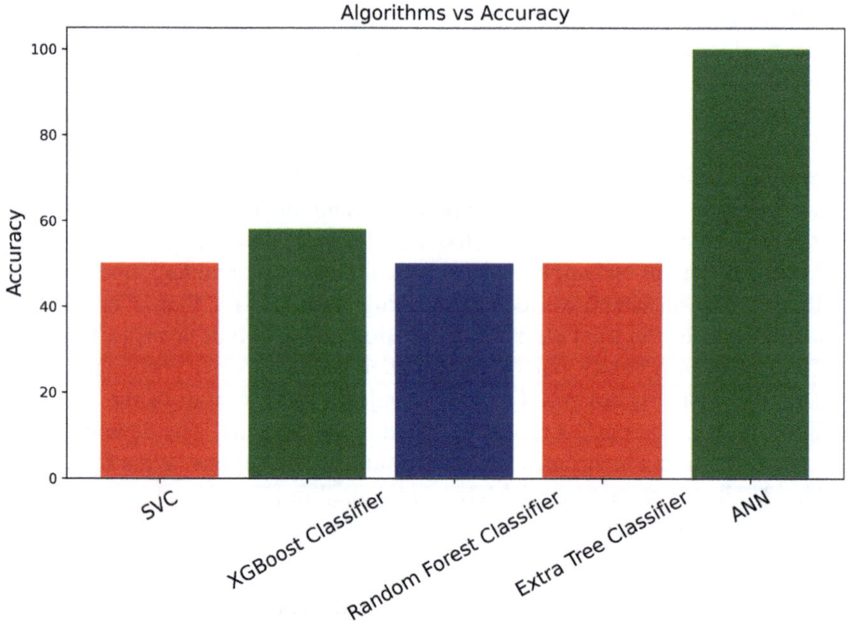

Fig. 7 Overall accuracy of each algorithm

4 Conclusion

In this article, we have analysed a variety of ML classification algorithms to determine the best algorithm for damage detection in a thin aluminium plate. Five of the most commonly used classification algorithms have been selected for this purpose. It is observed that the ANN is the best algorithm to serve the required purpose as it has achieved 100% accuracy against the unseen test samples. SVC, Random Forest Classifier, and Extra Tree Classifier have achieved an accuracy of 50%, whereas XGBoost classifier has achieved an accuracy of 58%. The comparison between the performances of these classification models has been shown in Fig. 7. The study can be further extended to perform damage diagnosis for more complex structures such as composite plates, plates with welds, rivets, ribs, etc.

References

1. Mitra M, Gopalakrishnan S (2016) Guided wave based structural health monitoring: a review. Smart Mater Struct 25(5):053001
2. Giurgiutiu V, Cuc A (2005) Embedded non-destructive evaluation for structural health monitoring, damage detection, and failure prevention. Shock Vibr Digest 37(2):83
3. Staszewski WJ (2005) Ultrasonic/guided waves for structural health monitoring. Key Eng Mater 293:49–62
4. Ying Y, Garrett JH Jr, Oppenheim IJ, Soibelman L, Harley JB, Shi J, Jin Y (2013) Toward data-driven structural health monitoring: application of machine learning and signal processing to damage detection. J Comput Civ Eng 27(6):667–680
5. Sen D, Nagarajaiah S (2018) Data-driven approach to structural health monitoring using statistical learning algorithms. In: Mechatronics for cultural heritage and civil engineering. Springer, pp 295–305
6. Neves AC, Gonzalez I, Leander J, Karoumi R (2017) Structural health monitoring of bridges: a model-free ANN-based approach to damage detection. J Civil Struct Health Monitor 7(5):689–702
7. Lu Y, Ye L, Su Z, Zhou L, Cheng L (2009) Artificial neural network (ANN)-based crack identification in aluminum plates with lamb wave signals. J Intell Mater Syst Struct 20(1):39–49
8. Nick W, Asamene K, Bullock G, Esterline A, Sundaresan M (2015) A study of machine learning techniques for detecting and classifying structural damage. Int J Mach Learn Comput 5(4):313
9. Nick W, Shelton J, Asamene K, Esterline AC (2015) A study of supervised machine learning techniques for structural health monitoring. MAICS 1353:36
10. Qian C, Ran Y, He J, Ren Y, Sun B, Zhang W, Wang R (2020) Application of artificial neural networks for quantitative damage detection in unidirectional composite structures based on lamb waves. Adv Mech Eng 12(3):1687814020914732
11. Rai A, Mitra M (2021) Lamb wave based damage detection in metallic plates using multi-headed 1-dimensional convolutional neural network. Smart Mater Struct 30(3):035010
12. Christ M, Braun N, Neuffer J, Kempa-Liehr AW (2018) Time series feature extraction on basis of scalable hypothesis tests (tsfresh-a python package). Neurocomputing 307:72–77
13. Prajapati KK, Rai A, Mitra M (2022) Lamb wave-based damage detection using artificial neural network and automated feature extraction. Trans Indian National Acad Eng: 1–8

Influence of Governing Parameters on the Molecular Dynamics Study of Boron Nitride Tube

Piyush Bhatt and Vikas Kukshal

Abstract In recent years, due to increasing computational power it's easier to run Molecular Dynamics (MD) simulations, which are used to figure out the mechanical and physical features of atoms. However, incorrect use or misuse of parameters may lead to inconsistent MD results in comparison to the experimental result. Parameters like interatomic potential, force field, cut-off distance, time increment, and many other parameters are needed to be chosen properly otherwise actual reality won't be captured due to the sensitivity of the results. Boron nitride tubes (BNT) attracted much research attention over carbon nanotubes as these possess high hardness, chemical and thermal stability, and excellent piezoelectric properties. In this paper, a thorough study has been conducted by choosing different values of these parameters and their contribution while simulating the Boron Nitride Tube (BNT). The investigation is done by taking different parameters and their effect on the properties of the composite. Various intermolecular potentials like TERSOFF, EXTEP, and other potentials are employed in MD to calculate properties with these potentials. Similarly, a different cut-off distance for Lennard–jones potential is deployed and crosschecked. The time step is also compared by deploying different values of time integration in MD simulation. In the end, a comparison of the present parameters is made with the existing experimental results and their closeness with the actual reality is done. So that those parameters can be directly used for further investigation in the future.

Keywords Molecular dynamics · Boron nitride tube

P. Bhatt (✉)
School of Mechanical and Materials Engineering, IIT Mandi, Himanchal 175005, India
e-mail: d22170@students.iitmandi.ac.in

V. Kukshal
Department of Mechanical Engineering, NIT Uttarakhand, Srinagar 246174, India
e-mail: vikaskukshal@nituk.ac.in

R. K. Ray et al. (eds.), *Advances in Theoretical and Applied Mechanics*, Lecture Notes in Mechanical Engineering, https://doi.org/10.1007/978-981-97-0418-7_17

1 Introduction

Carbon nanotube (CNTs) discovery by Iijima [1] was a landmark as CNT produces extraordinary thermal, mechanical, and electrical properties which encouraged other researchers to find other nanotubes one of which was Rubio et al. [2] who theoretically predicts Boron Nitride Tubes and later synthesized by Chopra et al. [3]. BNTs are more stable chemically and thermally compared to CNTs as suggested by many experimental and theoretical studies. BNTs also have comparable strength to that of CNTs (Young's modulus BNTs 1.1–1.3 TPa [4] vs. CNTs 1.3–1.8 TPa [5]). This makes BNTs a suitable material for many applications like composites, transistors, gas sensors, hydrogen storage, and many other applications. So many researchers are studying BNTs experimentally as well as through Molecular dynamics (MD). In recent times MD is used extensively to study nanomaterials [6] such as BNTs.

Properties calculated from MD are dependent on many parameters such as time step, inter-atomic potential, cut-off distance, time step, unit cell size, and strain rate [7]. Hence it is very important to choose these parameters very accurately otherwise the result obtained will be different from the actual reality. Several interatomic potentials are used by various researchers for studying BNTs, however harmonic functioned force fields like UVFF [8] force field does not permit breakage of bonds and can be used for small deformation while other reactive force fields such as TERSOFF can be used for large deformation and deformation as these allow bond breakage and formation. Cut-off radii also needed to be chosen as small cut-off radii won't capture full physics while a higher cutoff means no significant improvement but requires high computation. Choosing the time step is very crucial to obtain good results. The time step leads to compromise on result while smaller leads to high computational cost. Researchers have used time steps ranging from 0.1–15 fs to study BNTs. Similarly strain rate has to be chosen carefully as a high strain won't give enough time to equilibrate the molecule to the applied load while a smaller strain rate requires a high computational cost. Many researchers have done a study on strain rate influence but many of them are contradictory as some studies show significant change while some show no influence.

In this paper, a thorough study has been conducted by choosing different values of these parameters and their contribution while simulating the BNT. The investigation is done by taking different parameters and their effect on the properties. Various intermolecular potentials are employed in MD to calculate physical and mechanical properties. Similarly, a different cut-off distance for Lennard–jones potential is deployed and crosschecked. The time step is also compared by using different time integration values in MD simulation. Finally, the current parameters are compared to the existing experimental results to determine their closeness to reality. As a result, those parameters can be directly used for future research.

Table 1 List of parameters studied

Parameters	Used values
Potential	TERSOFF, DREIDING, UVFF, ReaxFF
Cut off distance	1.9–2.3 Angstrom
Strain rate	$2 \times 10^{-4}-16\ ns^{-1}$ (nanoseconds)
Time step	0.1–20 fs (femtoseconds)
BNT length	50–500 Å

2 Methodology

To study the influence of each parameter, every parameter is analyzed individually, and their effect is studied on the property of the BNT. This uniaxial tensile simulation of BNT is performed using individual parameters and their effect on young's modulus is studied how this result is near to the actual results. All MD simulations are studied through a Large-scale atomic/molecular massively parallel simulator (LAMMPS).

MD parameters studied in this paper are (Table 1).

3 Result and Discussion

In this section, results are presented and discussed for each parameter effect on the elastic modulus as well fracture property of BNT. Further results are compared with the results available in the literature and a set of parameters is suggested for future study so that one can directly use those parameters in their study.

3.1 Effect of Potential Function Type

Potential function defines the potential between two atoms mathematically by which we can calculate the total energy of the atomic model of the MD simulation.

Investigations were also done on how prospective function type would affect the expected mechanical characteristics of BNT. The tensile uniaxial stress-strain curves from MD simulations of a BNT utilizing various interatomic potentials are displayed in Table 2. The elastic and fracture parameters that were determined for each case considered are listed in Table 2. The findings show a moderate dependence of the property on the potential function, with Tersoff potential predicting the maximum elastic modulus of about 1100 GPa while Reax FF potential predict the lowest modulus of 847 GPa. There was a big impact of potential on the fracture property of BNTs. All potentials, except for CVFF force field, were also successful in capturing the fracture behaviour. Very similar outcomes are predicted by Tersoff

Table 2 BNT results for different potential function

Interatomic potential type	Young's modulus (GPa)	Fracture stress (GPa)	Fracture strain (%)
UVFF	921.9	–	–
ReaxFF	856.7	157.6	16.46
DREIDING	863.8	99.8	18.55
Tersoff	1101.4	109.3	17.6

potentials. ReaxFF force field predictions for fracture strain and fracture stress are 21% and 8% less than Tersoff potential results, respectively.

3.2 *Effect of Strain Rate*

With a 0.5 fs time step, uniaxial stress simulation of an armchair BNT with chirality (6, 6) and a length of 66 were performed in the TERSOFF potential. To choose the best loading strategy for MD simulations, both quasi-static as well as dynamic loading circumstances are examined. By moving one end of the BNT by 0.125 while leaving the other end fixed, the quasistatic loading scheme was applied. Followed by NVT ensemble for equilibration process [10 picoseconds (ps)] to give atoms to find equilibrium position after deformation. Up to the BNT's total fracture, this loading was applied repeatedly. In the dynamics loading method, atoms are brought into equilibrium during the loading process while strain rates are provided at the one end of the CNT. Different strain rates (0.016–16 ns^{-1}) uniaxial loading stress-strain curves are shown in Fig. 1. With an increase in strain rate, it was discovered that the elastic modulus decreased. According to Fig. 1, the strain rate has a substantial impact on the fracture stress. When rate dropped from 16 to 0.016 ns^{-1}, the fracture stress decreased from 131.8 to 88.8 GPa while fracture strain dropped from 18.8.4 to 15.5%. As atoms have more time to equilibrate at low strain rate than at high and are therefore more likely to cross the bond breaking barrier, higher strain rates result in increased fracture stress and strain. Additionally, the potential energy fluctuates widely and the BNT atoms cannot adequately equilibrate at strain rates greater than 10 ns^{-1}.

3.3 *Effect of Time Step*

An armchair BNT with a length of 66 and chirality (6, 6) was simulated under uniaxial stress using TRESOFF potential with time step ranging from 0.1 to 6 fs. In all simulations BNT was moved with constant head speed of 0.01 angstrom/ps in NVT ensemble at room temperature, with a strain rate of 0.16 ns^{-1}. The integration process for time step has a substantial impact on the stress-strain curves of the BNT.

Fig. 1 Stress–strain curve for different strain rate

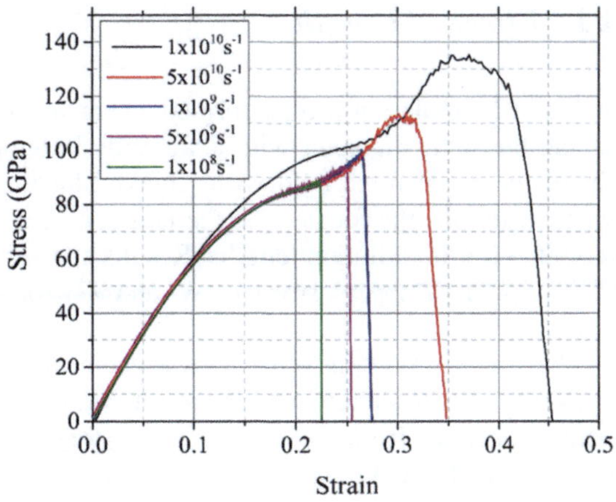

When the time step value was increased from 0.1 to 1 fs, all properties grew. In particular, the fracture stress increase by 15.9% while strain by 8.3%, Similarly Young's modulus show 41% increase from 794.7 GPa for 0.10 fs to 1120.2 GPa for 6.0 fs. The system becomes unstable at time steps greater than 6 fs, and the BNT does not react to the imposed load. According to these findings, the greatest time step for BNTs MD simulations is approximately 6 fs (Fig. 2).

Fig. 2 Stress–strain curve for different time step

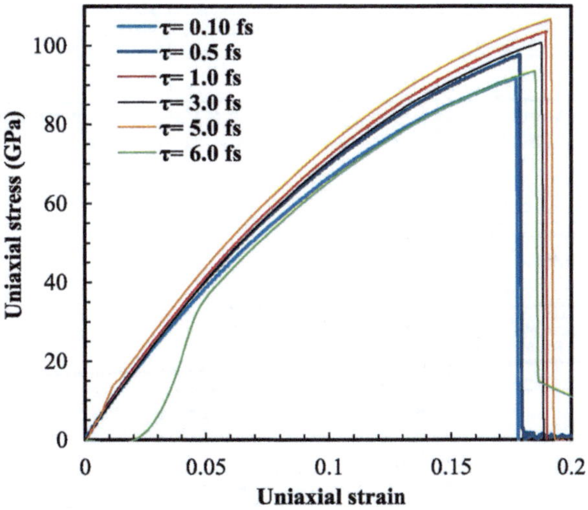

3.4 Effect of Length

In this MD simulation was performed on zigzag (17, 0) and armchair (10, 10) BNTs having the same diameter with a variation in their length as shown in Fig. 3 with Tersoff potential force field with Uniaxial loading MD simulation on BNTs until fracture. In this, the focus is on the length of the BNTs, and the stress-strain curve is plotted for the different lengths. Results show that the trend is the same up to the fracture point. The peak stress is 987 and 971 for armchair and zig zag respectively (Fig. 4).

The Young's modulus of BNTs was determined within the elastic limit in order to assess the effect of length, and the results were compiled in Table 3. Young's moduli were determined from the curve's linear stress-strain curves within an elastic limit according to Table 3.

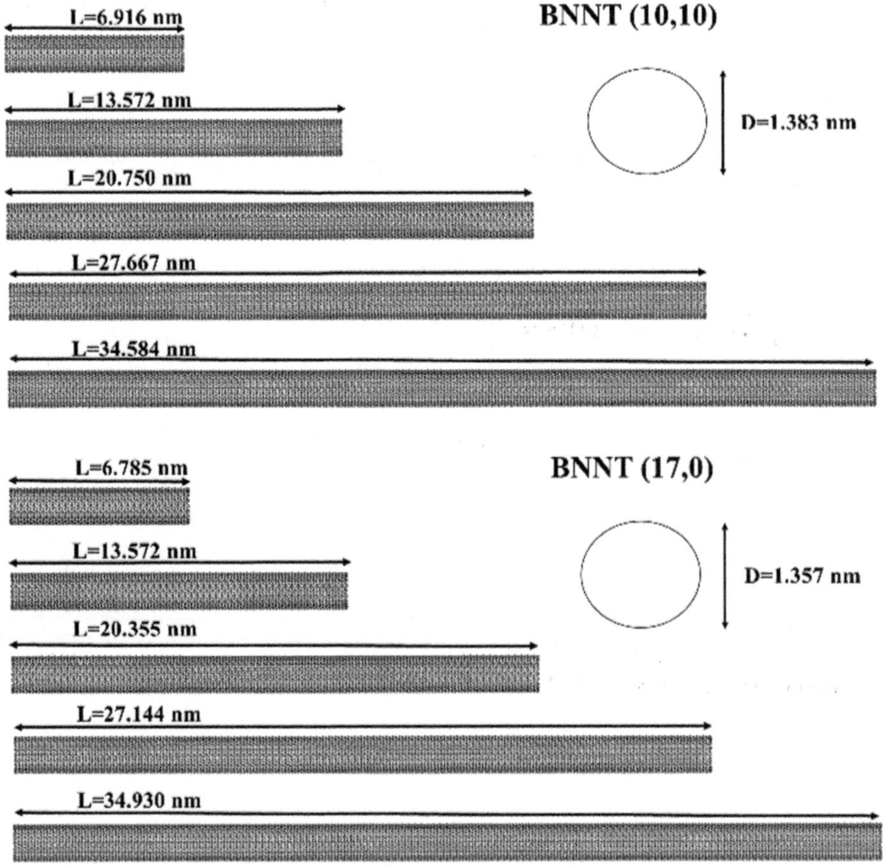

Fig. 3 Different BNTs lengths were used for MD simulation

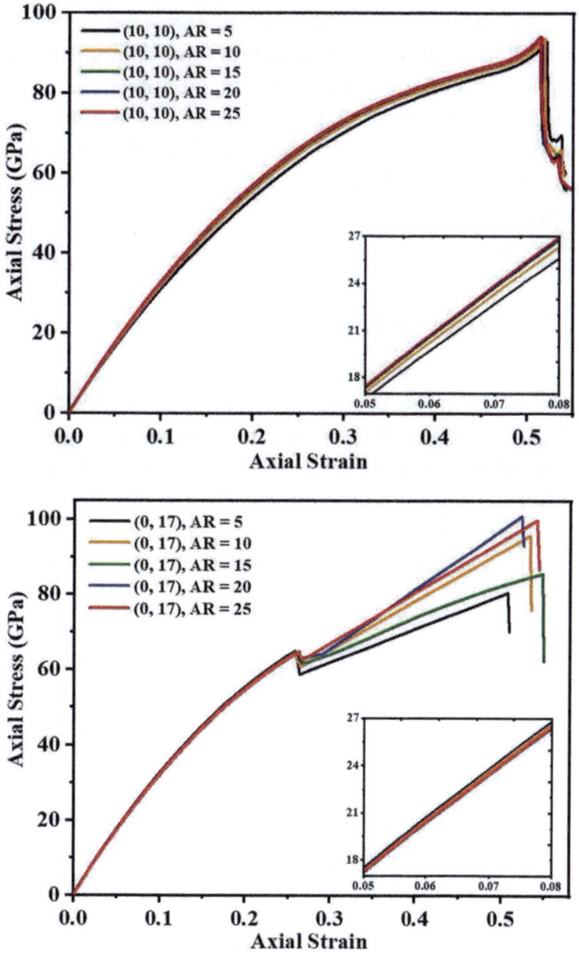

Fig. 4 Stress–strain curve for zig-zag (17, 10) and armchair (10, 10)

Table 3 Values of Young's modulus calculated through MD simulation

S. No.	Length (approx.) (nm)	E(TPa) (10, 10)	E(TPa) (17, 0)
1	7	0.8125	0.7998
2	13.5	0.9978	0.9789
3	20.5	1.0534	1.0658
4	27.5	1.0256	1.0341
5	34.5	1.0234	1.0278

Within the context of MD simulations, the Young's modulus of BNTs is examined, as well as the impacts of aspect ratio on Young's modulus. Our findings show that the aspect ratio of BNTs has a significant impact on its Young's modulus, which rises to a value of 15, after which it stabilises. For various aspect ratios, zigzag BNTs show a higher Young's modulus value than armchair BNTs. Additionally, MD simulations are used to describe the deformation characteristics of the BNTs under axial tensile loading.

3.5 Effect of Temperature

Due to the large kinetic energy of the atoms, BNT at higher temperatures fails at smaller stresses. The stress–strain curves at higher temperatures are found in Fig. 5 to be slightly lower than the curves at low temperatures, which suggests that the tube softens at higher temperatures. With strain, this effect steadily grows and becomes considerable towards the failure strain. The expansion of bonds at higher temperatures and the thermal contribution to vascular stress are to blame for this weakening of the tube.

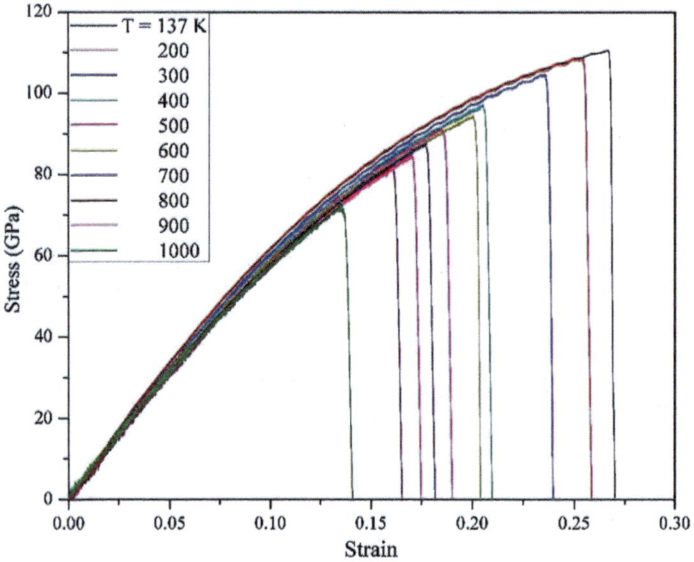

Fig. 5 Stress–strain curve for different strain rate

4 Conclusion

Currently, the literature on various atomistic simulations contains abnormalities and contradictions. This thorough investigation's goal is to identify the causes of these abnormalities and discrepancies in order to direct future atomistic research. To investigate the impact of strain rate, time step, length effect (from 6.8 to 35 nm of tube), and temperature effect (from 137 to 1000 K) on the elastic properties and fracture properties of BNTs, we specifically conducted extensive molecular dynamics studies. To investigate the impact of all these characteristics on the properties of the material under various loading circumstances, uniaxial tensile simulations were carried out. Our results are summarized as follows:

- The linear stress–strain behavior produced by the harmonic depiction of covalently bonded carbon atoms in the UVFF is devoid of any potential of bond rupture. The strain hardening at tensile strains, which is apparent, causes Tersoff force field to overestimate the fracture stress of BNT. ReaxFF does, however, forecast lower BNT fracture stress and strain.
- Nanostructure elastic properties are independent of strain rate less than $1 \times 10^9\ s^{-1}$. However, when strain rate increases, both the fracture stress and fracture strain grow exponentially. When subjected to strain rates greater than $2 \times 10^9\ s^{-1}$ during tensile MD simulations, the atomic system becomes unstable.
- All properties increase as the time step value was increased from 0.1 to 1 fs. Particularly, the fracture stress increased by 15.9% while strain by 8.3%, similarly Young's modulus increased by 41% from 794.7 GPa for 0.10 fs to 1120.2 GPa for = 6.0 fs. The mechanical properties then started to change as the time step was increased beyond 1 fs. Finally, the system becomes unstable at time steps greater than 6 fs, and the BNT does not react to the imposed load. Our findings show that the aspect ratio of BNTs has a significant impact on its Young's modulus, which rises up to a value of 15, after which it stabilises. For various aspect ratios, zigzag BNTs show a higher Young's modulus value than armchair BNTs.
- High atomic kinetic energy at higher temperatures causes BNT to fail at lower strains, which suggests that the tube softens. With strain, this effect steadily grows and becomes considerable towards the failure strain. The expansion of bonds at higher temperatures and the thermal contribution to vascular stress are to blame for this weakening of the tube.

Lastly, we think that the results of this study will help us understand the role of some atomistic parameters that lead to inconsistent and contradictory predictions among the published data, as well as help us find the right parameters for each different type of loading and material in future studies. Future studies like the one described here should look at how the factors that control MD simulations affect the mechanical properties of larger systems like advanced nanocomposites made of polymers and metals.

References

1. Iijima S (1991) Helical microtubules of graphitic carbon. Nature 354(6348):56–58
2. Rubio JL (1994) Theory of graphitic boron nitride nanotubes. Phys Rev B 49:5081–5084
3. Chopra NG (1995) Boron nitride nanotubes. Science 269:966–967
4. Communications SS (1998) Pergamon 105:297–330
5. Kundalwal SI, Choyal V (2018) Transversely isotropic elastic properties of carbon na-notubes containing vacancy defects using MD. Acta Mech: 1–22
6. Alian AR, Meguid SA (2016) Multiscale modeling of nanoreinforced composites. In: Meguid SA (ed) Advanced in Nanocomposites, Cham, pp 1–39
7. van Gunsteren WF, Berendsen HJC (1990) Computer simulation of molecular dynamics: methodology, applications, and perspectives in chemistry. Angew Chem Int Ed Engl 29:992–1023
8. Izadifar M, Abadi R, Jam AN, Rabczuk T (2017) Investigation into the effect of doping of boron and nitrogen atoms in the mechanical properties of single-layer polycrystalline graphene. Comput Mater Sci 138:435–447

The Study of Effect of Shear Wall on Building Resting on Hill Slopes

Chirag Nagar and Smita Kaloni

Abstract This paper examines the performance of buildings constructed on hilly terrain. Since most of the hilly region in India are earthquake prone therefore the stability of buildings in hilly slope become very important. This paper considers the study of behavior of building on hill slopes with and without shear walls. Shear walls are used to deal with lateral loads like earthquake effectively and efficiently. Shear walls are placed on different locations on different models in hill slope. In this study $G + 5$ floor building on approximate 20-degree hill slope angle have been modelled using structural analysis software Staad pro. The performance of shear wall placed on different locations across the slope in building on hill slope is also studied. The study emphasizes on suitability of orientation and location of structural wall on hill slope buildings.

Keywords Shear wall · Hill slope · Building on hill slopes · Staad pro · Earthquake force · Irregular building

1 Introduction

Due to asymmetry of plan and elevations, structures in hilly areas are more susceptible to earthquake forces. Due to increase in population in few decades and due to rapid urbanization, economic growth and climatic conditions, migration towards hilly areas is increasing day by day. In the Indian context hilly areas are more vulnerable to earthquakes. In such circumstances, it is very important for the researchers to examine the responses of structures subjected to any impact load such as earthquake.

In context of India, extensive field surveys in the Indian Himalayas regions have been carried out. The most common building configurations hill slopes were identified are step-back buildings, step back–set back buildings, and split foundation/vertical cuts buildings. These different configuration hill buildings vary on the position of center of stiffness and center of mass which results in a complex response

C. Nagar (✉) · S. Kaloni
Civil Engineering, NIT Uttarakhand, Srinagar, Garhwal, India
e-mail: chiragnagar010@gmail.com

R. K. Ray et al. (eds.), *Advances in Theoretical and Applied Mechanics*, Lecture Notes in Mechanical Engineering, https://doi.org/10.1007/978-981-97-0418-7_18

under seismic excitation. The analytical study on seismic behavior of building on hill slope was performed by Birajdar et al. [1], Pandey et al. [2], Singh et al. [3, 4], and Surana et al. [5]. Birajdar et al. [1] performed seismic analysis on buildings in different configurations i.e., set back, step back and step back and set back building concludes that step back and set back building are more convenient for the analysis on sloping ground. Singh et al. [4] performed an analytical study to compare the performance of buildings in hilly areas and general buildings and concluded that buildings more to road level are more vulnerable to earthquake than any other building. Dangi and Akhtar [6] investigated a $G + 6$ multistorey framed structure on different slopes of ground and results were compared based on periphery of shear walls and nodal displacements. Verma and Dubey [7] explored different type of shear wall in different panel condition and concluded that structure resting on sloping ground tends to displace orthogonally.

Mohammadi and Kumar [8] illustrate performance of shear wall placed on different locations subjected to seismic loads. Response spectrum analysis is done for the analysis and concludes that despite of different location of shear wall the efficiency of building increases with every time step.

In all of these studies the seismic behavior study was carried out on different types of the building configurations. It is observed that there has been very less study on structures resting on hilly areas. The irregularities and seismic proactivity in Indian context make it tough to design for hill slope buildings. Irregularity of any type in structure attracts greater damages and less stability. This study assess building resting on hill slope after providing shear wall/ RC Structural wall across the slope at different position and height at the outer periphery of the structure. The shear walls are very effective against lateral loads in multi storey buildings. Shear wall ensures the safety by neglecting the displacement. Shear walls brings solution to many complex problems by reducing the dynamic responses in case of earthquake. Staad Pro developed by Bentley Systems, a well-known and widely accredited Structural analysis tool is used for the modelling and primary analysis of the considered buildings. The software is easy to use because of its seamless user interface. It consists all the required tools, commands and latest IS codes. This software analyses models very fast with good efficiency.

2 Modeling and Analysis

In this study, step back building configurations are taken into account. The building is considered to be resting at hill slope consisting almost 20-degree slope. To study the behavior of shear wall on hill slope structure, four models are considered. Model A is building on hill slope without shear wall. On the rest three buildings, the shear wall is provided across the slope direction. Model B is building on hill slope with shear wall opposite to short columns with shear wall in two floors. Model C is building on hill slope with shear wall opposite to short columns with shear walls in all floors. Model D is building on hill slope with shear wall towards short columns. These four building models are examined for this paper. The model description is given in Table 1 and Fig. 1.

The current study directs mainly on effect of shear walls on building on hill terrain. For the analysis building models are used with different shear wall locations. A different load combination of dead load, imposed load and earthquake load are taken for the analysis. Building parameters are depicted in Table 1 shows the finite element model of $G + 5$ building.

Model Parameters

All the chosen models are rigid frame irregular buildings on almost 20-degree hill slope. Modelling and analysis are done using finite element software Staad pro. The modelled $G + 5$ buildings are consisting same building configuration, same properties of components like beam, column, slab, same material, support and loading. Seismic loading is applied according to IS 1893 2016, Zone IV—"This zone is called a high damage risk zone, zone factor is 0.24". Seismic Zone IV is considered for the wider range of usability and more practical consideration in physical utilization.

Shear wall has been discretized for the finite element analysis by generating a quadrilateral meshing of 1×1 in whole shear wall span and the same is also done for the Slabs. Depiction of this is shown in Fig. 2.

The details of the parameters considered in this study are given in Table 2.

The building models are analyzed for static and dynamic loading conditions. For getting the results and output of any building or analysis the main thing is to assume the behavior of applied forces. Based on tendency of force analysis are divided into two methods:

Table 1 Model description

Model ID	Description
Model A	Building on hill slope
Model B	Building on hill slope with shear wall at zero short height
Model C	Building on hill slope with shear wall at zero full height
Model D	Building on hill slope with shear wall at max

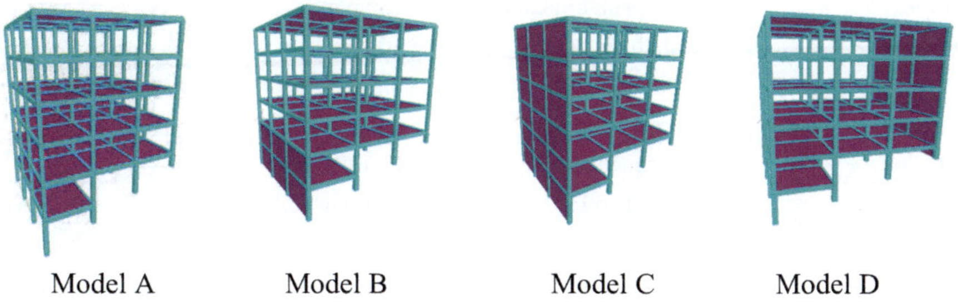

Fig. 1 Step back building models description (3D-represantation)

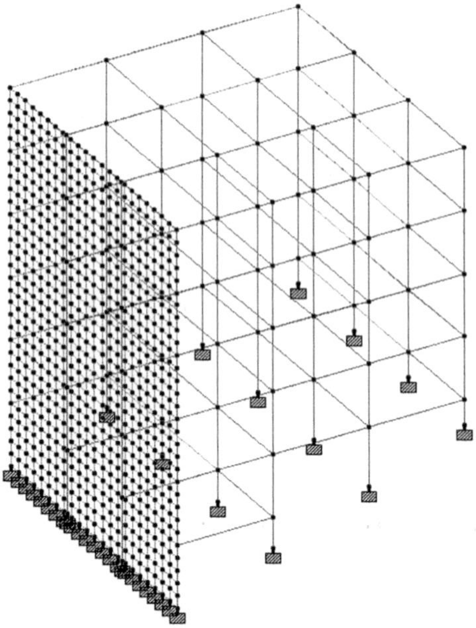

Fig. 2 Step back building model "C" (shear wall discretization)

Table 2 Building model parameters

S. No.	Parameters	Dimension
1.	Height max	30 m
2.	Height min	22.01 m
3.	Slope	~ 20°
4.	Length	22.5 m
5.	Width	22.5 m
6.	No. of bays	3
7.	Seismic zone	IV
8.	Reduction factor	5
9.	Importance factor	1
10.	Shear wall thickness	150 mm
11.	Slab thickness	100 mm
12.	Size of column	0.5 m × 0.5 m and 0.6 m × 0.6 m
13.	Size of beam	0.3 m × 0.6 m
14.	Earthquake load	IS-1893-2016
15.	Type of soil	Medium soil

I. Static method: in this method of analyzing structures all the forces applied are assume of fixed intensity at every time.
II. Dynamic method: in this method dynamic nature of forces are considered means real dynamic nature of time dependency of forces like earthquake are used to analyses the structures.

Although the main focus is centered on earthquake force on this study but the static analysis method is used because the core of study was to get the behavior of irregular structure after shear wall. All the models are analyzed with same method static analysis to get much realist and comparable results.

Three main steps are followed for every model:

I. Modelling: generating model by providing dimensions, column and beam size, properties and support.
II. Analysis: calculating deflection, displacement, forces and moment.
III. Post processing: getting the results.

After analyzing all the models' results are studied to find their behavior and the pattern of outputs. The building foundation is considered as fixed in all cases.

3 Result and Discussion

The analysis of step back building models was performed in Staad Pro software. The building model description is shown in Fig. 3. Column 1, 2, 3, and 4 are define in the same figure. All foundation is treated as fixed base. The ground level position is also defined in the figure.

Fig. 3 Step back building resting at 20° slope model

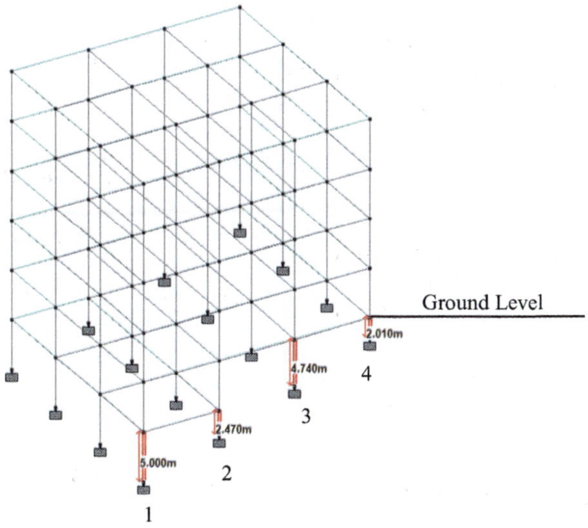

In this study, first static analysis is performed on all four considered model. Model A (without shear wall is compared with the other three models B, C, and D. From Table 3, it has been observed that the Model A (without shear wall) has larger value of shear force and bending moment. By providing shear wall in flexible side as shown in Model B and Model C, it has been observed that there is significant reduction in the value of shear force and bending moment (from the table). Shear wall's effectiveness across the slope direction can be observed and confirmed by static analysis of all the considered models

The following abbreviations are used in Table 3:

DL	Dead Load.
LL	Imposed Load.
EXTP	Earthquake Load in X-direction.
EZTP	Earthquake Load in Z-direction.

Further the response spectrum analysis (type of dynamic analysis) was performed for all considered models. Figure 4 shows the variation of time period of building (all models for different vibration modes). The frequency and period of the building all model obtained from are presented in Table 4. After comparing all the models, it is observed that building without shear wall have more time period than with shear wall at any location and height across the slope.

Peak storey shear: The variation of peak storey shear for all considered model can be observed using the Figs. 5 and 6 respectively. From the analysis result, it has been observed that the peak storey shear along and across the slope will be low in structure containing shear wall at any location. Shear walls reduces storey shear more effectively across its orientation. Shear wall reduces the shear in top floor very significantly. Further it is determined that the peak storey shear along the slope will be less when shear wall is placed near short columns. It is concluded from the study that the shear wall reduces the base shear in the structure.

Likewise in static analysis, it is observed here also that the Model B has the lowest value of peak shear. However, Model C and D are also effective in reducing the peak shear values in the structures. This shows the effectiveness of the proposed shear wall in the building resting on sloping areas. Due to the limitation of the analysis software, limited study is performed in this paper.

4 Conclusion

(1) A $G+5$ building in seismic zone IV in hilly slope of approximately 20° is taken for the analysis in finite element software.
(2) Shear wall was placed across the slope at different location on different models and analysis has been done.
(3) Effect of shear wall is considered and concluded that shear wall enhances the integrity of structure in case of earthquake.

Table 3 The variation of shear force and bending moment in along the slope direction of building

Column	Load case	Model A		Model B		Model C		Model D	
		Shear force	Bending moment	Shear force	Bending moment	Shear force	Bending moment	Shear force	Bending moment
1	EXTP	– 0.951	0.761	– 1.063	0.717	– 0.937	0.833	– 1.027	0.709
	EZTP	– 0.808	2.802	– 3.667	– 11.006	– 1.46	– 2.325	– 0.244	5.024
	DL	– 20.1	– 40.972	– 21.893	– 41.126	– 21.353	– 43.110	– 20.059	– 40.853
	LL	0.595	– 0.170	0.387	– 0.236	0.475	– 0.375	0.594	– 0.161
2	EXTP	– 0.471	– 10.589	0.578	– 9.164	– 0.440	– 10.744	– 0.604	– 11.123
	EZTP	8.324	– 2.420	– 20.215	– 14.523	– 1.436	– 2.673	11.254	– 2.710
	DL	17.914	64.259	23.647	63.600	14.567	58.748	18.208	64.574
	LL	– 4.552	– 3.541	– 3.774	– 3.301	– 4.873	– 4.071	– 4.497	– 3.486
3	EXTP	14.561	– 4.165	14.832	– 3.975	14.953	– 4.594	15.089	– 4.791
	EZTP	24.338	48.944	– 5.886	– 10.853	7.731	27.652	27.656	43.899
	DL	– 6.314	– 10.143	– 2.604	– 3.867	– 6.415	– 11.012	– 5.749	– 8.335
	LL	– 1.624	– 2.542	– 1.237	– 1.823	– 1.631	– 2.627	– 1.559	– 2.343
4	EXTP	201.167	28.400	208.338	31.463	207.642	29.053	210.229	29.574
	EZTP	180.574	99.531	– 46.135	– 22.936	22.461	42.633	239.02	102.488
	DL	11.403	69.033	1.635	58.140	12.261	67.449	12.117	70.060
	LL	4.614	20.789	4.084	19.982	4.735	20.641	4.616	20.740

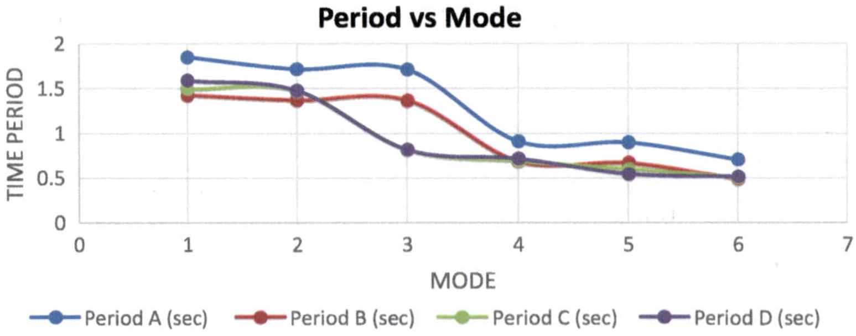

Fig. 4 Variation of time period with different modes of vibration

Table 4 Modal frequency and period of the building

Model A			Model B		Model C		Model D	
Mode	Frequency (cycles/s)	Period (s)	Frequency (cycles/s)	Period (s)	Frequency (cycles/s)	Period (s)	Frequency (cycles/s)	Period (s)
1	0.544	1.838	0.707	1.415	0.675	1.482	0.635	1.575
2	0.587	1.703	0.736	1.359	0.685	1.459	0.684	1.462
3	0.590	1.696	0.738	1.355	1.244	0.804	1.245	0.803
4	1.114	0.897	1.483	0.674	1.477	0.677	1.423	0.703
5	1.134	0.882	1.523	0.656	1.696	0.589	1.892	0.528
6	1.448	0.691	2.114	0.473	2.042	0.489	2.006	0.498

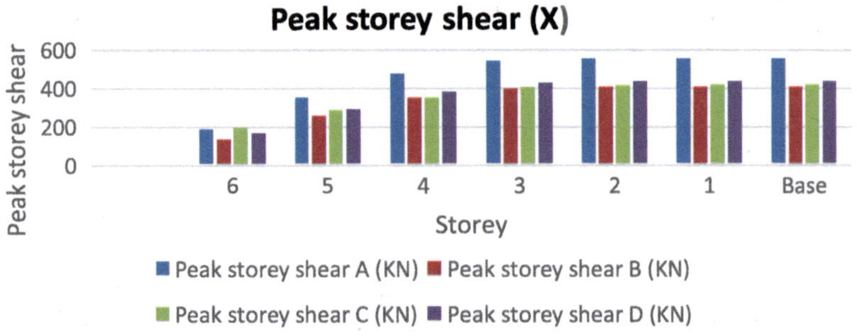

Fig. 5 Variation of peak storey shear of building with respect to height of building (*X* direction)

(4) From the analysis it was shown that the intensity of shear force and bending moments significantly reduces across the slope direction of buildings on hill slopes.
(5) Results show that the addition of shear walls reduces the base shear.
(6) In the building resting on hill slopes the major concern for structure designer is to reduce the short column effect because of column located at different levels. By providing the shear walls the short column effect can be minimized.

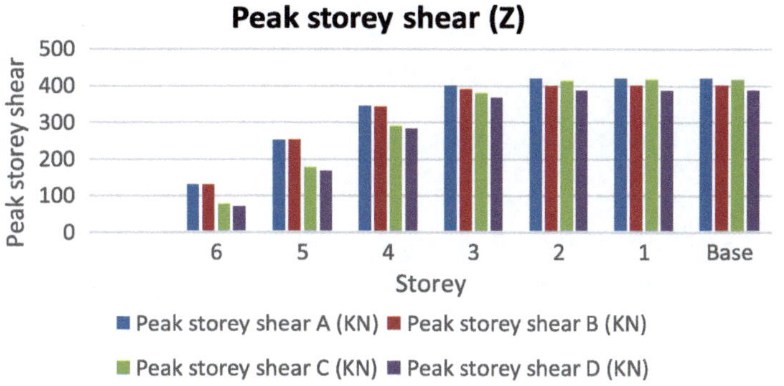

Fig. 6 Variation of peak storey shear of building with respect to height of building (Z direction)

In future, the study can further be extended considering the nonlinear analysis and the behavior of other building configuration like step back set back, split foundation, etc.

References

1. Birajdar BG, Nalawade SS (1995) Seismic analysis of buildings resting on sloping ground. In: 13th world conference on earthquake engineering. No. 1472; Gelman A, Carlin JB, Stern HS, Rubin DB (1995) Bayesian data analysis. Chapman and Hall, London
2. Pandey AD, Kumar P, Sharma S (2011) Seismic soil-structure interaction of buildings on hill slopes. Int J Civ Struct Eng 2(2):544–555
3. Singh Y et al (2012) Seismic behavior of buildings located on slopes-an analytical study and some observations from Sikkim earthquake of September 18, 211. In: 15th world conference on earthquake engineering journal; Gull SF (1989) Developments in maximum-entropy data analysis. In: Skilling J (ed) Maximum entropy and Bayesian methods. Kluwer Academic, Dordrecht, pp 53–71
4. Singh Y, Gade P, Lang DH, Erduran E (2012) Seismic behavior of buildings located on slopes-an analytical study and some observations from Sikkim earthquake of September 18, 2011. In World conference in earthquake engineering
5. Surana M, Singh Y, Lang DH (2015) Seismic fragility analysis of hill buildings in Indian Himalayas. In: Proceedings of SECED—an international conference on earthquake risk and engineering towards resilient world
6. Dangi SK, Akhtar S (2019) Seismic analysis of a RC building on sloping ground with shear wall at different positions. AIP Conf Proc 2158(1)
7. Verma SK, Dubey H (2021) Seismic performance of buildings with various configurations in hilly regions. Civ Eng Archit 9(7):2205–2236
8. Mohammadi MH, Kumar P (2022) A review on effect of the positioning of shear wall for earthquake resistance Multi-Storey building

Free Vibration Analysis of Thin Membranes Using EFGM

Sheetal and Anup Ghosh

Abstract Vibration of thin membranes under tension is a topic of very high physical significance. One of its most common application is in a typical western percussion instrument where a uniform membrane under tension is used to produce sound. Some classical Indian percussion instruments use a variable density membrane to generate complex modes of vibrations. Present work is aimed at developing an efficient computer program based on Galerkin scheme, a particular case of Mesh Free Methods, to solve the basic wave propagation equation for the specific case of a thin rectangular membrane fixed along the perimeter. A MATLAB code is written to extract its first few modes of vibration. Starting from the basic wave equation, a weak form for the governing differential equation is formulated, followed by the formulation of discrete equations, as prescribed by Element Free Galerkin Method (EFGM). It is observed that the code can produce accurate mode shapes associated with the vibration. A convergence study is performed to determine the minimum number of nodes required to produce accurate results. It was observed that the results converged in a relatively low number of nodes in comparison to a traditional FEM technique. The error plots produced by EFGM method also proved that the results obtained from the meshfree method were much more reliable than the traditional FEM results. The work has future scope to be developed into a full-fledged 3-D capable tool to analyses complex geometries and composite cavities.

Keywords EFGM · Thin membrane modal analysis · Vibration analysis · Helmholtz equation · Meshless method

Sheetal (✉) · A. Ghosh
Department of Aerospace Engineering, IIT Kharagpur, Kharagpur, India
e-mail: sheetalgodiyal011@gmail.com

A. Ghosh
e-mail: anup@aero.iitkgp.ac.in

R. K. Ray et al. (eds.), *Advances in Theoretical and Applied Mechanics*, Lecture Notes in Mechanical Engineering, https://doi.org/10.1007/978-981-97-0418-7_19

1 Introduction

Numerical simulations are powerful tools for research when it comes to solving complex differential equations. Researchers have used several techniques to arrive at a solution numerically, Finite Element Method (FEM) being one of the most popular and effective. FEM has advanced exponentially over time, but it brings in severe inconvenience and inaccuracy in results when it comes to problems involving re-meshing. This is where Mesh-free methods come in handy. Unlike FEM, mesh free methods does not depend on a systematically defined mesh or connectivity matrix, rather it takes in individual node positions for calculation. Each node is assigned with its own domain of influence and affects properties in its respective domains. The influence of a node on a particular point on the domain is controlled by the choice of an appropriate weight function. Mesh free methods are relatively new and developing in comparison to traditional finite element methods. Present study focuses on the normal modes of vibrations of a uniform rectangular membrane. The membrane is fixed along the perimeter, obeys a zero pressure Dirchilet boundary condition along the four edges. 2-D membrane under tension supports transverse vibrations. There are numerous works on membrane vibrations that rely on finite element methods for extracting natural modes [1, 2]. Most of these methods are invariably accurate for results involving low wave numbers, but for high wave numbers the Helmholtz equation suffers the so called "pollution effect" [3]. This "pollution effect" leads to a significant deviation of the results from the actual values. This calls for thinking about some other method that behaves more robustly with respect to wave number. As rightly pointed out by Suleau and Bouillard [4], mesh free methods are less affected by the pollution and dispersion effects and is capable of producing much more accurate results at high wave numbers. Meshfree methods have been explored extensively by researchers like Liu [5] and Dolbow and Belytschko [6], who developed proper systematic methods to approach a problem using mesh free methods. Their works are usually seen as the foundation for mesh free methods. Moving least square approximations, a prerequisite for EFGM, were introduced by Lancaster and Salkauskas [7]. Vibration of membranes or more specifically the Helmholtz equation was solved using mesh free methods in 1997 by Bouillard and Suleaub [8].

2 Formulation

2.1 Strong Form of Wave Equation

Acoustic wave equation is derived from the basic conservation laws of mass and momentum along with a thermodynamic relation. The fluctuation in all physical quantities are assumed to be small and the equation is linearized. The resulting 3D wave equation takes the following form.

$$\nabla^2 = \frac{1}{c^2}\frac{\partial^2 \tilde{P}}{\partial t^2} \tag{1}$$

To obtain the harmonic solutions of the equation, pressure perturbation is assumed to be harmonic in nature.

$$\tilde{P} = Pe^{j\omega t} \tag{2}$$

Substituting Eq. (2) in (1) we arrive at the strong form of the wave equation or Helmholtz equation.

$$\nabla^2 P - k^2 P = 0 \tag{3}$$

where $\tilde{P}$ is pressure perturbation, 'c' is speed of sound in dry air at 20 °C, 'ω' is angular frequency and 'k' is wave number.

3 Variational Formulation of Helmholtz Equation

Strong form of wave equation involves the evaluation of second order differentials of the primary variable. This might pose numerical difficulties when the differentiability of the primary variable is limited. Weak form of differential equation are used to obtain integral equations which are much easier to work with and imposes lesser constraints on the differentiability of the primary variable.

$$\int_{\Omega} \omega\left[\nabla^2 P - k^2 P\right]\mathrm{d}\Omega = 0 \tag{4}$$

$$\int_{\Omega} \left[\nabla\omega\nabla P - k^2\omega P\right]\mathrm{d}\Omega = 0 \tag{5}$$

Equation (5) is a weak formulation of Helmholtz equation and is derived for a case where the acoustic damping and excitation are set to zero. In an EFGM formulation, the weight functions ω is chosen to be same as the shape functions.

4 Discrete EFGM Formulation of Acoustic Equation

Substituting the weight function, and appropriate discrete forms of the primary variable and its gradient, leads to the discrete system of equations. Pressure is the only degree of freedom in the system, and it is assumed to take the following form.

$$p^h = \sum_{i=1}^{n} \Phi_i P_i = \mathbf{\Phi P} \tag{6}$$

Substituting Eq. (6) and the value of ω into Eq. (5) the following relationship is obtained.

$$\int_{\Omega} \left[\nabla\mathbf{\Phi} \sum_{\mathrm{i}=1}^{\mathrm{n}} \nabla\Phi_i P_i - k^2 \mathbf{\Phi} \sum_{\mathrm{i}=1}^{\mathrm{n}} \Phi_i P_i \right] \delta P = 0 \tag{7}$$

Above equation can be re-written as

$$[K - k^2 M]P = 0 \tag{8}$$

where p^h is pressure approximation, Φ_i is EFGM shape function, K is stiffness matrix and M is mass matrix.

5 Moving Least Square Approximation and Shape Functions

Moving least square approximations or MLS is an efficient data fitting technique. It tries to approximate a function that best fits all the given discrete data points. MLS methods were first looked at by Lancaster and Salkauskas [7].

The domain be represented by several discrete nodes. Each node has an associated quantity called nodal parameter p_i. It is to be noted that the nodal parameters are not equal to the nodal values of the function. $p_i \neq p(x_i)$.

$$p^h(\mathbf{x}) = \sum_{j=0}^{j=m} f_j(x) a_j(x) = \mathbf{f^T}(\mathbf{x})\mathbf{a}(\mathbf{x}) \tag{9}$$

$f(x)$ represents an appropriate basis set, and $a(x)$ is the matrix of non-constant coefficients. These non-constant coefficients are called MLS approximants. The choice of $f(x)$ depends on the problem under consideration. For the specific case of Helmholtz equation 2-D linear basis is good enough to approximate the final pressure distribution.

$$\mathbf{f^T}(\mathbf{x}) = [1xy] \tag{10}$$

Each node is associated with its own domain of influence as illustrated in Fig. 1. The radius of the domain of influence is determined by the local nodal spacing. The properties at any point in the domain will be influenced by all the nodes whose domains of influence overlap at that point. Coefficient matrix **a(x)** is determined by

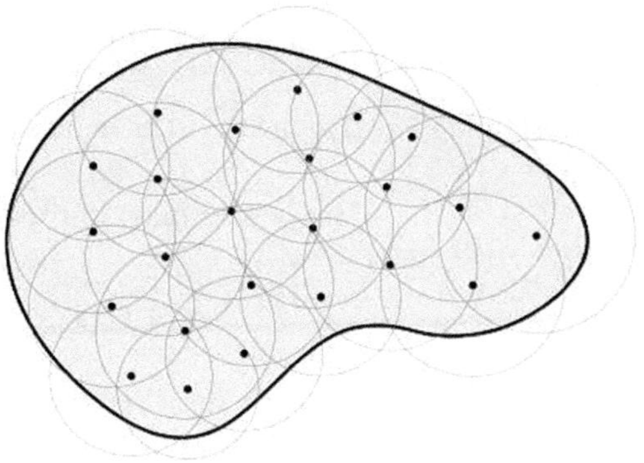

Fig. 1 Domain of influence of each node

minimizing the weighted discrete L_2 norm.

$$J = \sum_i W(\mathbf{x} - \mathbf{x_i})\left[\mathbf{f}^{\mathbf{T}}(\mathbf{x_i})\mathbf{a}(\mathbf{x}) - p_i\right]^2; i = 1, 2, 3, \ldots .n \tag{11}$$

where $\mathbf{x}$ denotes the position vector at the point of evaluation, and $\mathbf{x_i}$ denotes the position vector of the neighbouring nodes that influence the point. W denotes the weighting function which depends on the distance between the node and the point of evaluation. Most of the times, the weight functions are chosen such that they smoothly tend to zero as the distance is increased. Cubic spline weight function was found to be most suitable for this work (Table 1).

To minimize J with respect to $\mathbf{a}$, the first derivative of J with respect to $\mathbf{a}$ is set equal to zero.

$$\frac{\partial J}{\partial \mathbf{a}} = 0 \tag{12}$$

$$\sum_i W(\mathbf{x} - \mathbf{x_i})\mathbf{f}(\mathbf{x_i})\left[\mathbf{f}^{\mathbf{T}}(\mathbf{x_i})\mathbf{a}(\mathbf{x}) - p_i\right] = 0; i = 1, 2, 3, \ldots .n \tag{13}$$

Equation 13 can be rewritten as

$$\mathbf{A}(\mathbf{x})\mathbf{a}(\mathbf{x}) = \mathbf{C}(\mathbf{x}) \tag{14}$$

Table 1 Model specifications

Property	Value
Membrane length (L)	0.16 m
Membrane breadth (B)	0.14 m
Applied pre-tension (T)	100 N/m
Material density (ρ)	1390 kg/m^3
Membrane thickness	2.54 m e^{-4}

$$\mathbf{a(x)} = \mathbf{A}^{-1}\mathbf{(x)C(x)} = \mathbf{A}^{-1}\mathbf{(x)C_p p} \quad (15)$$

Substituting Eq. (15) in Eq. (9) an expression for $p^h(\mathbf{x})$ is obtained.

$$p^h(\mathbf{x}) = \mathbf{f^T(x)a(x)} = \mathbf{f^T(x)A}^{-1}\mathbf{(x)C_p p} = \boldsymbol{\Phi}\mathbf{p} \quad (16)$$

$$\boldsymbol{\Phi} = \mathbf{f^T(x)A}^{-1}\mathbf{(x)C_p} \quad (17)$$

As can be observed from Eq. (17), the shape function calculations involve the evaluation of inverse of **A(x)** which can be computationally expensive. An efficient method for evaluating the shape functions were introduced by Dolbow and Belytschko [6]. This method relies on LU decomposition of **A(x)** rather than a direct inverse calculation.

6 Weight Function

There are several different weight functions that can be used for the purpose. For the specific case of uniform node distribution in a rectangular membrane, it was found that a cubic spline weight function yielded accurate results. Cubic spline weight functions are one of the most commonly used weight functions in EFGM. It is defined as below

$$d_I = ||\mathbf{x} - \mathbf{x_I}|| \quad (18)$$

where $\mathbf{x}$ denotes the position of the point of evaluation and $\mathbf{x_I}$ denotes the position of Ith node.

$$r = \frac{d_I}{d_{mI}} = \frac{d_I}{d_{\max} C_I} \quad (19)$$

where d_{mI} stands for the radius of domain of influence of the Ith node. C_I stands for a radius around the Ith node, which encloses a minimum number of neighbouring nodes in it. $d_{\max}$ is a scaling parameter whose values usually lie between 2.0 and 4.0 [6]. For present study $d_{\max}$ is considered as 2.

The weight is a function of the parameter r and is defined as follows.

$$W(r) = \begin{cases} \frac{2}{3} - 4r^2 + 4r^3 & \text{if } r \le \frac{1}{2} \\ \frac{4}{3} - 4r + 4r^2 - \frac{4}{3}r^3 & \text{if } \frac{1}{2} < r \le 1 \\ 0 & \text{if } r > 1 \end{cases} \quad (20)$$

7 Boundary Conditions

As discussed earlier, the nodal parameters p_i are not equal to the functional values at the nodes $p(x_i)$. This very property makes it difficult to apply boundary conditions directly on to the assembled stiffness matrix. To resolve this issue a scaled transformation method is applied. In this method the nodal values at the boundary nodes are forced to be equal to the nodal parameters. To apply this transformation, all the boundary nodes and corresponding boundary conditions are identified first. A scaling matrix Λ is generated, by populating the rows corresponding to boundary nodes with the shape functions evaluated at that node. The remaining positions are filled in with δ, with unity at the positions along the diagonal.

$$\Lambda = \begin{Bmatrix} \Phi_1^T \\ \Phi_2^T \\ \cdots \\ \delta \\ \Phi_N^T \\ \cdots \end{Bmatrix} \tag{21}$$

N stands for the number of boundary nodes, and δ is the Dirac delta function. Dimension of the Λ matrix is same as that of the assembled K matrix. Thus scaled K and M matrices are

$$K^{'} = \Lambda^{-1T} K \Lambda^{-1} \tag{22}$$

$$M^{'} = \Lambda^{-1T} M \Lambda^{-1} \tag{23}$$

Thus final equation or eigen value equation to extract natural modes can be written as

$$\left[K^{'} - k^2 M^{'}\right] p = 0 \tag{24}$$

8 Results

There are several factors that decide the accuracy of the predicted final pressure. The value of the parameter $d_{\max}$, the number of nodes used, and the integration scheme used being some of the most important ones. A convergence study was performed to find the optimum number of nodes to obtain reasonably accurate results. The results obtained by using the optimum value of number of nodes were compared with analytical and commercial FEM package results. Variation of first five modes by varying number of nodes are presented in Fig. 2. Modal frequencies and their comparison with analytical and FEM results are tabulated in Table 2 and mode shapes of first five modes are shown in Fig. 3. Natural frequencies of a thin isotropic rectangular membrane under free vibration without damping can be calculated analytically [9] from the equation given below.

$$\lambda_{nm}^2 = \frac{\pi^2}{C^2}\left(\frac{n^2}{L^2} + \frac{m^2}{B^2}\right) \tag{1}$$

where $c = \sqrt{(T/\rho)}$.

Fig. 2 Variation of 1st five modal frequencies (in Hz) with number of nodes

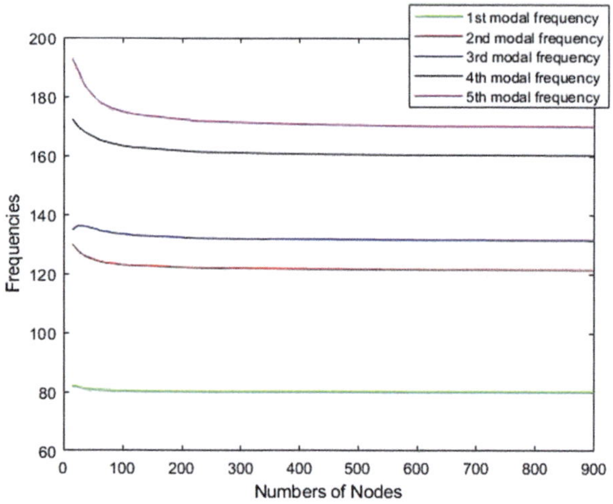

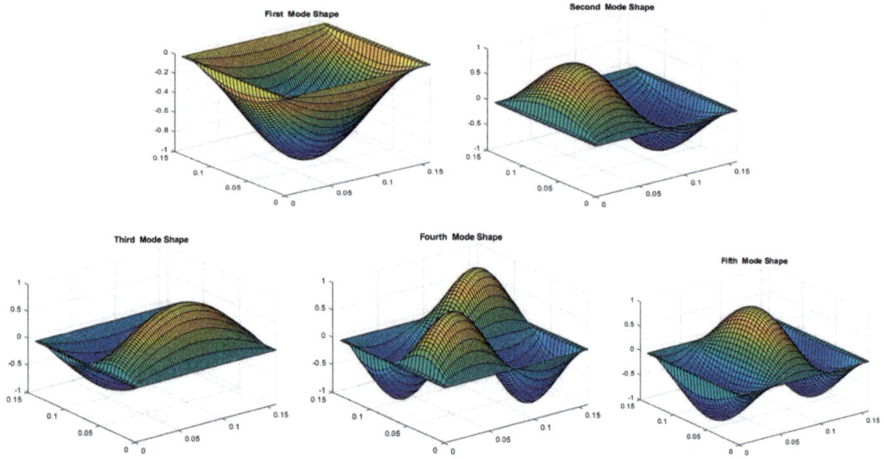

Fig. 3 First five mode shapes of the membrane

Table 2 Modal frequency data comparison with FEM and analytical solution

Mode No.	Analytical frequency (in Hz)	FEM (400 nodes) (in Hz)	EFGM (400 nodes) (in Hz)	Error (FEM)	Error (EFGM)
1	79.86	79.869	79.757	0.002602395	0.137208912
2	121.14	121.32	121.185	0.142402073	0.03095039
3	131.21	131.57	131.140	0.271897125	0.055517131
4	159.73	159.74	159.832	0.003854478	0.061810038
5	168.83	169.84	169.153	0.59272402	0.185661347
6	187.83	189.48	188.195	0.877809715	0.194019788
10	239.60	234.58	240.528	2.095471043	0.387105995
20	318.84	319.28	321.32	0.135284288	0.77632817
30	375.66	369.35	380.13	1.680338246	1.189357198
40	435.76	427.08	442.97	1.993807523	1.653259451
50	484.58	474.37	490.26	2.108986005	1.171925387

9 Conclusion

EFGM formulation has been done for the stretched thin membrane. A MATLAB code is written for extracting the natural modal frequencies and corresponding mode shapes. The work was successful at extracting the natural modes and corresponding frequencies of vibration of the membrane using a mesh-free scheme. There were several remarkable conclusions that could be drawn from the study. Observing the trends from the convergence plots, it can be concluded that EFGM results attain a very high degree of convergence with fairly low number of nodes. It was also observed that the errors produced by EFGM are much less and consistent in comparison to the FEM.

References

1. Harari I, Grosh K, Hughes TJR, Malhotra M, Pinsky PM, Stewart JR, Thompson LL (1996) Recent developments in finite element methods for structural acoustics. Arch Comput Methods Eng 3(2):131–309
2. Ihlenburg F, Babuška I (1995) Finite element solution of the helmholtz equation with high wave number part I: the h-version of the fem. Comput Math Appl 30(9):9–37
3. Deraemaeker A, Babuška I, Bouillard P (1999) Dispersion and pollution of the fem solution for the helmholtz equation in one, two and three dimensions. Int J Numer Meth Eng 46(4):471–499
4. Suleau S, Bouillard P (2000) One-dimensional dispersion analysis for the element-free galerkin method for the helmholtz equation. Int J Numer Meth Eng 47(6):1169–1188
5. Liu GR, Gu YT (2005) An introduction to meshfree methods and their programming, 1edn. Springer
6. Dolbow J, Belytschko T (1998) An introduction to programming the meshless element f reegalerkin method. Arch Comput Methods Eng 5(3):207–241

7. Lancaster P, Salkauskas K (1981) Surface generated by moving least square methods. 37:141
8. Bouillard P, Suleaub S (1998) Element-free Galerkin solutions for Helmholtz problems: formulation and numerical assessment of the pollution effect. Comput Methods Appl Mech Eng 162(1–4):317–335
9. Weaver Jr W, Timoshenko SP, Young DH (1991) Vibration problems in engineering. Wiley, pp 484–488

An Efficient Enrichment Approach for Fracture Simulations Using Element Free Galerkin Method

Ayush Awasthi and Mohit Pant

Abstract In the current work an approach to enhance the computational efficiency of element free Galerkin method (EFGM) has been proposed by modifying the usage of enrichment functions along with regional nodal density. Effective use of linear basis and fully enriched basis function has been proposed such that area near to tip utilizes fully enriched basis function which subsequently changes to linear basis function in a well-defined proposed way in moving away from the crack tip. Similarly, the regional near crack is discretised for high nodal density than the remaining region. This newly proposed scheme has been validated by evaluating the stress intensity factor for crack problems under thermoelastic loads. Obtained results from proposed scheme show good agreement with the results from literature and relatively took 58% less computational time thereby enhancing its computational efficiency. Also disadvantage of conventional EFGM in evaluating field variables inaccurately for non-convex domains has been overcome by the proposed criteria.

Keywords EFGM · Fracture · Crack · MLS basis function

1 Introduction:

Advancement in materials and its applications has increased the importance of fracture mechanics to produce damage tolerant design. All materials contain flaws or discontinuity which act as the region for high concentration during their service life leading to failure. Researchers have been trying to limit these kinds of failures by proposing and enhancing different computational methods. Evaluation of Stress intensity factor becomes important in determining the safe design of a component. Over the time various analytical and computational methods have been developed to determine the SIFs in a component. Finite element method has been widely developed method used in simulations of various such problems [1]. But this method has some

A. Awasthi (✉) · M. Pant
Department of Mechanical Engineering, National Institute of Technology Hamirpur, Hamirpur, India
e-mail: ayush@nith.ac.in

R. K. Ray et al. (eds.), *Advances in Theoretical and Applied Mechanics*, Lecture Notes in Mechanical Engineering, https://doi.org/10.1007/978-981-97-0418-7_20

limitations like in crack propagation where remeshing is required. Extended finite element method is another important method developed in handling fracture problems [2, 3]. Significant research in mesh-based methods led to the development of mesh free methods which are very effective in handling crack problems [4]. Some of such methods include Smoothed particle hydrodynamics (SPH) [5], Diffuse element method (DEM) [6], Element-free Galerkin method (EFGM) [7], Reproducing kernel particle method (RKPM) [8] etc. Out of these meshless methods, the element free Galerkin method (EFGM) proposed by Beystscho et al. is widely accepted method for crack problems [7]. In EFGM only nodal data is required and discretizes the domain by inserting the gauss points of some order in between the nodes for the interpolation. Specific enrichment basis functions are used in this EFGM to approximate the result. Element free galerkin method has gained importance in handling fracture problems due to its advantage in its approximation procedure based only on nodal data [9]. In EFGM, domain is discretized by generation of gauss points of some order between these nodal positions and field variables are approximated using the moving least square (MLS) shape function [10]. Though this EFGM is quite efficient in handling crack problems but there is always a need to enhance its computational efficiency. One of the primary drawbacks of the Element-Free Galerkin Method (EFGM) is the significant computational overhead required to achieve accurate results and inability to provide accurate results in case of non-convex domains. In the current work an approach to enhance the computational efficiency of EFGM for handling crack problems has been proposed. Algorithm developed is in form of modifications in use of basis function along with variation in nodal density by development of generic MATLAB codes. Present study proposes algorithm for the effective use of linear basis and fully enriched basis function such that area near to tip utilizes fully enriched basis function which subsequently changes to linear basis function in a well-defined proposed way in moving away from the crack tip. High nodal density is provided to region near the crack than the rest of domain. This scheme has been developed on a conclusion that only the area near to tip is susceptible to high stresses which require accurate approximations but area far from it is relatively less predominant to stress so the computational cost can be saved in this case by using simple basis function for approximation. Discretization of domain has been done by division of problem domain into specific regions and then incorporating the varying nodal density along with effective use of basis functions for approximation of results. Primary focus of current study is development of efficient enriched EFG algorithm in terms of computational efficiency and solution accuracy. This has been done to overcome certain limitations of conventional EFGM in handling problems in non-convex domains along with optimization of computational costs. The developed algorithm has been tested for simulation of cracks under thermoelastic loads by evaluating stress intensity factors and obtained results have been compared with reference results of conventional EFGM.

2 Element Free Galerkin Algorithm Formulation

Two-dimensional domain of body bounded by $\boldsymbol{\Gamma}$ subjected to various forces have been shown in Fig. 1 where $\boldsymbol{u}$ is the displacement vector, $\bar{\boldsymbol{t}}$ represents traction force, and $\bar{\boldsymbol{n}}$ is the unit normal.

$$\nabla.\boldsymbol{\sigma} + \boldsymbol{b} = 0 \,\text{over}\, \Omega \quad \text{(governing Equation)} \tag{1}$$

where $\boldsymbol{\sigma}$ is stress sensor given as $\boldsymbol{\sigma} = \boldsymbol{D}(\boldsymbol{x})[\boldsymbol{\varepsilon}]$ and $\mathbf{D}(\mathbf{x})$ is the material matrix,

$$\boldsymbol{D} = E/(1-\nu)\begin{bmatrix} 1 \; \nu & 0 \\ \nu \; 1 & 0 \\ 0 \; 0 & (1-\nu)/2 \end{bmatrix} \quad \text{(For plane stress)}$$

$$= E/(1+\nu)(1-2\nu)\begin{bmatrix} 1-\nu & \nu & 0 \\ \nu & 1-\nu & 0 \\ 0 & 0 & (1-2\nu)/2 \end{bmatrix} \quad \text{(For plane strain)}$$

$$\boldsymbol{u} = \bar{\boldsymbol{u}} \,\text{over}\, \Gamma_u \quad \text{(Essential boundary condition)} \tag{2}$$

$$\boldsymbol{\sigma}.\bar{\boldsymbol{n}} = \bar{\boldsymbol{t}} \,\text{over}\, \Gamma_t \quad \text{(Natural boundary condition)} \tag{3}$$

Now applying langrage multiplier method enforce boundary conditions, the discrete equation formed from Equation is as [7]:

$$\begin{bmatrix} \mathrm{K} & \mathrm{G} \\ \mathrm{G^T} & 0 \end{bmatrix}\begin{Bmatrix} \mathrm{u} \\ \lambda \end{Bmatrix} = \begin{Bmatrix} \mathrm{f} \\ \mathrm{q} \end{Bmatrix} \tag{4}$$

where

$$K_{IJ} = \int_{\Omega} \boldsymbol{B}_I^T \boldsymbol{D} \boldsymbol{B}_I \mathrm{d}\Omega,\; f_I = \int_{\Gamma_t} \bar{\boldsymbol{t}} \Phi_I \mathrm{d}\Gamma_t,\; q_K = -\int_{\Gamma_u} \boldsymbol{N}_K \bar{u} \mathrm{d}\Gamma_u,$$

here, K is the global stiffness matrix, f is the global force vector, G and q are the terms related to boundary conditions linked to the langrage multiplier, Φ_{I} denotes

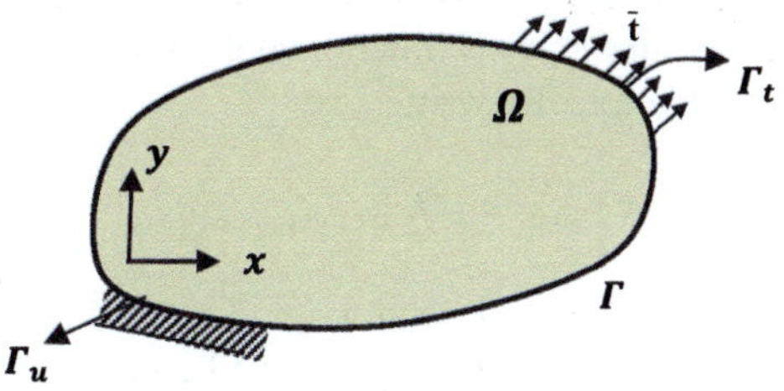

Fig. 1 Domain representation with boundary conditions

the meshfree shape function B_I is the differential operator and N_K is 1D Lagrange interpolant given as:

$$B_I = \begin{bmatrix} \Phi_{I,x} & 0 \\ 0 & \Phi_{I,y} \\ \Phi_{I,y} & \Phi_{I,x} \end{bmatrix}, N_K \begin{bmatrix} N_K & 0 \\ 0 & N_K \end{bmatrix}$$

3 Mixed Enrichment Criterion

For incorporation of current logic specific mathematical logics are developed in MATLAB. First step of current algorithm is division of whole domain into different regions as shown in Fig. 2. First step is to mark region between $c * a$ ($c = 1.5$) vertical distance above and below the crack and name it as region 1. Now mark the remaining domain above the region 1 as region 2 and similarly mark the remining region below the region 1 as region 3. Now as per EFGM formulation the domain is discretized by generation of nodes. Region 1 in which crack is located is given high nodal density while the region 2 and region 3 are provided with lower nodal density. Now subdivide the region 1 by drawing circles around the crack. The first circle is drawn taking radius (r_1) equal to $a/1.5$ having centre at the crack tip and mark that as region 4. Draw the second circle of radius (r_2) $c * r_1$ having centre at crack tip and mark that as region 5. Now depending upon the location of evaluation point in these regions the subsequent enrichment function is employed for its contribution in forming global stiffness matrix $[K]$. As the global stiffness matrix is calculated by sum of all local stiffness matrices of all gauss points in a domain so depending upon the location of evaluation point in these regions subsequent enrichment function is used in its approximation as shown in Fig. 3. If the point lies in region 1 and region 2 having low nodal density, simple basis function is employed for contribution in forming global stiffness matrix. Now for gauss points lying in region 4 full intrinsic enrichment basis function is employed in formation of global stiffness matrix. In case gauss point lies inside region 5 the blending of full intrinsic enriched function and basis function is used such that the point lying in this region which are near to region 4 will have higher contribution of fully enriched basis function and least contribution from simple basis function and similarly for points in this region near to remaining area region 1 excluding region 4 and region 5 will have higher contribution from simple basis function and least from fully enriched basis function. This can be mathematically written as:

$$[K]_{\text{Total}} = \left([K]_{\text{Fully enriched basis}} \times Q\right) + \left([K]_{\text{Fully enriched basis}} \times (1 - Q)\right) \quad (5)$$

where $Q = 1$ for region 4 and $Q = 0$ for region 2, region 3 and remaining area of region 1 excluding region 4 and region 5.

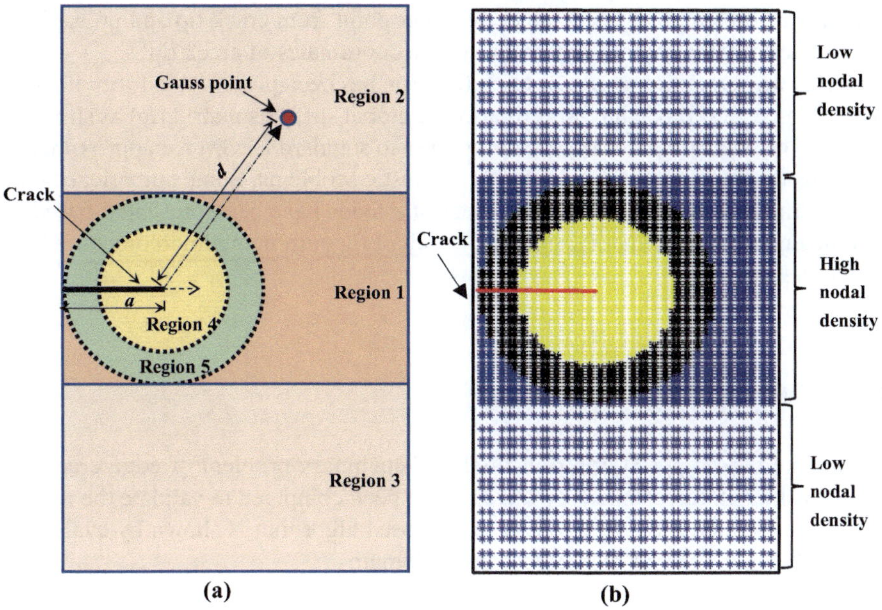

Fig. 2 Mixed enrichment criterion **a** regional discretization, **b** nodal density variation

Fig. 3 Flowchart for mixed enrichment criterion

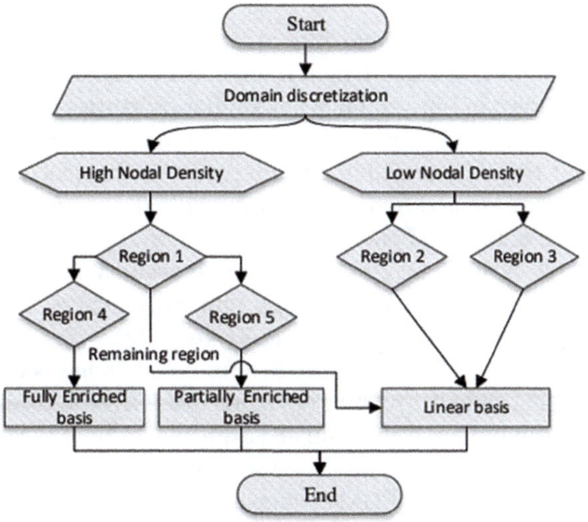

$$Q = \frac{c * (a/1.5) - d_g}{(c-1) * \left({}^{a}/_{1.5}\right)} \text{ and } d_g = \sqrt{(pt(x) - \text{Tip}(x))^2 + (pt(y) - \text{Tip}(y))^2}$$

Here a is crack length, d_g is distance of gauss point from crack tip and pt(x, y) are coordinates of gauss points and Tip(x, y) are coordinates of crack tip.

In this way an optimum use of basis function has been proposed in form of modification in general procedure in formation of global stiffness matrix $[K]$ as shown in Eq. (5). This algorithm has been incorporated to standard EFGM for approximation of results in simulation of cracks in thermoelastic problems. Some numerical examples of cracked structures under thermoelastic loads have been presented based on current algorithm in the later sections. Flowchart for current methodology have been shown in Fig. 3.

4 Results and Discussions

To validate the proposed criterion firstly a benchmark problem of edge crack has been considered. Stress intensity factors have been computed to validate the results. In the later section the effectiveness of proposed algorithm is shown by evaluating thermal fields in circular disk (non-convex domain).

4.1 Rectangular Plate with an Edge Crack

In the current example edge crack in rectangular plate having dimensions $L = 1$, $W = 0.5$ having crack length a subjected to thermoelastic loads have been considered as shown in Fig. 4a. Plate has similar properties as reference problem [11]. Thermal boundary conditions include imposition of temperature of – 1000 °C on the right edge and 1000 °C on the left edge of plate. Mechanical constraints include the restriction of both top and bottom edges in vertical direction and restriction of extreme right node of top and bottom edge in both horizontal and vertical directions. Temperature fields evaluated are used an input to evaluate the displacements and stresses which are required in evaluating stress intensity factors. SIFs has been evaluated using M integral and normalized using $\mathrm{K_I}/(\alpha * 100 * \sqrt{\pi * \mathrm{a}} * \frac{\mathrm{E}}{1-\mathrm{v}})$. . For comparison with reference result [11], firstly SIF for $a = 0.25$ has been evaluated and it showed good agreement with a relative deviation of 1.1%. Now Stress intensity factors for different a/W ratios have been evaluated using both proposed scheme and conventional EFGM. Results obtained by proposed scheme show good agreement with the results obtained by conventional EFGM as shown in Fig. 4b. Newly proposed criterion relatively took nearly 58% less computational time as compared to conventional EFGM.

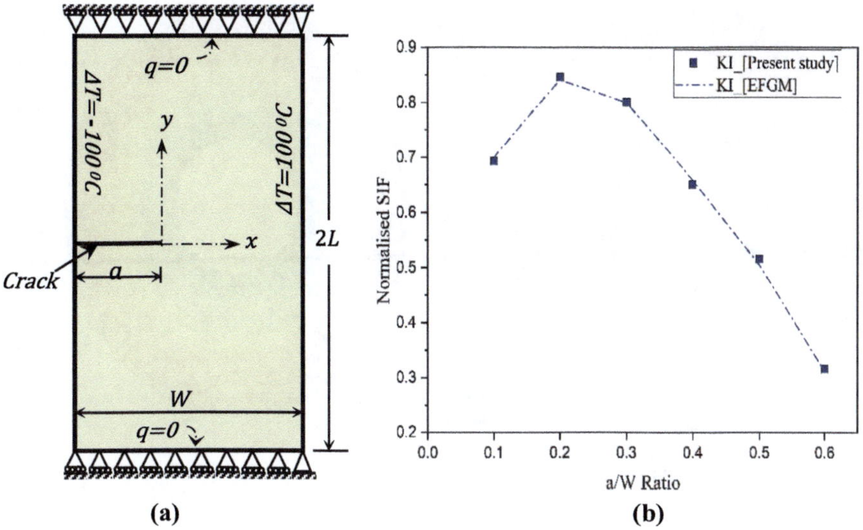

Fig. 4 **a** Problem geometry, **b** normalised stress intensity factor (KI)

4.2 Non-convex Domain Problem

In case of thermoelastic problems temperature fields are used as an input to evaluate stress fields so its correct evaluation is very important. In this example a circular disk subjected to thermal loads have been considered as shown in Fig. 5. Circular disk consists of inner radius (ri) = 10 mm, outer radius (ro) = 30 mm and crack length (a) = 10 mm crack inclination angle (θ) = 30°. Thermal boundary conditions include temperature of 0 °C on the inner diameter and 100 °C on the outer diameter. Thermal fields are evaluated using both conventional EFGM and by using proposed criteria as shown in Fig. 6.

Fig. 5 Problem geometry with boundary conditions

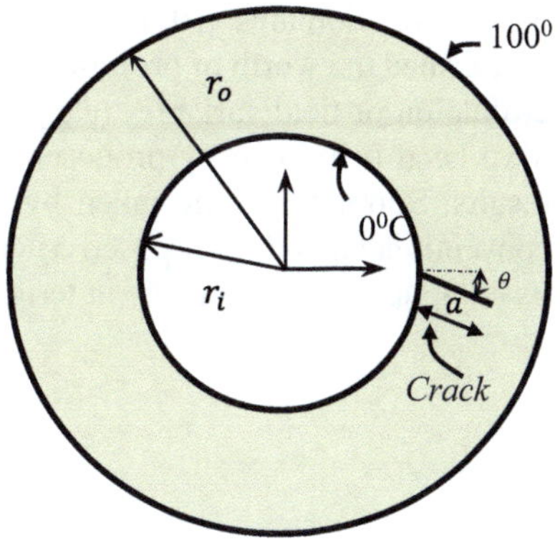

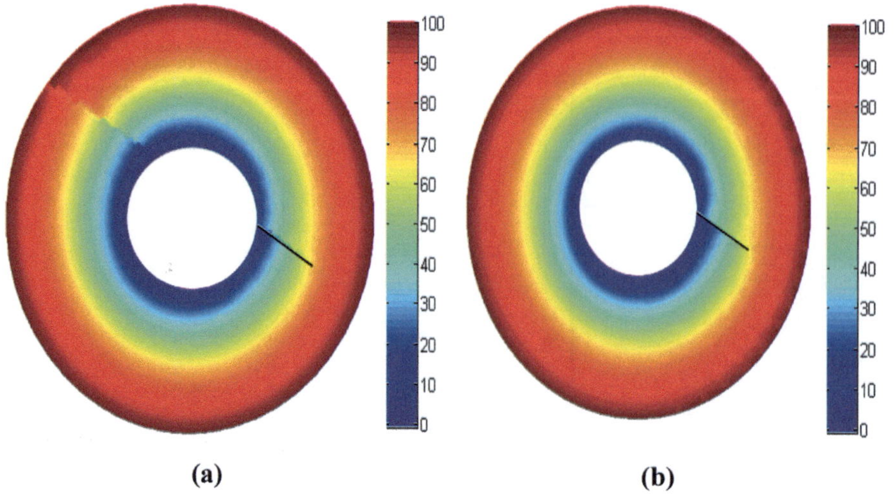

Fig. 6 Temperature distribution **a** traditional EFGM, **b** enriched EFGM

From Fig. 6 it has been observed that Temperature fields observed by conventional EFGM are not accurate as discontinuity in temperature fields due to crack is observed on opposite side of the crack. Whereas temperature fields observed using the proposed criterion are accurate as clear discontinuity in thermal fields are observed due to the presence of crack.

5 Conclusions

In the current work the efficient enrichment approach has been proposed to enhance the computational efficiency of EFGM by developing generic MATLAB codes. The proposed criteria have been tested for thermoelastic problems in both convex and non-convex domains. SIF results validated from literature and conventional EFGM, established the worth of proposed criteria. Also, limitation of conventional EFGM in evaluation of field variables (e.g. Temperature etc.) in case of non-convex domains have been overcome by proposed criteria which have been proved from obtained results. Simulation time taken by current criterion is nearly 58% lower than the conventional EFGM. Proposed criterion enhances conventional EFGM by increasing its computational efficiency in terms of both simulation time and result accuracy.

References

1. Ortiz M, Leroy Y, Needleman A (1987) A finite element method for localized failure analysis. Comput Methods Appl Mech Eng 61(2):189–214
2. Pathak H, Singh A, Singh IV (2013) Fatigue crack growth simulations of bi-material interfacial cracks under thermo-elastic loading by extended finite element method. Euro J Comput Mech 22(1):79–104
3. Pathak H (2020) Crack interaction study in functionally graded materials (FGMs) using XFEM under thermal and mechanical loading environment. Mech Adv Mater Struct 27(11):903–926
4. Garg S, Pant M (2018) Meshfree methods: a comprehensive review of applications. Int J Comput Methods 15:4
5. Xu X, Yu P (2017) Modeling and simulation of injection molding process of polymer melt by a robust SPH method. Appl Math Model 48:384–409
6. Nayroles B, Touzot G, Villon P (1992) Generalizing the finite element method: diffuse approximation and diffuse elements. Comput Mech 10:307–318
7. Belytschko T, Gu L, Lu YY (1994) Fracture and crack growth by element free Galerkin methods. Modell Simul Mater Sci Eng 2(3):519–534
8. Khoei AR, Samimi M, Azami AR (2007) Reproducing Kernel particle method in plasticity of pressure-sensitive material with reference to powder forming process. Comput Mech 39(3):247–270
9. Singh IV, Sandeep K, Prakash R (2003) Heat transfer analysis of two-dimensional fins using meshless element free Galerkin method. Numer Heat Transf Part A Appl 44(1):73–84
10. Ju SH, Hsu HH (2014) Solving numerical difficulties for element-free Galerkin analyses. Comput Mech 53(2):273–281
11. Duflot M (2008) The extended finite element method in thermoelastic fracture mechanics. Int J Numer Meth Eng 74(5):827–847

Simulation of Functionally Graded Thermal Barrier Coating (TBC) Over Al–Si Piston Using Finite Element Method

Ankit Sharma and Mohit Pant

Abstract The piston is an essential part of an internal combustion engine which transfers force from gas expansion in cylinder to the crankshaft. Owing to its functionality, piston is subjected to high thermal gradients due to expansion of gases which often leads to development of high thermal stresses. Use of coating on engine components has been found to yield the efficiency of the engine as it entraps unwanted heat transfer to the engine components. Over the years functionally graded materials are gaining importance as thermal barrier coating owing to its spatial variation of constituent. FGM consists of metallic and ceramic phases which vary in composition from one end to other in a well-defined way. The effort to increase the performance of piston using thermal barrier coatings in form of functionally graded materials have been the main aim of current work. Spatial variation of constituents in FGM has enabled it to be used as thermal barrier coatings as its ceramic constituent exposed to high-temperature acts as a barrier to heat transfer due to its lower thermal conductivity. Functionally graded thermal barrier coating enhances high-temperature durability of the components by reducing heat transfer between the underlying piston metal. Current work focuses on finite element analysis of Al–Si piston using ANSYS workbench having different type of FGM coatings ($MgZrO_3$ − NiCrAl, YSZ − NiCoCrAlY, NiCrAlY − YSZ, $La_2Zr_2O_7$ − NiCoCrAlY). The major aim of this study is to predict the most suitable FGM coating after comparing the outcomes of the analysis performed using the Finite element method.

Keywords Functionally graded material · Finite element method · Piston · Coating

A. Sharma (✉) · M. Pant
Department of Mechanical Engineering, National Institute of Technology Hamirpur, Hamirpur, India
e-mail: 21mme101@nith.ac.in

M. Pant
e-mail: mohitfme@nith.ac.in

R. K. Ray et al. (eds.), *Advances in Theoretical and Applied Mechanics*, Lecture Notes in Mechanical Engineering, https://doi.org/10.1007/978-981-97-0418-7_21

1 Introduction

Internal combustion engines are used in a variety of industries including transportation, agriculture, and sports. A piston is a moving component of an engine that is susceptible to high levels of inertia force as well as external forces. To achieve greater efficiency in terms of fuel consumption, it is necessary to make the engine support structure out of lightweight material. On the other hand, the material in question should have excellent casting and machining qualities. In addition, resistance to wear and heat conductivity are also desirable features that must be taken into consideration during the process of selecting materials for pistons. As the piston is subjected to very high temperature, to reduce heat transfer inside the piston it is necessary to coat it with different types of coating. Researchers are very interested in improving the efficiency and performance of engines by selecting aluminium-based alloy and cast-iron pistons. TBC (thermal Barrier coating) has been used for decades to increase the performance and efficiency of many machine parts. These coatings enable greater operating temperatures while limiting heat exposure to structural components, hence extending part life, and in the engine, it also reduces the knocking [1]. TBCs of YSZ (yttria-stabilized zirconia) over the marine diesel engine increase the thermal efficiency of the engine [2]. Coating of ceramic is widely used in aircraft and turbine blades to protect superalloy parts from oxidation and high temperature which increases the life of the blade [3]. Zirconia coating over the piston increases the brake thermal efficiency by 5.89% and indicated thermal efficiency by 11.14% in comparison to the uncoated piston [4].

TBC protects against failure fatigue when working at high temperatures during analysis and reduces heat rejection from the engine. TBC protects the piston against thermal stress, corrosion assaults, and excessive heat emissions while also reducing fuel consumption and the piston's total heat flux. It is used to improve the efficiency and performance of IC engines, SI engines, and gas turbines by applying it to surfaces where heat transfer occurs or locations where temperatures are very high [5]. TBCs were researched for "adiabatic" diesel engines owing to first law thermodynamic predictions of improved fuel efficiency, less heat rejection, and enhanced power density. In Internal Combustion Engine, due to the expansion of gases heat is generated and absorbed by the piston. To control the thermal stresses and deformation due to thermal stresses, it is very important to calculate and minimize the temperature variation at the crown [6]. By coating these components with functionally graded thermal barrier coatings (TBCs), the damaging effects of wear, friction, heating, corrosion, and oxidation reduces. In certain experiments, engine coating has been shown to boost engine power, reduce specific fuel consumption, significantly reduce exhaust gas emissions, and reduce smoke density when compared to an uncoated engine. The coated pistons have been used to keep the combustion chamber's temperature at the desired level. This will lessen the piston's heat loss [4]. So, to operate Piston at a higher operating temperature functionally graded TBCs of the ceramic and metallic substrate are applied in a specific manner that improves

Table 1 Material properties of piston, ring, ceramics and metallic substrates

Material	Elasticity E (GPa)	Conductivity, k (W/m°C)	Diffusivity, α * $10^{(-6)}$ (1/°C)	Density, ρ(kg/m^3)	Poisson's ratio (μ)
Al–Si piston [1]	80	155	21	2700	0.28
Steel [2]	200	79	12.2	7300	0.3
$MgZrO_3$ [9]	46	0.8	0.80	5600	0.20
NiCrAl [9]	90	16.1	12	7870	0.27
YSZ [10]	40	2.12	10.7	6010	0.22
NiCoCrAlY [11]	86	16.1	17.5	7870	0.8
$La_2Zr_2O_7$ [11]	63	1.2	9.1	6300	0.3
NiCrAlY [10]	186	6.4	24	7320	0.3
Oil ring [10]	160	25–42	10	7200	0.29
Compression ring [6]	110	46–59	10	7300	0.3

the durability, reliability, and efficiency of the IC engine. During Power stroke, high-temperature gases strike over the piston top surface and crown, and to reduce the heat rejection between piston and cylinder functionally grade TBCs applied over the surface of the piston that also protects the surface from thermal fatigues, due to power and exhaust strokes in IC engine process [7].

In this paper aluminium alloyed Al–Si piston has been selected because of its compositions like Magnesium, Copper, Nickel provide good strength, and less deformation and stress as compared to Aluminium piston [8]. The primary focus is on the thermal behaviour of functionally graded coatings generated by employing a commercial code, specifically ANSYS 2022 R1, on Al–Si piston surfaces. Thermal Analysis employed to metallic, ceramic and cremate functionally graded thermal barrier coating of **$MgZrO_3$ + NiCrAl, YSZ + NiCoCrAlY, YSZ + NiCrAlY, $La_2Zr_2O_7$ + NiCoCrAlY** whose properties with citation are shown in table 1, are applied over the Aluminium Alloy piston. It has been get that functionally graded coating of **$MgZrO_3$ + NiCrAl** have 56.6% temperature higher than uncoated piston, and there also minimum temperatures et al.-Si crown in this FG thermal barrier coating.

2 Functionally Graded Material

There are different types of combinations for generating FGM like metal–metal, metal-ceramic, ceramic-ceramic and ceramic-polymer [12]. This progressive transition of thermomechanical characteristics between the ceramic and the metallic coat eliminates the harmful interfacial tensions that are often caused by thermal expansion mismatch [6].

For comparative analysis FGM of $\mathbf{MgZrO_3 - NiCrAl}$, $\mathbf{YSZ - NiCoCrAlY}$, $\mathbf{YSZ - NiCrAlY}$, $\mathbf{La_2Zr_2O_7 - NiCoCrAlY}$ have been selected, and the Al–Si piston surface have been coated with selected different FGMs. A metallic substrate have coated over the surface of the piston and at the topmost, there have a ceramic substrate. In between ceramic and metallic substrates selected FGM have been present throughout whose properties varies as given below equation. Metallic substrate, FGM and topmost ceramic substrate plays important role in the reduction of internal stresses. Ceramics coating's thermal expansion coefficient, elastic modulus, and thermal conductivity coefficient determine how well it can withstand thermal shock [13].

$$E_{\text{effective}} = [(E_{\text{ceramic}} * V) + (E_{\text{metallic}} * (1 - V))] \tag{1}$$

$$\alpha_{\text{effective}} = [(\alpha_{\text{ceramic}} * V) + (\alpha_{\text{metallic}} * (1 - V))] \tag{2}$$

$$\mu_{\text{effective}} = [(\mu_{\text{ceramic}} * V) + (\mu_{\text{metalic}} * (1 - V))] \tag{3}$$

$$\rho_{\text{effective}} = [(\rho_{\text{ceramic}} * V) + (\rho_{\text{metalic}} * (1 - V))] \tag{4}$$

$$K_{\text{effective}} = \frac{(K_{\text{ceramic}} * K_{\text{metallic}})}{K_{\text{ceramic}} * (1 - V) + K_{\text{metallic}} * V} \tag{5}$$

where V is volume fraction. Equations 1–5 are varies linearly, variation in the FGM can be quadratic or polynomial depends upon the how properties we have to varies.

3 Steady State Finite Element Heat Transfer Analysis of Functionally Graded Material

Numerical Simulation analysis of the Al–Si diesel engine piston performed using Ansys 2022 R1. Three-dimension finite element thermal analysis carried out over uncoated and FGM coated piston. Meshed model for full model Al–Si piston have solid linear tetrahedron elements with mid-side nodes. Total number of elements were 366,176. Mesh model of the full piston model is shown in Fig. 1. This model contains contact between surface of piston ring and groove for the ring and these contacts are meshed as contact mesh in computational code. Steady state thermal was performed in Ansys 2022 R1 for complete piston model. Piston consists thermal boundary conditions of ring land and skirt thermal boundary condition, underside thermal boundary condition, combustion side thermal boundary condition. Thermal boundary conditions, convective heat transfer coefficient and ambience temperature shown in Fig. 1b [9]. To represent heat transmission in the ring land and skirt zone, the thermal circuit approach is applied as shown in Fig. 2. Esfahanian et al. [14] with following assumptions.

- It is assumed that the influence of piston motion on heat transmission is negligible.
- The rings as well as the skirt are completely submerged in oil, and there are no cavitations present.
- The rings are not twisted in any way.
- Oil film conductive heat transfer is ignored.

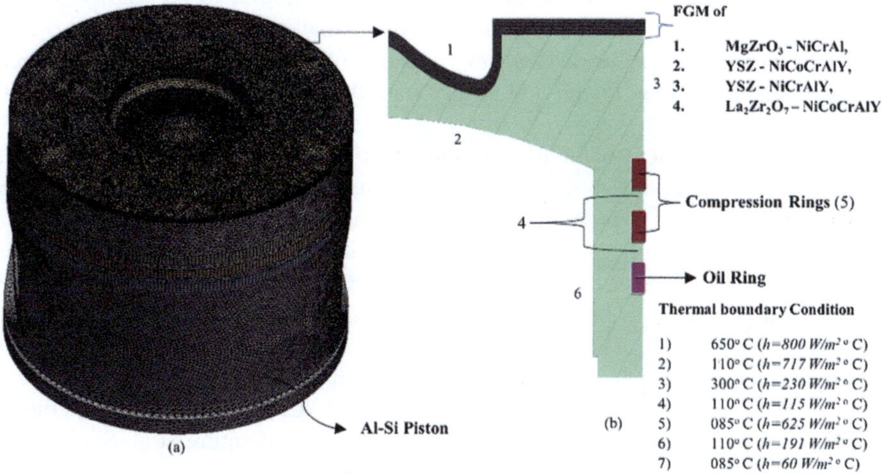

Fig. 1 **a** Finite element Meshed geometry of the diesel engine Al–Si Piston and FGM coat over the piston, **b** sectional view and applied piston boundary condition

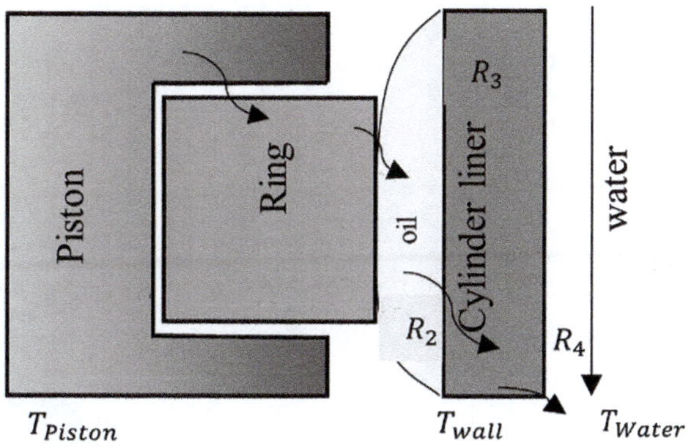

Fig. 2 Thermal circuit resistance model for heat transfer between piston and water. Conductive resistance of the ring R_1. Conductive resistance of the oil film R_2. Conductive resistance of liner R_3. Convective resistance between liner and cooling water R_4 [14]

4 Results and Discussion

Heat conduction is generally more dominant than radiation for TBCs in an engine. Hence, to estimate temperature distributions and heat flows the thermal conductivity of the material is very important [15]. Several numerical analyses are carried out in Ansys 2022 R1 to evaluate temperature gradient over the uncoated Al–Si and different selected TBCs Al–Si Piston.

Temperature gradient of standard Al–Si piston is shown in Fig. 3. The maximum crown temperature value of uncoated Al–Si piston is found to be 363.75 °C, the minimum temperature is 196.11 °C and average temperature is 241.69 °C.

To reduce the temperature at the crown of piston four different functionally graded TBCs Selected over the Piston and steady state thermal analysis has been done in simulating Code.

Temperature gradient of firstly selected functionally Graded TBC of **$MgZrO_3$ – NiCrAl** Al–Si piston is shown in Fig. 4a. The maximum temperature value of FG coated Al–Si piston is found to be 569.9 °C, the minimum temperature is 163.62 °C and average temperature is 292.22 °C. Temperature et al.-Si crown is found as 284.51 °C which is lesser than the standard (Uncoated) piston crown. Steady state thermal analysis also has been done over the secondly selected functionally graded thermal barrier coating of **YSZ – NiCoCrAlY**, temperature gradient of selected FGM TBC has shown in Fig. 4b, and it has been observed that the maximum temperature value of FG coated Al–Si piston is found to be 486.1 °C, the minimum temperature is 179.07 °C and average temperature is 301.85 °C. Temperature et al.-Si crown is found as 330.61 °C.

Further steady state thermal analysis has been done over the selected functionally grade coating of **YSZ – NiCrAlY** during the simulation in computational code, it has been getting that the maximum temperature value of FG coated Al–Si piston is found

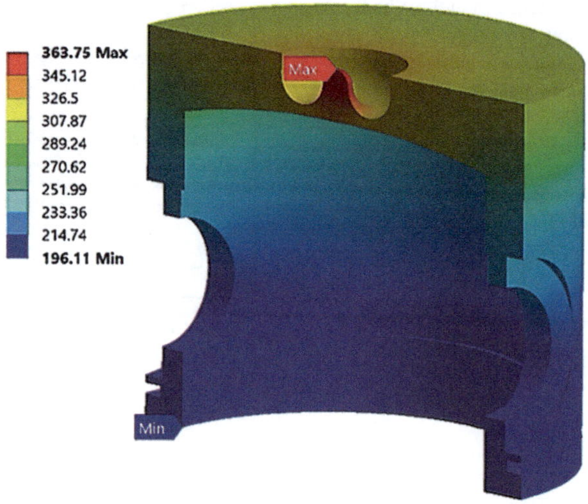

Fig. 3 Temperature distribution of uncoated Al–Si piston

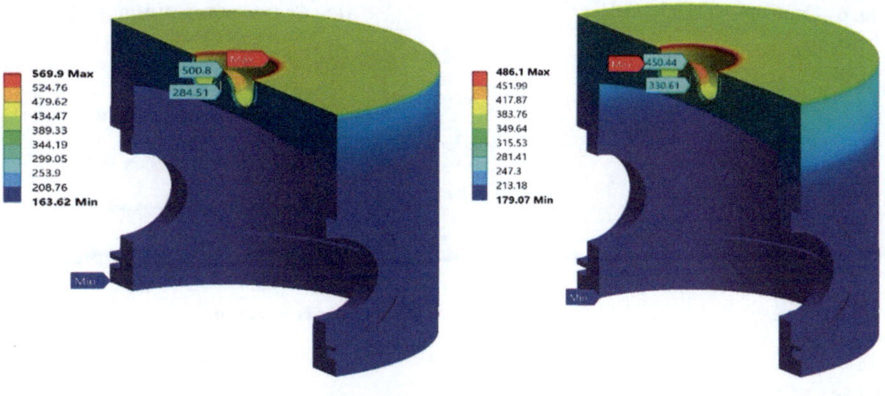

Fig. 4 **a** Temperature distribution of $MgZrO_3$ – NiCrAl, **b** temperature distribution of YSZ – NiCoCrAlY functionally graded coated Al–SI piston

to be 498.2 °C, the minimum temperature is 176.71 °C and average temperature is 305.24 °C. Temperature et al.-Si crown is found as 324.86 °C as shown in Fig. 5a.

Steady state thermal analysis has been done over selected **$La_2Zr_2O_7$ – NiCoCrAlY** functionally graded thermal barrier coating and it has been got that maximum temperature is 533.65 °C and, minimum temperature is 170.94 °C and average temperature is 296.8 °C. Temperature et al.-Si crown is found as 311.03 °C as shown in Fig. 5b, and comparative graphical analysis is shown in Fig. 6.

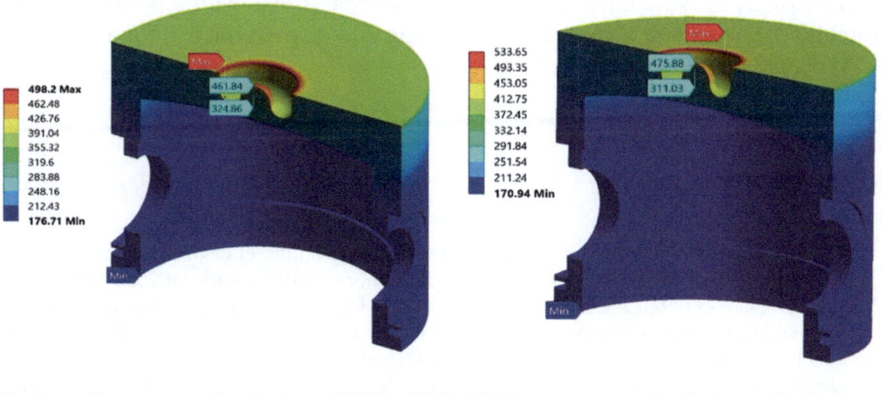

Fig. 5 **a**. Temperature distribution of YSZ – NiCrAlY, **b** temperature distribution of $LaZr_2O_7$ – NiCoCrAlY functionally graded coated Al–SI piston

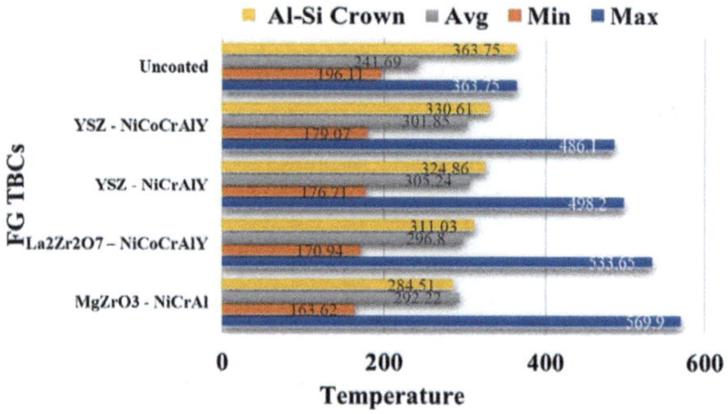

Fig. 6 Comparative analysis of FG TBCs

5 Conclusion

A comparative Analysis was made between the temperature gradient of the uncoated Al–Si piston surface and different selected FGM coated Al–Si Piston. From the numerical simulations, it has been revealed that the surface temperature of the $MgZrO_3$ – NiCrAl FGM coated Al–Si piston was higher than by 56.67% that of the uncoated Al–Si piston. From simulation it has been also observes that surface temperature of YSZ – NiCoCrAlY, YSZ – NiCrAlY, $La_2Zr_2O_7$ – NiCoCrAlY was higher than by 33.63%, 36.96%, 46.70% that of standard Al–Si piston respectively. It has been revealed that Al–Si piston crown temperatures in functionally graded thermal barrier coated piston is reduced as comparison to standard (uncoated) piston.

According to the numerical simulations performed in this work, it has been determined that the use of FGM coating on Al–Si pistons raises the combustion chamber temperature and thermal strength of the base metal. These coatings reduces the thermal stress and strains over the Al–Si Pistons. These FGM coatings helps to improves the efficiency and effectiveness of the engines as compared to uncoated pistons. It also helps to improve the life of the piston in automobile engines. In Future Different FGM coatings can be selected over the Al–Si or steel piston. Transient thermal and structural analysis for the power stroke can be performed with suitable FGM coatings. Experimental work can also be performed in future work for FGM coated piston assembly.

References

1. Taymaz I (2007) The effect of thermal barrier coatings on diesel engine performance. Surf Coatings Technol 201(9–11):5249–5252. https://doi.org/10.1016/j.surfcoat.2006.07.123
2. Fei CG, Qian ZQ, Ren J, Zhou XJ, Zhu SW (2021) Numerical and experimental research on thermal insulation performance of marine diesel engine piston based on YSZ thermal barrier coating. Coatings 11(7). https://doi.org/10.3390/coatings11070765
3. Darolia R (2013) Thermal barrier coatings technology: critical review, progress update, remaining challenges and prospects, vol 58, no 6. https://doi.org/10.1179/1743280413Y.0000000019

4. Sathyamoorthi S, Prabhakaran M, Muhammed Abraar SA (2016) Numerical investigation of ceramic coating on piston crown using finite element analysis. Int J Sci Eng Appl Sci 2:258–263 [online]. Available www.ijseas.com
5. Sharma JK, Raj R, Kumar S, Jain RK, Pandey M (2021) Finite element modelling of lanthanum cerate ($La_2Ce_2O_7$) coated piston used in a diesel engine. Case Stud Therm Eng 25:100865. https://doi.org/10.1016/j.csite.2021.100865
6. Buyukkaya E, Cerit M (2007) Thermal analysis of a ceramic coating diesel engine piston using 3-D finite element method. Surf Coatings Technol 202(2):398–402. https://doi.org/10.1016/j.surfcoat.2007.06.006
7. Varshney N, Mishra D, Agrahri PK, Gupta S, Chandan BB, Kumar A (2018) Thermal analysis of ceramic coated piston head in diesel engine, pp 3770–3779
8. Francis S, Umesh V, Shivakumar S (2021) Materials today: proceedings design and analysis of vortex bladeless wind turbine. Mater Today Proc 2021. https://doi.org/10.1016/j.matpr.2022.07.211
9. Buyukkaya E (2008) Thermal analysis of functionally graded coating AlSi alloy and steel pistons. Surf Coatings Technol 202(16):3856–3865. https://doi.org/10.1016/j.surfcoat.2008.01.034
10. Ganapathy Pandian S, Puzhuthivakkam Rengarajan S, Babu TP, Natarajan V, Kanagasabesan H (2015) Thermal and structural analysis of functionally graded NiCrAlY/YSZ/Al_2O_3 coated piston. SAE Int J Mater Manuf 8(2):578–585. https://doi.org/10.4271/2015-01-9081
11. Fei C, Lei T, Qian Z, Shu Z (2022) Piston thermal analysis of heavy commercial vehicle diesel engine using lanthanum zirconate thermal-barrier coating. Energies 15(12):4225. https://doi.org/10.3390/en15124225
12. El-Galy IM, Saleh BI, Ahmed MH (2019) Functionally graded materials classifications and development trends from industrial point of view. SN Appl Sci 1(11):1–23. https://doi.org/10.1007/s42452-019-1413-4
13. Sarikaya O, Islamoglu Y, Celik E (2005) Finite element modeling of the effect of the ceramic coatings on heat transfer characteristics in thermal barrier applications. Mater Des 26(4):357–362. https://doi.org/10.1016/j.matdes.2004.06.020
14. Esfahanian V, Javaheri A, Ghaffarpour M (2006) Thermal analysis of an SI engine piston using different combustion boundary condition treatments. Appl Therm Eng 26(2–3):277–287. https://doi.org/10.1016/j.applthermaleng.2005.05.002
15. Heywood JB (1988) Internal combustion engine fundamentals. McGraw-Hill, New York

Effect of the Internal Length Scale Parameter in Gradient Damage Mechanics

Sourabh Puranik and Sitikantha Roy

Abstract Gradient enhanced continuum damage mechanics is one of the best methods to remedy the mesh sensitivity and mathematically infeasible results usually obtained in the numerical implementation of continuum damage mechanics theories. It involves the introduction of internal length scales through nonlocal averaging. We evaluate the effects of the internal length scale parameter on structural response in this study. A half model of a single-edge notched specimen is analyzed with large and small length scale parameters using an in-house MATLAB code assuming isotropic damage and plane strain conditions. A scalar damage variable with exponential evolution law is considered. The results demonstrate that with increasing internal length scale parameter, the damage spread increases and its maximum value decreases; hence, its selection should be made considering the experimental responses of a material. A detailed description of finite element formulations and implementation algorithms for solving the nonlinear problem is elucidated.

Keywords Gradient damage mechanics · Internal length parameter · Nonlinear finite element methods

1 Introduction and Objective

Continuum damage mechanics is a widely pursued field in the analysis of structures. Many researchers have contributed significantly to this field [1, 2]. The numerical implementation of continuum damage mechanics leads to spurious mesh sensitivity

S. Puranik (✉) · S. Roy
Department of Applied Mechanics, Indian Institute of Technology Delhi,
New Delhi 110016, India
e-mail: sourabh2396@gmail.com

S. Roy
Yardi School of Artificial Intelligence, Indian Institute of Technology Delhi,
New Delhi 110016, India

R. K. Ray et al. (eds.), *Advances in Theoretical and Applied Mechanics*, Lecture Notes in Mechanical Engineering, https://doi.org/10.1007/978-981-97-0418-7_22

and inaccurate results because the damaged material exhibits strain-softening behaviors and localization. Moreover, due to damage evolution, the governing differential equations lose ellipticity after certain loading [3]. As a solution to the problem discussed above, an approach involving the inclusion of an internal length scale parameter that can control the localization zone width through gradients is introduced by Peerling et al. [4]. This length is considered dependent on the material's crystal structure. Since it is difficult to measure the internal length parameter experimentally, it is considered only an input parameter to the model. It can be calculated theoretically through inverse analysis and parameter identification techniques [5]. This paper presents the internal length parameter's effect on a material's structural response. A comparison of the spread of the damage and the maximum damage value is shown.

We solved the gradient damage problem on the single-edged notched specimen (half model), assuming two different internal length scales: small and large. Linearized governing equations are solved incrementally using the Newton Raphson algorithm in the context of finite element methods.

2 Gradient Damage Mechanics Model

We represent all the microcavities and microcracks by a field variable D, which is isotropic and scalar and can give insight into the material condition at any time. By using concepts of effective stress, the stress–strain relation can be given as

$$\boldsymbol{\sigma} = (1 - D)\boldsymbol{E} : \boldsymbol{\varepsilon} \tag{1}$$

where $\boldsymbol{E}$ is the stiffness tensor, and Damage evolution law is guided by a scalar strain history parameter (κ) describing the most severe deformation a material point has undergone and is expressed by an equivalent strain (ε) in strain-based formulations. The scalar strain history parameter (κ) is defined as

$$\kappa(t) = \max\{\varepsilon | 0 \leq \tau \leq t\} \tag{2}$$

where τ and t denote instantaneous and total time, respectively,

For the damage evolution, exponential law has been used given in Eq. (3) where κ_i is the initial damage threshold, and κ is a history parameter that keeps track of the maximum nonlocal strain.

$$D(\kappa) = \begin{cases} 0, & \text{if}\, \kappa < \kappa_i \\ 1 - \frac{\kappa_i}{\kappa}\{1 - \alpha + \alpha \exp[-\beta(\kappa_i - \kappa)]\}, & \text{otherwise} \end{cases} \tag{3}$$

α, β are material parameters.

2.1 Equivalent Strain

We have used the von mises equivalent strain, which is given as

$$\varepsilon = \frac{k-1}{2k(1-2\nu)}I_1 + \frac{1}{2k}\sqrt{\frac{(k-1)^2}{(1-2\nu)^2}I_1^2 + +\frac{2k}{(1-\nu)^2}J_2} \tag{4}$$

where I_1 represents the first invariant of the strain tensor and J_2 the second invariant of the deviatoric strain tensor, ν is the poisons ratio. Material parameter k is defined as the ratio of compressive strength (f_c) and tensile strength (f_t).

2.2 Nonlocal Averaging

To avoid mesh sensitivity, we introduce an internal length parameter l through nonlocal averaging. We define nonlocal equivalent strains by the weighted integral of local equivalent strain over the domain V.

$$\overline{\varepsilon}(x) = \frac{\int_V W(\boldsymbol{\xi}).\varepsilon(\boldsymbol{x}+\boldsymbol{\xi})\,\mathrm{d}V}{\int_V W(\boldsymbol{\xi})\mathrm{d}V} \tag{5}$$

where $W(\boldsymbol{\xi})$ is the gaussian weight function, and it depends on neighborhood $\boldsymbol{\xi}$ as

$$W(\boldsymbol{\xi}) = \frac{1}{\sqrt{2\pi}l}e^{-\frac{\xi^2}{2l^2}} \tag{6}$$

Substituting Eq. (6) in Eq. (5), we finally get an implicit gradient form for nonlocal strain as

$$\overline{\varepsilon} - c\frac{\partial^2\overline{\varepsilon}}{\partial x^2} = \varepsilon \tag{7}$$

where $c = l^2$.

3 Finite Element Formulations and Solution Algorithm

We solve the linear momentum balance equation and Eq. (7) through the nonlinear finite element method. Without the presence of body and traction forces, the linear momentum balance equation can be written as follows:

$$\nabla.\sigma = 0 \tag{8}$$

By utilizing Eq. (1), we obtain the resulting weak form of Eqs. (7) and (8):

$$\int_{\Omega} \nabla(v)^T : \boldsymbol{\sigma} \mathrm{d}\Omega = 0 \tag{9}$$

$$\int_{\Omega} (w\overline{\varepsilon} + \nabla w.c\nabla\overline{\varepsilon})\mathrm{d}\Omega = \int_{\Omega} w\varepsilon \mathrm{d}\Omega \tag{10}$$

where $\boldsymbol{v}$ and w represent the weight functions used.

We discretize our elements using serendipity (8 noded) elements for displacements ($\boldsymbol{u}$) and quadratic elements (4 noded) for nonlocal strain as:

$$\begin{aligned} \boldsymbol{u} &= N\underline{u};\ \boldsymbol{v} = N\underline{v} \\ \overline{\epsilon} &= \overline{N}\underline{\overline{\epsilon}};\ w = \overline{N}\underline{w} \end{aligned}. \tag{11}$$

Substitutions of the discretization mentioned above in Eqs. (9) and (10)

$$\int_{\Omega} B^T \boldsymbol{\sigma} \mathrm{d}\Omega = 0 \tag{12}$$

$$\int_{\Omega} \left(\overline{N}^T \overline{N} + \overline{B}^T c\overline{B}\right)\mathrm{d}\Omega = \int_{\Omega} \overline{N}^T \varepsilon \mathrm{d}\Omega \tag{13}$$

The above two equations (Eqs. 12 and 13) are nonlinear coupled equations. Linearizing them, we get

$$K_{i-1}^{uu}\delta\underline{u}\epsilon_i + K_{i-1}^{u}\delta\underline{\overline{\epsilon}}_i = -\boldsymbol{f}_{i-1}^{\boldsymbol{u}} \tag{14}$$

$$K_{i-1}^{\epsilon u}\delta\underline{u}_i + K_{i-1}^{\epsilon\epsilon}\delta\underline{\overline{\epsilon}}_i = \boldsymbol{f}_{i-1}^{\epsilon} - K^{\epsilon\epsilon}\underline{\overline{\epsilon}}_{i-1} \tag{15}$$

where

$$\begin{aligned} K_{i-1}^{\epsilon u} &= \int_{\Omega} \overline{N}^T \underline{s}_{i-1}^T B \mathrm{d}\Omega \\ K_{i-1}^{u\epsilon} &= -\int_{\Omega} B^T H\epsilon_{i-1} q_{i-1} \overline{N} \mathrm{d}\Omega \\ K_{i-1}^{uu} &= \int_{\Omega} B^T (1 - D_{i-1}) H B \mathrm{d}\Omega \\ K^{\epsilon\epsilon} &= \int_{\Omega} \left(\overline{N}^T \overline{N} + \overline{B}^T c\overline{B}\right)\mathrm{d}\Omega \end{aligned}$$

$$\underline{s}_{i-1} = \frac{\delta \varepsilon}{\delta \underline{\varepsilon}_{i-1}}$$

and the damage evolution rate is given as

$$q_{i-1} = \begin{cases} \frac{\delta D}{\delta \kappa_{i-1}} & \text{if} \quad \overline{\varepsilon}_{i-1} > \kappa_0 \\ 0 & \text{if} \quad \overline{\varepsilon}_{i-1} \leq \kappa_0 \end{cases}$$

The Newton Raphson method is used to solve Eqs. (14) and (15) incrementally, and the algorithm applied is presented in Fig. 2.

4 Results and Discussions

On account of symmetry, half model of single edge notched specimen assuming linearly elastic plane strain and tensile loading condition keeping the material properties $E = 1000$ MPa, $\mu = 0.2$, $\alpha = 0.99$, $k = 10$, and damage initiation threshold $\kappa_i = 0.002$ constant is considered. Internal length scale and softening parameters c, β have been taken to be 1.56 mm^2 and 8, and 400 mm^2 and 205, respectively, for small and large internal length scale cases.

In the present study, isotropic damage with scaler damage variable and exponential damage evolution law has been used, and gradients have been introduced by performing nonlocal weighted averaging considering the Gaussian weight function. We represented the crack in the model using appropriate boundary conditions, and a displacement was applied, as shown in Fig. 1b.

Fig. 1 **a** Illustration of the single-edge notched specimen in tension, **b** prescribed boundary conditions of half model

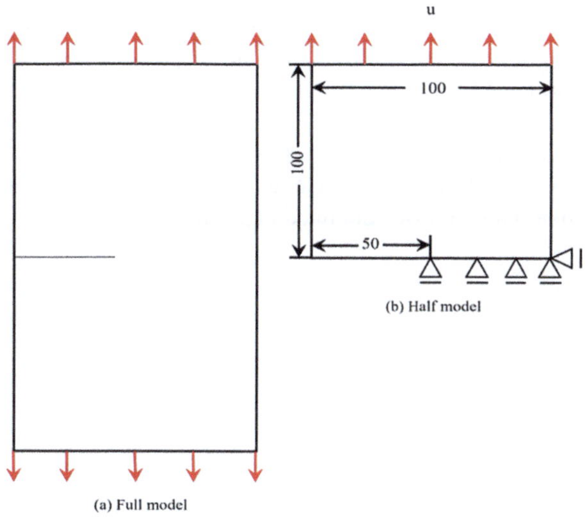

1. Initialization:
   Initialize Nodal displacements $\underline{u}$,nonlocal strains $\underline{\bar{\epsilon}}$ and history parameter κ.
2. Set iteration counter=0
3. Displacement steps loop starts:
   - Calculate displacements for current step
   - Newton Raphson Loop starts (until converged)
     - ❖ Elemental loop starts
       - ➢ Integration Point loop starts
         Compute κ, D and q
         Compute $K^{\epsilon u}$, K^{uu}, $K^{\epsilon\epsilon}$, $K^{\epsilon u}$, f^{u}, f^{ϵ}
         Integration point loop ends

       Elemental loop ends

     - ❖ Assemble K, F and apply Boundary conditions
     - ❖ Solve for δu and $\delta\underline{\bar{\epsilon}}$
     - ❖ Update $\underline{u}$ and $\underline{\bar{\epsilon}}$
     - ❖ Check convergence
   - Calculate Reaction forces, stresses, and strains with converged value
   - Update history parameter κ

   Displacement steps loop ends

Fig. 2 Algorithm used for the incremental solution

The results from the half model are shown in Figs. 3 and 4. It is also cross verified with the UEL implementation presented in [6], and it is evident that the structural response depends on the internal length scale parameter. As we increase the parameter, the maximum value obtained of damage decreases, and the spread increases because the strains localize in larger volumes. The selection of the internal length scale, which explains the response better, can be done by optimizing the parameter using the simulation along with the experimental data for the different materials.

It was found that the internal length predominantly affects the structural response, and the optimum internal length scale should be carefully selected based on the actual material responses obtained experimentally.

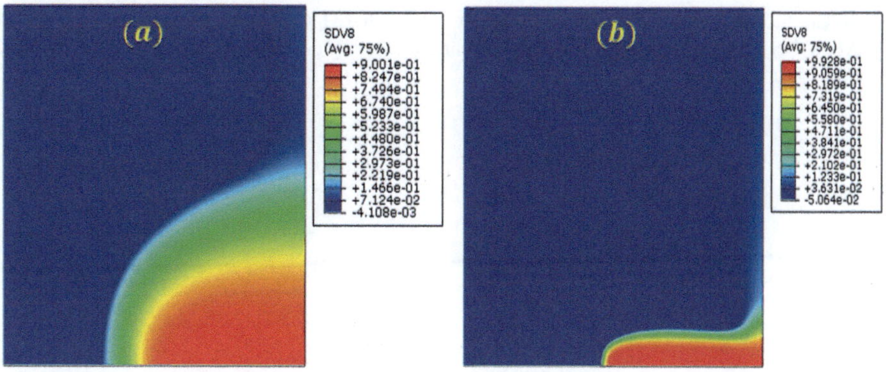

Fig. 3 Damage profiles were obtained by considering **a** large internal length scale parameter and **b** small internal length parameter

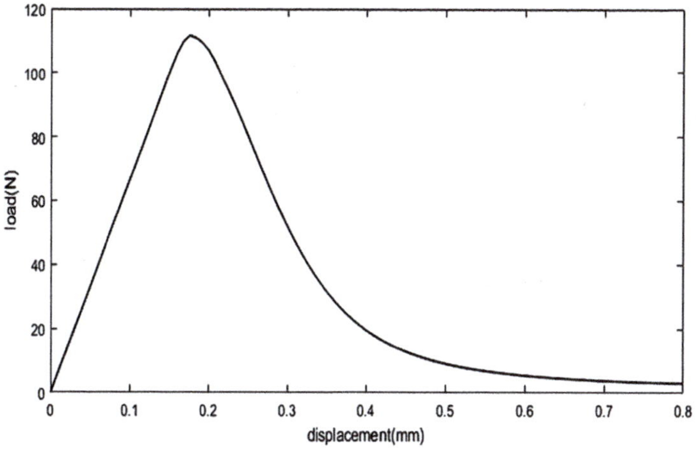

Fig. 4 Load-displacement curve corresponding to c = 400 mm^2

References

1. Krajcinovic D, Lemaitre J (eds) (1987) Continuum damage mechanics: theory and applications. Springer-Verlag, New York
2. Murakami S (2012) Continuum damage mechanics: a continuum mechanics approach to the analysis of damage and fracture, vol 185. Springer Science & Business Media
3. Lasry D, Belytschko T (1988) Localization limiters in transient problems. Int J Solids Struct 24(6):581–597
4. Peerlings RHJ, René de Borst WA, Brekelmans M, de Vree JHP (1996) Gradient enhanced damage for quasi-brittle materials. Int J Numer Methods Eng 39(19):3391–3403
5. Mahnken R, Kuhl E (1999) Parameter identification of gradient enhanced damage models with the finite element method. Euro J Mech-A/Sol 18(5):819–835

6. Sarkar S, Singh IV, Mishra BK, Shedbale AS, Poh LH (2019) A comparative study and ABAQUS implementation of conventional and localizing gradient enhanced damage models. Finite Elements Anal Des 160:1–31

FFT-Homogenization of Elastic and Thermo-elastic Properties of Concrete

Maninder Pal Singh, Subhamoy Sen, Gaurav Arora, Himanshu Pathak, Ajay Bhushan Dogra, and Kundan Singh

Abstract Meso-scale homogenization of elastic and thermo-elastic properties of concrete is carried out using a semi-random sequential adsorption algorithm taking basis on properties of aggregate randomly distributed in a homogenized matrix (cement mortar). The aggregates are generated by the sequential reduction of a cuboid into a polyhedron with multiple faces. The original Moulinec-Suquet (1994) FFT-based algorithm was used for the homogenization of elastic properties subjected to quasi-static loading with periodic boundary conditions. For the homogenization of thermo-elastic properties, the original algorithm is modified for temperature loading. The model is validated by comparing the homogenization results with the commercial DIGIMAT software. The results were verified for spherical inclusion by varying the properties of the constituents (matrix and inclusion) and the volume fraction. The volume fraction of the inclusion was however kept within 30% for comparison due to the limitation of DIGIMAT. A good agreement between the numerical and DIGIMAT results was obtained. Later on, the same algorithm is applied to a computationally generated realistic concrete mesostructure RVE with a higher volume fraction to obtain the homogenized properties of the concrete mix.

Keywords Concrete mesostructure · FFT-homogenization · RVE · DIGIMAT

M. P. Singh · S. Sen (✉)
i4S Laboratory, Indian Institute of Technology Mandi, Kamand, India
e-mail: subhamoy@iitmandi.ac.in

G. Arora
Chandigarh University, Chandigarh, India

H. Pathak
Indian Institute of Technology Mandi, Kamand, India

A. B. Dogra · K. Singh
Satluj Jal Vidyut Nigam (SJVN) Limited, Shimla, India

R. K. Ray et al. (eds.), *Advances in Theoretical and Applied Mechanics*, Lecture Notes in Mechanical Engineering, https://doi.org/10.1007/978-981-97-0418-7_23

1 Introduction

Concrete is a complex material for which the properties are very much dependent on the ingredients, their mix proportion, and the final microstructure post-casting. Eventually, reproducing its properties exactly is hardly possible since unlike other composites, the geometric and material properties of its constituents are far too random. Moreover, the same quality control cannot be expected to be maintained for every casting. It should further be noted that the ambient conditions (temperature and humidity) during casting can also greatly affect the behavior of concrete. Eventually, the micro-scale model, while possible with great computational expense, is not that practical for concrete. Instead, a homogenized mesostructure can be a handy solution that can replicate the behavior of the mix being within sufficient accuracy limits.

The FFT-homogenization technique was first introduced by Moulinec and Suquet in 1994 [1] in order to define homogenized properties of a mix, taking basis on its constituents. This homogenization approach uses the Lippmann-Schwinger formulation and relies on non-conforming Fourier-Galerkin discretization to generate a fast and robust solution. The approach has been extensively employed in several engineering problems wherein the mesoscale properties of a mix are necessary to model a full-scale domain. Modeling concrete structures can be identified as a good application case to demonstrate such challenges. Meshing a 3-dimensional microstructure of a composite material like concrete is highly challenging for finite element formulations [2]. To counter this problem, the FFT-based homogenization yields the homogenized properties of the mix so that the mix can be replicated with a representative volume element (RVE) providing an alternate way to perform simulations on a regular grid. The FFT-based methods are considerably faster compared to finite element solvers since no mesh has to be generated. The curse of dimensionality is also overcome by an iterative method that does not require the formation of the stiffness matrix [1], thereby leaving a smaller memory footprint. The 2D/3D images can also be utilized in this approach as an input to solve the micro-mechanics of the material [3].

The structure of the paper is as follows: Sect. 2 presents a detailed explanation of the methodology employed for the FFT iterative scheme used in elastic and thermo-elastic homogenization. In Sect. 3, the validity of the proposed methodology is assessed by comparing it with the commercial DIGIMAT software, with the inclusion limit set at 30%. Additionally, concrete mesostructures are generated using the semi-random sequential adsorption (SRSA) algorithm, and the homogenized properties of the concrete mix are calculated. Finally, Sect. 4 provides a comprehensive conclusion summarizing the findings.

2 Methodology

2.1 Problem Statement

A concrete mix can be idealized through a parallelopiped with dimensions L_1, L_2, and L_3 (Fig. 1). This 3-dimensional material domain (Ω) can further be discretized into several subdomains as $\Omega_1, \Omega_2, \ldots, \Omega_n$, each defining an independent elastic phase. The problem is to find the constitutive relationship (stress-strain) in Ω under mechanical or thermal loading conditions considering that the compatibility conditions and the linear momentum balance are satisfied under the periodic boundary conditions. Moreover, it is necessary for the volume-averaged strain field to match the given macroscopic strain in order to determine the homogenized mechanical properties of the mixture volume. This connection can be leveraged to establish the definition of homogenized mechanical properties.

2.2 Elastic Homogenization

The effective elastic material properties are computed on the basis of the following governing equations:

$$\begin{aligned} \nabla\sigma(x) = 0, \quad \sigma(x) = \mathbb{C}(x) : \varepsilon(x) \\ \varepsilon(x) = \tilde{\varepsilon}(x) + \bar{\varepsilon}, \quad \langle\varepsilon(x)\rangle = \bar{\varepsilon} \end{aligned} \tag{1}$$

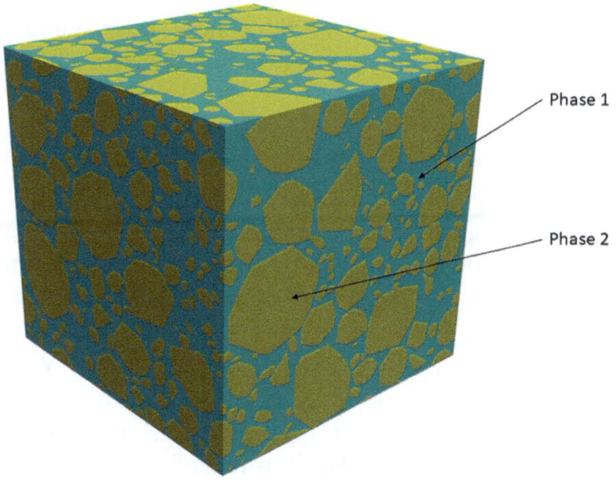

Fig. 1 3D representation of a two-phase material

$\nabla\varepsilon(x)$ and $\sigma(x)\cdot\vec{n}(x)$ are periodic and anti-periodic respectively; with, $\sigma(x)$ denotes the local Cauchy stress tensor, $\mathbb{C}$ being the fourth-order stiffness tensor, $n(\vec{x})$ being the local normal unit vector on the boundary of the periodic Ω, and $\varepsilon(x)$ being the local strain field which can be decomposed into two parts: (a) average macroscopic strain $\bar{\varepsilon}$, (b) periodic fluctuating strain field $\tilde{\varepsilon}(x)$ such that $\langle\tilde{\varepsilon}(x)\rangle$ is zero. The symbol $\langle\bullet\rangle$ represents a spatial average of the quantity $\bullet$, and x represents the location index of the variable in the 3-dimensional space. An average macroscopic strain $\bar{\varepsilon}$ is employed as the external load on the subdomain Ω.

The subsequent step involves establishing the auxiliary problem by employing a homogeneous reference material characterized by the stiffness tensor $\mathbb{C}_0$. By incorporating and deducting the influence of the reference medium from the heterogeneous problem's local stress tensor as defined in Eq. (1), the following results are obtained:

$$\begin{aligned}\sigma(x) &= [\mathbb{C}_0 + \mathbb{C}(x) - \mathbb{C}_0] : (\tilde{\varepsilon}(x) + \bar{\varepsilon}) \\ \sigma(x) &= \mathbb{C}_0 : (\tilde{\varepsilon}(x) + \bar{\varepsilon}) + [\mathbb{C}(x) - \mathbb{C}_0] : (\tilde{\varepsilon}(x) + \bar{\varepsilon})\end{aligned} \tag{2}$$

One can further establish the definition of the stress polarization tensor, represented as $\tau(x)$, as follows:

$$\tau(x) = [\mathbb{C}(x) - \mathbb{C}_0] : (\tilde{\varepsilon}(x) + \bar{\varepsilon}) \tag{3}$$

From the linear momentum balance equation in (1), the following equation can then be arrived at:

$$\begin{aligned}\nabla(\mathbb{C}_0 : (\tilde{\varepsilon}(x) + \bar{\varepsilon})) &= -\nabla(\tau(x)) \\ \text{Since,}\quad \nabla(\bar{\varepsilon}) &= 0 \\ \text{therefore,}\quad \nabla(\mathbb{C}_0 : (\tilde{\varepsilon}(x))) &= -\nabla(\tau(x))\end{aligned} \tag{4}$$

Moreover, the compatibility condition between microscopic strain and displacement can be formally expressed under the assumption of small strains as follows:

$$\begin{aligned}\varepsilon(x) &= \nabla^s u(x) \\ \tilde{\varepsilon}(x) &= \nabla^s \tilde{u}(x)\end{aligned} \tag{5}$$

In this equation, ∇^s represents the symmetric gradient operator, while $u(x)$ and $\tilde{u}(x)$ correspond to the local and fluctuating displacement field vectors, respectively. By incorporating the term $\tilde{u}(x)$ into Eq. (4), we obtain the following expression:

$$\nabla(\mathbb{C}_0 : \nabla^s \tilde{u}(x)) = -\nabla(\tau(x)) \tag{6}$$

The quantity $-\nabla(\tau(x))$ can be understood as the source term in the elastic partial differential equation (PDE) formulated for a homogeneous material. Solving Eq. (6) involves the utilization of Green's function. Green's function is an integral kernel employed to solve non-homogeneous PDEs [3].

The solution of (1) is expressed in real space by a periodic Green operator Γ_0 associated with $\mathbb{C}_0$ as,

$$\varepsilon(\tilde{u}(x)) = -\Gamma_0 * \tau(x) + \bar{\varepsilon} \quad \forall\, x \in V \tag{7}$$

The same can be expressed in Fourier's space as,

$$\hat{\varepsilon}(\xi) = -\hat{\Gamma}_0 * \hat{\tau}(\xi) \quad \forall\, \xi \neq 0, \quad \hat{\varepsilon}(0) = \bar{\varepsilon} \tag{8}$$

wherein $(\hat{\bullet})$ is representation of $(\bullet)$ in Fourier space, ξ is frequency vector associated with the x in Fourier space, and (*) is the convolution operator.

The operator Γ_0 for the isotropic reference material [1] with Lame's constants λ_0 and μ_0 where each symbol has their usual meanings can be expressed as follows:

$$\hat{\Gamma}_0^{ijkh} = \frac{1}{4\mu_0\xi^2}\left(\delta_{ki}\xi_h\xi_j + \delta_{hi}\xi_k\xi_j + \delta_{kj}\xi_h\xi_i + \delta_{hj}\xi_k\xi_i\right) - \frac{\lambda_0 + \mu_0}{(\lambda_0 + 2\mu_0)\,\mu_0}\,\frac{\xi_i\xi_j\xi_k\xi_h}{\xi^4} \tag{9}$$

Here, the function δ represents Kronecker's delta, which is defined as follows:

$$\begin{aligned} \delta_{mn} &= 1 \quad \text{for} \quad m = n \\ \delta_{mn} &= 0 \quad \text{for} \quad m \neq n \end{aligned} \tag{10}$$

2.3 The Algorithm

The algorithm operates based on the principle of employing Eqs. (3), (4), (7), and (8) alternately in real and Fourier space. This iterative scheme is utilized to solve (1). For a detailed demonstration of the iterative scheme, please refer to Algorithm 1.

Algorithm 1 Proposed iteration scheme

Initialization: $\varepsilon_0(x) = \bar{\varepsilon}, \quad \forall x \in V$
$\sigma_0(x) = \mathbb{C}(x) : \varepsilon_0(x), \quad \forall x \in V$ ▷ cf. Equation (1)
Iterate $i + 1$: ε_i and σ_i being known
(a): $\tau_i(x) = \sigma_i(x) - \mathbb{C}_0 : \varepsilon_i(x)$,
(b): $\hat{\tau}_i = \mathcal{FFT}(\tau_i)$ ▷ $\mathcal{FFT}$ denotes the fast Fourier transform
(c): Convergence test
(d): $\hat{\varepsilon}_{i+1}(\xi) = -\hat{\Gamma}_0 : \hat{\tau}(\xi) \quad \forall \xi \neq 0$ and $\hat{\varepsilon}_{i+1}(0) = \bar{\varepsilon}$ (8)
(e): $\varepsilon_{i+1} = \mathcal{FFT}^{-1}(\hat{\varepsilon}_{i+1})$ ▷ $\mathcal{FFT}^{-1}$ denotes the inverse fast Fourier transform
(f): $\sigma_{i+1}(x) = \mathbb{C}(x) : \varepsilon_{i+1}(x), \quad \forall x \in V$ (1)

The convergence is reached when σ_{i+1} is in equilibrium. The error between two successive iterations should be smaller than a prescribed tolerance value (10^{-8} in our case).

Check for error: $\frac{\langle(\mathrm{div}(\sigma_i)^2)\rangle^{1/2}}{\langle\sigma_i\rangle} = \frac{\langle(\xi\hat{\sigma}_i(\xi)^2)\rangle^{1/2}}{\hat{\sigma}_i(0)} <= \text{tolerance}$

2.4 Thermo-elastic Homogenization

Similar to the homogenization procedure demonstrated for mechanical properties of the adopted domain Ω, the thermo-elastic homogenization requires the stress-strain relationships to be estimated for temperature loading. For this case, the total strain will however be the sum of the mechanical and the thermal strain. The pertinent governing equation is modified for mechanical stress and strain as below

$$\nabla \sigma_m = 0, \quad \sigma_m = \mathbb{C} : (\varepsilon - \varepsilon_T), \quad \varepsilon_T = \alpha \Delta T \tag{11}$$

In this context, σ_m represents the local mechanical stress tensor, ε_T denotes the thermal strain, α is the coefficient of thermal expansion, and ΔT signifies the change in temperature. The external loads are applied to the domain Ω through thermal loading, specified in terms of temperature changes.

The coefficient of thermal expansion is homogenized by considering a fully constrained specimen such that no deformations are allowed. Under such constraints, the specimen is subjected only to thermal loading and the induced stresses are measured. The hydro-static condition is achieved leading to only principal stresses to exist. For this state, the volumetric coefficient of thermal expansion (α_{vol}) may be defined as

$$\alpha_{\text{vol}} = \frac{\sigma_h}{K \cdot \Delta T} \tag{12}$$

where σ_h is the hydro-static stress state and K is the bulk modulus of elasticity of the concrete.

Subsequently, for small-strain, the volumetric strain can be considered as the sum of the linear strains in each direction. Similarly, the volumetric coefficient of thermal expansion can be considered as the sum of linear thermal coefficients in each direction. With an assumption of isotropic and homogeneous material imposed on the concrete volume, the associated coefficient of thermal expansion can be obtained as

$$\alpha_{\text{hom}} = \frac{\alpha_{\text{vol}}}{3} \tag{13}$$

3 Results and Discussion

In order to investigate the level of accuracy that can be achieved with the proposed homogenization approach, its performance is compared against the output of DIGIMAT software. Two identical mixes with spherical inclusion are treated in a similar manner with the DIGIMAT software and the adopted approach. Nevertheless, the volume fraction of the inclusion had to be restricted within 30% since DIGIMAT can not handle beyond that level. The homogenized properties obtained through both the approaches are compared in Tables 1 (elastic properties) and 2 (thermo-elastic prop-

Table 1 Comparison of elastic homogenization results with DIGIMAT

E_m (GPa)	E_a (GPa)	ν_m	ν_a	V_a	E^D_{hom} (GPa)	ν^D_{hom} (GPa)	$E^{\text{FFT}}_{\text{hom}}$ (GPa)	$\nu^{\text{FFT}}_{\text{hom}}$	ϵ_E (%)
19	62	0.2	0.2	0.2	23.8453	0.2007	23.875	0.2006	0.12
19	62	0.2	0.2	0.25	25.2048	0.2008	25.3089	0.1995	0.41
19	62	0.2	0.2	0.3	26.6347	0.2002	26.6367	0.2002	0.01
25	72	0.2	0.2	0.2	30.6673	0.2005	31.2528	0.1962	1.91
19	57	0.2	0.2	0.25	24.7329	0.2006	24.8267	0.1995	0.38
16	67	0.2	0.2	0.3	23.94	0.2016	24.5436	0.1984	2.52
22	62	0.2	0.2	0.25	28.235	0.2005	28.2211	0.1985	0.05

Subscript m and a denote matrix and aggregate, E and ν correspond to elasticity and poisson's ratio, superscript FFT and D signify the association of the corresponding entity to FFT-based/DIGIMAT-based analyser, subscript hom stands for homogenized/effective property, and ϵ is the relative error in %

erties) wherein it can be verified that the adopted approach is capable of achieving accuracy levels similar to DIGIMAT. In both observations, RVEs with a 75^3 number of voxels were considered for comparison.

3.1 Concrete Mesostructure

After validation of the adopted algorithm's performance in homogenizing elastic and thermo-elastic properties, the same is employed for preparing the much desired RVE for the mass concrete under consideration. For the computation of elastic and thermo-elastic properties of the mass concrete, a mesostructure is developed based on the mix proportions of concrete shown in Table 3. A virtual mesostructure is generated using the computational tool called as Concrete Mesostructure Generator (CMG) based on Python, available on https://pycmg.readthedocs.io/en/latest [4]. Nevertheless, the available CMG was limited to a two-phase system only which includes matrix (cement mortar) and aggregates as inclusions. Since air-entraining agents are usually used in order to protect the concrete from freezing and thawing action under extreme weathering conditions, the CMG is modified to deal with a three-phase system by including voids into the system. About 4% air is entrained during the concrete casting, which is thereby included in the microstructure through the modified CMG (Fig. 2).

The same level of detail and information is maintained to generate a realistic mesostructure of the concrete mix under consideration. Two virtual mesostructures, with dimensions 251 mm × 251 mm × 251 mm each, are generated, one with two phases, viz., matrix (cement mortar includes cement, sand, and water) and aggregates of size ranging from 5mm to 80 mm as primary inclusion as shown in Fig. 3. The other mix is also modeled with three phases with voids as the secondary inclusion as shown in Fig. 4. The resolution for each mesostructure was kept as 1 mm^3 per voxel.

Table 2 Comparison of thermo-elastic homogenization results with DIGIMAT

E_m (GPa)	E_a (GPa)	ν_m	ν_a	α_m	α_a	V_a	E_{eq}^{D} (GPa)	ν_{eq}^{D}	α_{eq}^{D}	E_{eq}^{FFT} (GPa)	ν_{eq}^{FFT}	α_{eq}^{FFT}	ϵ_E	$\epsilon_\alpha(\%)$
19	62	0.2	0.2	12.8	7.5	0.2	23.845	0.2008	1.1233	23.7552	0.2006	1.12346	0.38	0.014
19	57	0.2	0.2	13.3	7.5	0.25	23.733	0.1925	1.127	24.8095	0.1991	1.1282	4.54	0.106
16	67	0.2	0.2	12.8	8	0.3	23.94	0.2016	1.0687	24.256	0.1988	1.0671	1.32	0.150
22	62	0.2	0.2	12.8	8.5	0.25	28.235	0.2005	1.1319	28.5677	0.2	1.12538	1.18	0.576
16	62	0.2	0.2	13.8	8.5	0.3	23.464	0.2013	1.1507	23.4516	0.1987	1.15286	0.05	0.188
25	72	0.2	0.2	13.3	7.5	0.2	30.67	0.2005	1.1646	31.31	0.1971	1.1568	2.09	0.670

α denotes coefficient ($\times 10^{-}6/°C$) of thermal expansion of the entities

Table 3 Mix design (20% fly ash replacement)

Material	Quantity (kg/m^3)
Cement	205.92
Fly ash	51.48
Water	123
Coarse aggregate	1432
Fine aggregate	508.95
Water reducing agent (WRA)	2.574
Air-entraining agent (AEA)	0.36

To generate the mesostructure, a semi-random sequential algorithm (SRSA) was used which works well for volume fraction more than 30% and is almost 4 times faster compared to the traditional random sequential algorithm (RSA) which implements the periodic boundary conditions. The concrete mesostructure is generated in two steps as follows:

1. Inclusions: Generating the realistic aggregates using irregular polyhedrons with concave depressions as primary inclusion and voids as other inclusion.
2. Packing: The generated aggregates and voids are packed into the host material (cementitious mortar).

3.1.1 Data for Mesostructure Generation

Two different mesostructures are generated based on the concrete mixes shown in Table 3. In the first mesostructure (two-phase), void spaces are completely ignored which is an approximate estimate of the selected concrete mix, while the second one (three-phase) is the more closer replica of the actual concrete mix. The volume fraction of the individual constituent is listed in Table 4.

The grading of the aggregates conforms to IS 383:2016. The distribution of the aggregates sizes can be seen in Fig. 2.

3.2 Elastic Homogenization

The material properties considered for the elastic homogenization are taken from the work of [4] as shown in Table 5. A two-step homogenization procedure is employed to determine the homogenized elastic properties. First elastic properties of cementitious mortar are determined using the FFT-homogenization scheme described in Algorithm 1 and then the overall properties of the concrete using the same algorithm. The elastic properties of the cementitious mortar are determined by considering fine aggregates as spherical hard inclusions in the host cement paste matrix.

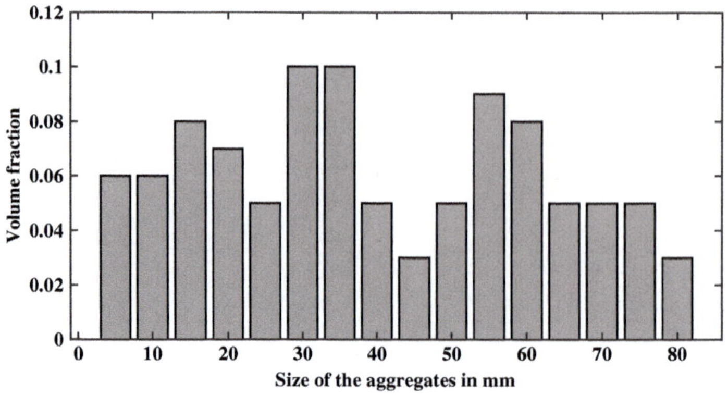

Fig. 2 Distribution of aggregates in concrete mix

Fig. 3 Two-phase concrete mesostructure

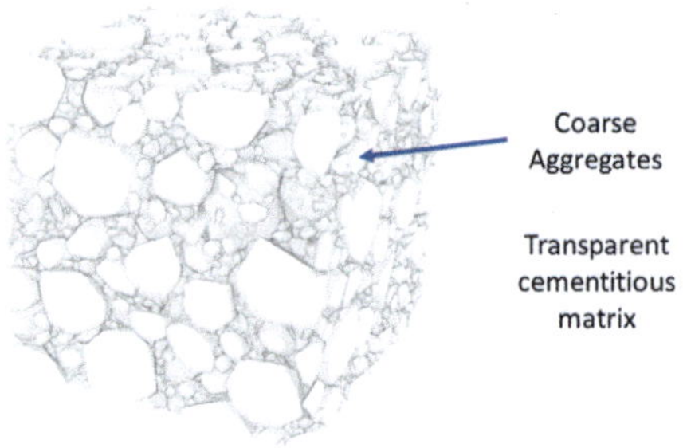

Fig. 4 Three-phase concrete mesostructure

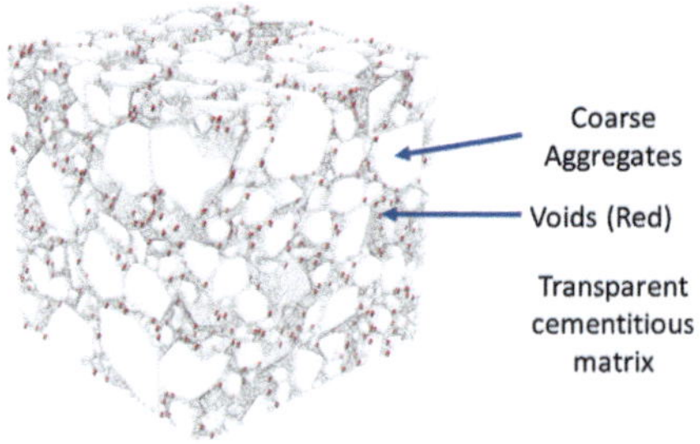

Table 4 Volume fraction of the constituents

Material	Cement paste (%)	Coarse aggregates (%)	Fine aggregates (%)	Voids (%)
Two phase	22.6	56.66	20.5	Nil
Three phase	22.6	53.9	19.5	4

Table 5 Elastic properties of the constituents

Material	Density (kg/m^3)	Young's modulus (GPa)	Poisson's ratio
Cement paste	1898	18.7	0.24
Aggregate (quartzite)	2560	84.6	0.12

Table 6 Homogenized elastic properties of concrete

Material	Young's modulus (GPa)	Poisson's ratio
Concrete (two phase)	60.10	0.15
Concrete (three phase)	59.70	0.15

Table 7 Thermo-elastic properties of the constituents

Material	Coeff. of thermal expansion ($10^{-6}/°C$)	Young 's modulus (GPa)	Poisson's ratio
Cement paste	10.50	18.7	0.24
Aggregate (quartzite)	11.75	84.6	0.12

The modulus of elasticity and Poisson's ratio of the homogeneous cement mortar from the FFT-homogenization scheme were determined as 38.927 GPa and 0.188 respectively. RVE size of 75^3 is considered for homogenization of cementitious mortar. The corresponding values of homogenized elastic properties of the concrete are given in Table 6.

3.3 *Thermo-elastic Homogenization*

The coefficient of thermal expansion for cement paste and the aggregates (quartzite) reported by [5, 6] is listed in Table 7, while the same elastic properties listed in Table 5 are considered for thermo-elastic homogenization.

The thermal and elastic characteristics of cementitious mortar and concrete obtained from the FFT-homogenization scheme were determined and were presented in Table 8.

Table 8 Thermo-elastic properties of the constituents

Material	Coeff. of thermal expansion ($10^{-6}/°C$)	Young's modulus (GPa)	Bulk modulus (GPa)	Poisson's ratio
Cementitious mortar	11.285	38.899	20.81	0.188
Concrete (two phase)	11.581	60.084	28.82	0.15
Concrete (three phase)	11.615	59.70	28.54	0.15

4 Conclusion

FFT-homogenization scheme is utilized for elastic and thermo-elastic homogenization of concrete. The results of the adopted algorithm were compared with the DIGIMAT software with a constraint on the inclusion levels, and a good agreement between the results was obtained. The proposed FFT-based homogenization scheme is perceived to be an efficient computational solver because of its advantages over conventional FEM-based solvers. The benefits originate from the aspects of faster convergence, lower memory footprints, a mesh-independent approach, and the practicality of employing 2D/3D images of the microstructure for homogenization. However, FFT-based homogenization is not reliable if a locally fine discretization is desirable and also not suitable for domains with non-periodic boundary conditions. For better convergence, it is also desirable that there should not exist a large contrast in the properties between the phases which however was not the case with our problem. Still, various efforts have been made in recent times to improve the convergence and to incorporate non-periodic boundary conditions which can be investigated in our future research, and the relative cost–benefit analysis can be drawn taking basis on real experimental data.

Acknowledgements This study is funded by Satluj Jal Vidyut Nigam (SJVN) Limited, Shimla, India, under project no. IITM/SJVNL/SUS/338.

References

1. Moulinec H, Suquet P (1998) A numerical method for computing the overall response of nonlinear composites with complex microstructure. Comput Methods Appl Mech Eng 157(1–2):69–94
2. Kabel M, Merkert D, Schneider M (2015) Use of composite voxels in FFT-based homogenization. Comput Methods Appl Mech Eng 294:168–188
3. Lucarini S, Upadhyay MV, Segurado J (2021) FFT based approaches in micromechanics: fundamentals, methods and applications. Model Simul Mater Sci Eng

4. Holla V, Vu G, Timothy JJ, Diewald F, Gehlen C, Meschke G (2021) Computational generation of virtual concrete mesostructures. Materials 14(14):3782
5. Sakyi-Bekoe K (2008) Assessment of the coefficient of thermal expansion of Alabama concrete. PhD thesis
6. Wang H, Mang H, Yuan Y, Pichler BL (2019) Multiscale thermoelastic analysis of the thermal expansion coefficient and of microscopic thermal stresses of mature concrete. Materials 12(17):2689

Assessment of Intra-Laminar Macro-Damage in Textile Composite Using Gradient-Enhanced Continuum Damage Mechanics Method

Aditya S Deshpande and Bhrigu Nath Singh

Abstract Although textile composites have been popular over laminated composites, their complex fiber architecture poses hurdles in modeling and predicting material behavior under various static and dynamic conditions. The development of an effective methodology for capturing the material response to macro-damage onset and evolution is even more challenging. The present work discusses the intra-laminar macro-damage initiation and evolution in the plain woven textile composite. A two-step homogenization approach is employed to deduce the effective properties of the composite material which are then used in developing the strain-damage coupled constitutive equations. Gradient-enhanced continuum mechanics technique is used to tackle the localization issues due to mesh sensitivity in otherwise vastly efficient and popular continuum damage mechanics method. The numerical solution is evaluated with the help of ABAQUS user element subroutine (UEL) via nonlinear finite element solution techniques. The material response to macro-damage initiation and evolution is observed to be consistent with the experimental data thus validating the efficiency and accuracy of the present work.

Keywords Macro-damage · Gradient-enhanced continuum damage mechanics · Regularization · Textile composite · Two-step homogenization · Nonlinear finite element method

1 Introduction

Over the last few decades, textile composites have become a popular choice of composite materials as they are more economical, offer better mechanical properties, and are convenient for manufacturing complex structural components. Unlike unidirectional composites, which have very good properties in in-plane directions

A. S Deshpande (✉) · B. N. Singh
Department of Aerospace Engineering, Indian Institute of Technology Kharagpur, Kharagpur, West Bengal, India
e-mail: adityadeshpande@iitkgp.ac.in

R. K. Ray et al. (eds.), *Advances in Theoretical and Applied Mechanics*, Lecture Notes in Mechanical Engineering, https://doi.org/10.1007/978-981-97-0418-7_24

only, textile composites made up of interwoven fabric reinforcements have demonstrated excellent out-of-plane and transverse properties, better impact resistance, etc. [2, 8]. They also exhibit higher fracture toughness and damage tolerance as the macroscopic cracks experience greater resistance at yarn intersections than unidirectional composites [9]. Due to these advantages, textile composites have found their applications in aerospace, automobile, naval and structural engineering, etc. [10]. On the other hand, due to the complex architecture of textile composites, their properties are sensitive to fiber orientation and weave pattern, etc.; and, hence, it is tedious to develop constitutive material models capable of predicting effective properties with better accuracy. Due to these aforementioned difficulties, studying the material response for textile composite has been predominantly restricted to experimental methods.

To understand the behavior of textile composites, an accurate estimation of their properties is critical. Many analytical, numerical, and homogenization-based models have been proposed to evaluate material properties for complex textile architectures. However, it was reported that analytical models are not efficient in predicting out-of-plane properties for textile composites. Numerical models based on finite element methods have been widely used to study textile composite and provide better estimation for material properties compared to analytical methods. But complex fiber arrangements and large textile models are difficult to model with numerical methods [2]. TexGen, a multi-functional pre-processor open source code, provides a comprehensive tool to develop a numerical model for complex textile architecture for its further modeling and numerical analysis [1]. An add-on feature to TexGen called TexGen4SC is based on the mechanics of structure genome (MSG). It uses a semi-analytical approach to model and analyzes heterogeneous composite structures by homogenizing them and estimating the equivalent material properties with great computational accuracy and minimum loss of geometry and information [2].

In structures, micro-defects in the form of micro-cracks, voids, etc. are spread over the entire region. These micro-defects experience local stress concentration regions and as a result, these defects coalesce and grow into macro-defects. Computationally, the continuum damage mechanics (CDM) technique has been proven to be versatile in modeling and analyzing damage initiation and evolution problems from micro-to-macro scale [3]. But, due to its high mesh sensitivity and localization issues, its use has been limited. To tackle these difficulties, the regularization method in the form of gradient enhancement is applied to obtain the nonlocal strain fields, which helps in reducing the local sense of damage evolution equations [4].

In the present work the two-step homogenization technique [2] for textile composites in combination with gradient-enhanced CDM approach is used to study the macro-damage response of the composite structure. This approach enables in obtaining the strain-damage coupled constitutive equations for the textile. The effective properties of plain woven carbon fiber epoxy composite material are obtained by a two-step homogenization method and the homogenized material is numerically analyzed for the intra-laminar macro damage initiation and evolution. The results obtained are in agreement with the experimental values thus, validating the effectiveness of the approach and the accuracy of the numerical solution.

2 Mathematical Formulation and Composite Homogenization

2.1 Gradient-Enhanced Continuum Damage Mechanics Method

CDM formulation incorporates the state of material damage in the constitutive equation and the global structural damage response is obtained by solving the equilibrium equations. The constitutive stress-stress relation incorporating the damage variable is given by Eq. (1)

$$\sigma = (\mathbf{1}-D)\, C : \boldsymbol{\varepsilon} \tag{1}$$

where $\boldsymbol{\sigma}$: Cauchy stress tensor; C: stiffness tensor; ε: linear strain tensor, and D: scalar damage variable. The evolution of the scalar damage variable, D, is defined by the scalar history parameter, κ, which is a function of equivalent strain ε_{eq}. The history parameter represents the most severe deformation that which material has experienced and facilitates growth of the damage variable, $D(\kappa)$. The evolution of the damage variable is analyzed by employing the damage evolution law as given in Eq. (2); where, κ_i and κ_c are initial and critical history parameter values respectively.

$$D(\kappa) = \begin{cases} \frac{\kappa_c}{\kappa} \cdot \frac{\kappa - \kappa_i}{\kappa_c - \kappa_i} & \text{if} \quad \kappa_i \le \kappa \le \kappa_c \\ 1 & \text{if} \quad \kappa \ge \kappa_c \end{cases} \tag{2}$$

The state of damage is dependent on the equivalent local strain; thus, making CDM a local method. This local sense of scalar damage variable can be corrected by redefining it in terms of nonlocal equivalent strain. Using the principle of nonlocality with the Eq. (3), the corresponding equivalent gradient form of the nonlocal equivalent strain is obtained as Eq. (4), where, c is the gradient parameter.

$$\overline{\varepsilon_{\text{eq}}}(x) = \frac{1}{V} \int_V W(\xi)\, \varepsilon_{\text{eq}}(x+\xi) \mathrm{d}V \tag{3}$$

$$\overline{\varepsilon_{\text{eq}}} - c\nabla^2 \overline{\varepsilon_{\text{eq}}} = \varepsilon_{\text{eq}}; \ \text{subjected to BC: } \nabla \overline{\varepsilon_{\text{eq}}} . \vec{n} = 0 \tag{4}$$

The gradient equivalent strain equation, Eq. (4), is solved along with the equations of equilibrium to evaluate the macro-damage response of the structure.

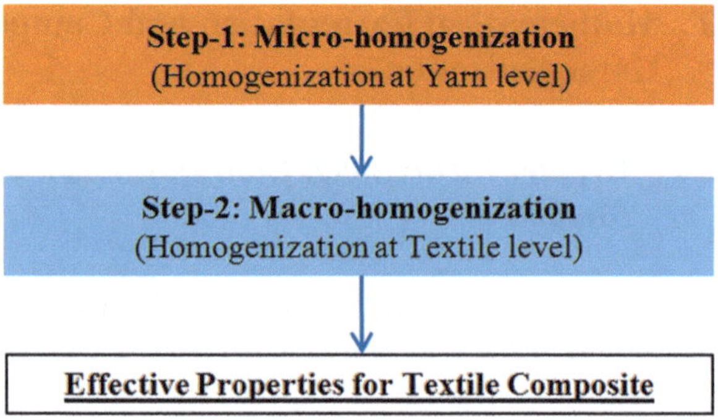

Fig. 1 Two-step homogenization approach using MSG

2.2 *Two-Step Homogenization of Plain Woven Composite*

The present work employs the open-source code TexGen4SC to estimate the homogenized effective material properties for plain woven textile composite. It uses MSG based two-step homogenization procedure as shown in Fig. 1. Micro homogenization at the yarn level gives effective properties from fiber and matrix; and by using these effective yarn properties, the macro homogenization gives overall homogenized material properties of the composite [2].

In this work, a plain woven textile composite with carbon fiber and epoxy matrix is homogenized by following the two-step homogenization approach to deduce the equivalent material properties. For the given plain-woven textile composite as shown in Fig. 2, a 2D structural genome, which is the smallest geometric domain of the structure to be analyzed, is considered. Fibers and matrix arrangement in the yarn at the micro-level are modeled as a hexagonal pack for the representative volume element (RVE) and individual constituent properties are used to evaluate the effective yarn properties by micro-homogenization. These yarn properties are then used for the genome arrangement at the macro-level to obtain the homogenized textile properties. TexGen4SC efficiently resolves computational difficulties in evaluating the effective homogenized properties and accurately estimates the material properties for complex textile composites.

3 Numerical Implementation

The intra-laminar macro-damage response of plain textile woven composite is determined by implementing gradient-enhanced damage formulation. Here, the gradient form of nonlocal equivalent strains as given in Eq. (4), and structural equilibrium equations are solved simultaneously. Along with the global structural displacements, the nonlocal equivalent strains are also considered as independent variables. The numerical solution is evaluated by a mixed Finite Element (FE) nonlinear solver using Netwon-Raphson nonlinear solution method in ABAQUS FE solver with the

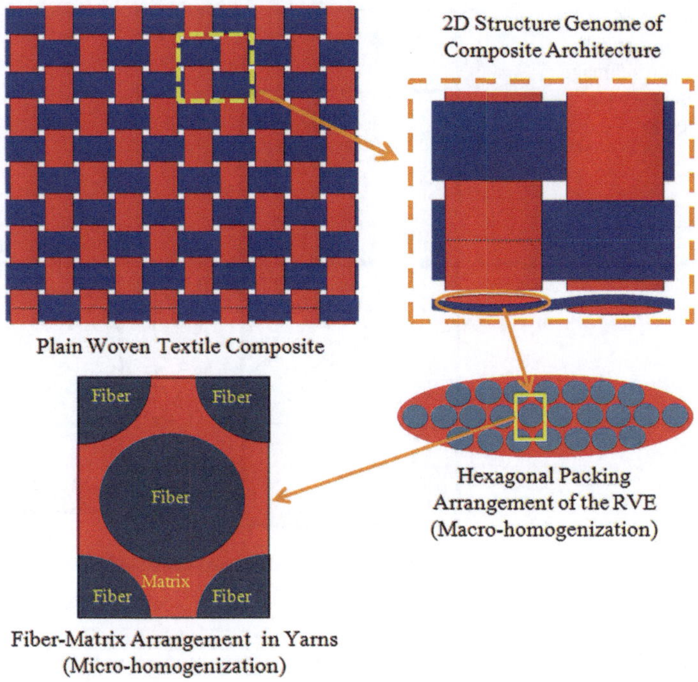

Fig. 2 Implementation of two-step homogenization for the plain woven textile composite

help of a User Element subroutine (UEL) [6, 7]. A schematic flowchart for the numerical implementation of gradient-enhanced CDM via ABAQUS UEL is given in Appendix 1.

4 Results and Discussion

To study the macro-damage response of plain woven textile composite, a single edge notch specimen with macro-damage in the form of edge crack is modeled as shown in Fig. 3. In the present work, the textile composite is modeled with carbon fiber T300 and epoxy matrix 3601. The constituents' material properties are given in Table 1. The volume fraction of fibers in yarns is 80%; while the fiber volume fraction in the weave is 45%. Effective homogeneous properties for the composite are evaluated following the two-step homogenization method. The homogenized material properties are compared with the experimental data [11] and the results of the analysis are furnished in Table 2.

From the obtained results, it can be inferred that the present approach of homogenizing the composite properties via MSG-based two-step homogenization evaluates the effective material properties accurately and with relative ease. The variation in

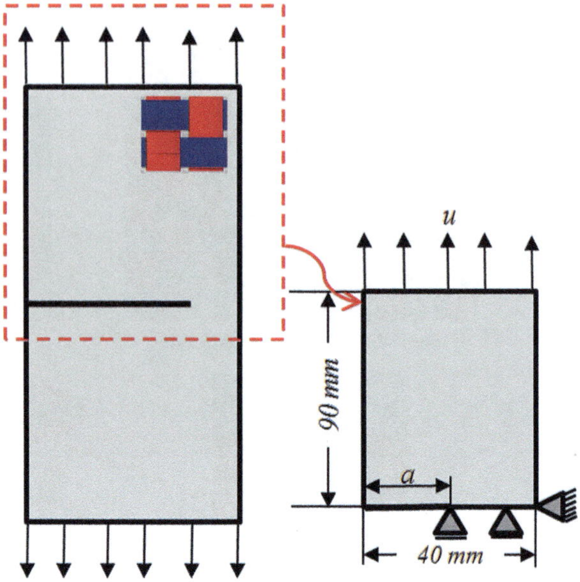

Fig. 3 Modeling of geometry for numerical analysis of macro-damage analysis in plain woven textile composite

the evaluated homogenized properties and the experimental values is due to the geometric and numerical approximations considered in the genome modeling and its two-step homogenization process. The open-source code, TexGen4SC proves to be an efficient tool for evaluating the material properties of textile composites which can be further used to study the behavior and response of composite structures in various applications.

The macro-damage response of the plain woven textile composite is studied by considering the material properties obtained from the MSG-based two-step homogenization approach. The effective homogeneous composite properties are assumed to be transversely isotropic and used to model the equivalent material. A single edge notch specimen with macro-damage in the form of edge crack is modeled as half-model and is analyzed under uniaxial tension (Mode-I) to study the macro-damage response of plain woven textile composites as shown in Fig. 3. The gradient-enhanced CDM analysis is carried out considering the scalar damage variable to predict the macro-damage response of the structure. Parameters considered for this analysis are: $c = 100\ \text{mm}^2$; $\kappa_i = 1 \times 10^{-4}$ and $\kappa_c = 0.125$ [5]. The analysis is performed for two different edge crack lengths $a = 4$ mm and $a = 8$ mm.

The macro-damage response of the textile composite is captured as shown in Fig. 4. It is observed from the response curve that the material responds linearly to the applied displacements up to the apex points. This gradual response is indicative of the steady increase in the material's damage state under the action of external loading. The maxima of the curves signify the attainment of a critical damage state in the material. After reaching the maxima on the Force–Displacement curves, the curves suddenly drop indicating the level of critical damage growth leading to final failure. This sudden drop in the curve is also known as the softening region in the damage response curve. The nature and response of the material damage curves, as indicated in Fig. 4, agree well with the experimental results [11] as shown in Fig. 5. This establishes the validity of the present numerical solution determined using the combined approach with two-step homogenization along with the gradient-enhanced CDM formulation in studying the macro-damage response of the plain textile composites.

In numerical simulations, the softening response with the sudden drop in the damage response curve cannot be captured as effectively as the experimental damage response curves [12]. The results presented in this work match well with the experimental results. But, more deliberation and refinement in softening laws to exactly capture the experimental response in the numerical solutions is a topic of ongoing research.

Contour plots as shown in Fig. 6 indicate the distribution of the nonlocal equivalent strain which governs the growth of damage in the material and in turn the macro-damage response. Figure 7 shows the contour plots for the growth of the scalar damage variable in the material for the numerical solution. From Fig. 7, it can be

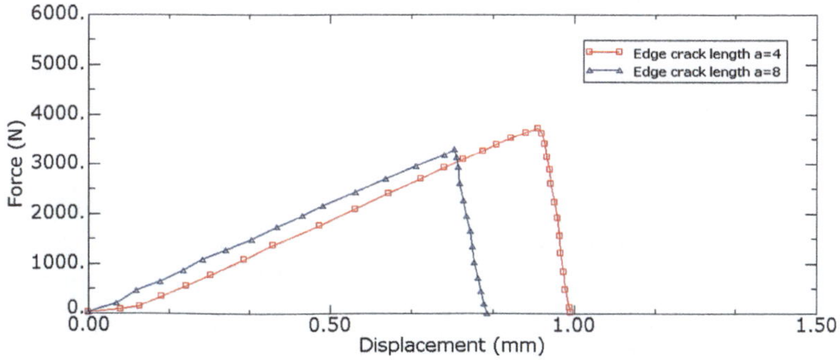

Fig. 4 Force–displacement response curve for homogenized plain-woven textile composite

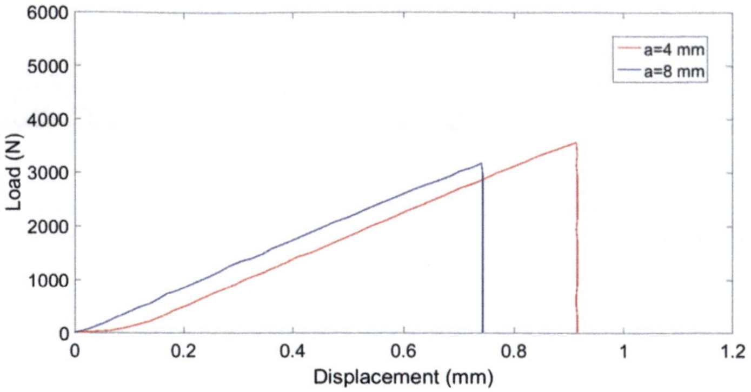

Fig. 5 Load–displacement curve for specimen with edge cracks: $a = 4$ mm and $a = 8$ mm from experimental results [11]

seen that the critical region of the damage variable and its direction of evolution is around the edge crack's tip. This captures the phenomenon of the fracture process zone in the vicinity of the crack tip for Mode-I fracture crack and its propagation.

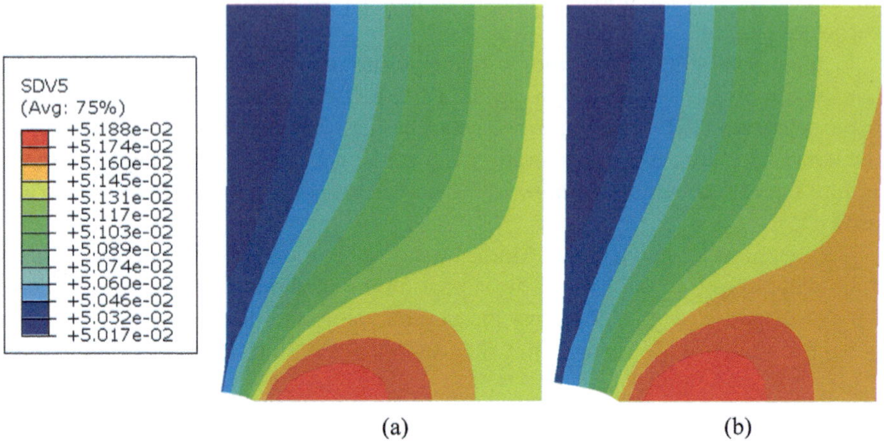

Fig. 6 Contour plots showing nonlocal equivalent strains in the plain-woven textile composite for edge crack lengths **a** $a = 4$ mm and **b** $a = 8$ mm

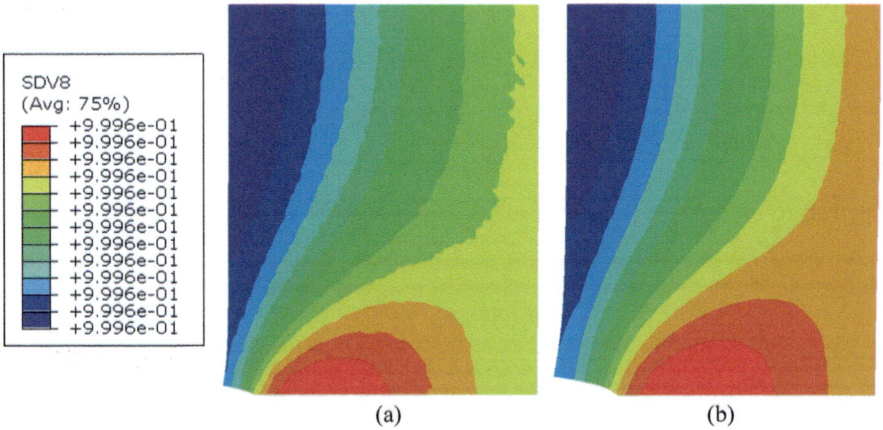

Fig. 7 Contour plots showing macro-damage response in the plain-woven textile composite for edge crack lengths **a** $a = 4$ mm and **b** $a = 8$ mm

Table 1 Material properties of carbon fiber (T-300) and epoxy-3601 matrix [10]

T300 carbon fibre				Epoxy 3601	
E_1 (GPa)	230	G_{23} (GPa)	14.3	E_1 (GPa)	3.5
$E_2 = E_3$ (GPa)	40	$\nu_{12} = \nu_{13}$	0.26	$G_{12} = G_{13}$ (GPa)	1.3
$G_{12} = G_{13}$ (GPa)	24	ν_{23}	0.44	$\nu_{12} = \nu_{13}$	0.35

Table 2 Comparison of present results of homogenized composite properties with experimental results [11]

Property	Present	[11]	Diff. %
E_1 (GPa)	28.7	26.7	7.49
E_2 (GPa)	28.7	26.7	7.49
E_3 (GPa)	9.16	9.36	2.13
G_{12} (GPa)	2.2	2.4	8.34
G_{13} (GPa)	2.2	2.4	8.34
G_{23} (GPa)	2.36	2.64	10.6
ν_{12}	0.13	0.13	0.00
ν_{13}	0.35	0.36	2.77
ν_{23}	0.35	0.36	2.77

5 Conclusions

The present work has been successful in studying the macro-damage response of plain woven textile composites with macro-damage in the form of an edge crack under uniaxial tension. Following conclusions can be noted:

- MSG-based two-step homogenization approach accurately estimates the effective properties of woven textile composites.
- Gradient-enhanced damage method in combination with MSG-based two-step homogenization approach is an efficient and computationally economical technique to predict the damage response of textile composite.
- The present study finds its applicability in analyzing more complex textile composite architectures and, more future work is planned on these lines.

Declarations

Conflict of Interest The authors declare that there is no conflict of interest.

Appendix 1: Flow Chart for Numerical Implementation

See Fig. 8.

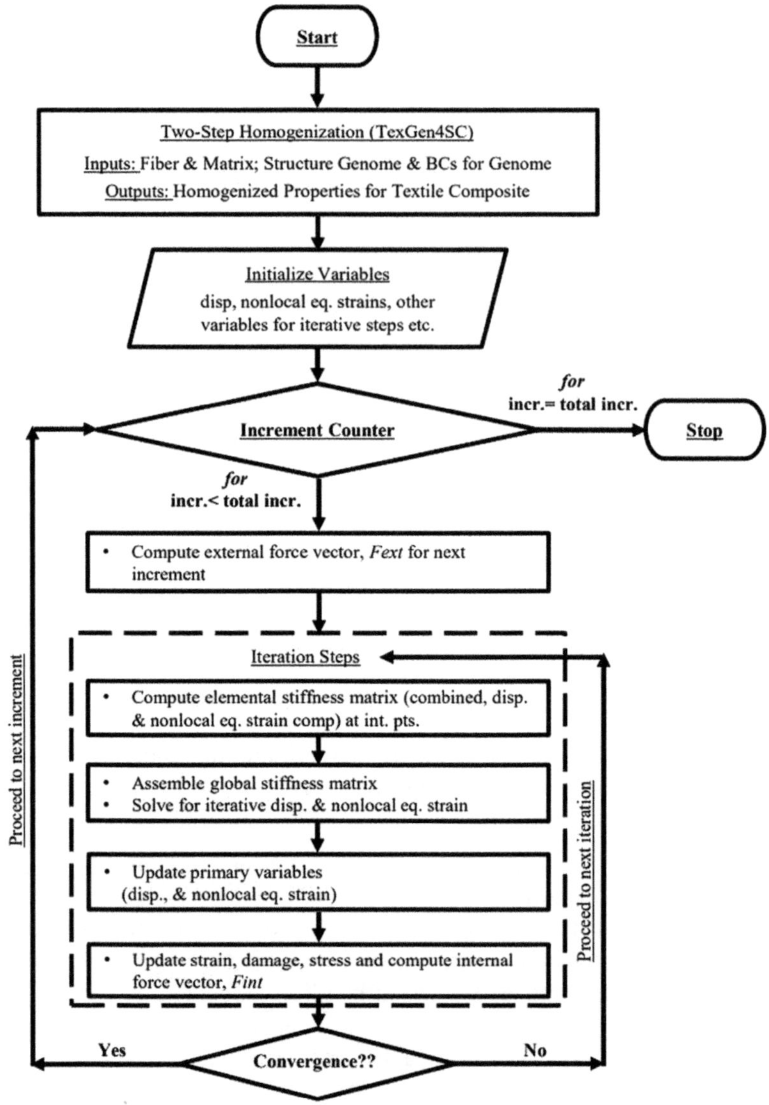

Fig. 8 Flowchart for the numerical implementation of gradient-enhanced CDM via ABAQUS UEL

References

1. Lin H, Brown LP, Long AC (2011) Modelling and simulating textile structures using TexGen. Adv Mater Res 331:44–47. https://doi.org/10.4028/www.scientific.net/AMR.331.44
2. Liu X, Rouf K, Peng B, Yu W (2017) Two-step homogenization of textile composites using mechanics of structure genome. Compos Struct 171:252–262. https://doi.org/10.1016/j.compstruct.2017.03.029

3. Lorentz E, Godard V (2011) Gradient damage models: Toward full-scale computations. Comput Methods Appl Mech Eng 200:1927–1944. https://doi.org/10.1016/j.cma.2010.06.025
4. Peerlings RHJ, De Borst R, Brekelmans WAM, De Vree JHP (1996) Gradient enhanced damage for quasi-brittle materials. Int J Numer Methods Eng 39:3391–3403. https://doi.org/10.1002/(SICI)1097-0207(19961015)39:19%3c3391::AID-NME7%3e3.0.CO;2-D
5. Peerlings RHJ, De Borst R, Brekelmans WAM, De Vree JHP (1995) Computational modelling of gradient-enhanced damage for fracture and fatigue problems. Comput Plast Fundam Appl Proc 4th Int Conf: 975–986
6. Sarkar S, Singh IV, Mishra BK et al (2019) A comparative study and ABAQUS implementation of conventional and localizing gradient enhanced damage models. Finite Elem Anal Des 160:1–31. https://doi.org/10.1016/j.finel.2019.04.001
7. Sarkar S, Singh IV, Mishra BK et al (2019) Source codes and simulation data for the finite element implementation of the conventional and localizing gradient damage methods in ABAQUS. Data Br 26:104533. https://doi.org/10.1016/j.dib.2019.104533
8. Udhayaraman R, Mulay SS (2017) Multi-scale approach based constitutive modelling of plain woven textile composites. Mech Mater 112:172–192. https://doi.org/10.1016/j.mechmat.2017.06.007
9. Wang C, Roy A, Silberschmidt VV, Chen Z (2017a) Modelling of damage evolution in braided composites: recent developments. Mech Adv Mater Mod Process 3. https://doi.org/10.1186/s40759-017-0030-4
10. Wang L, Wu J, Chen C et al (2017) Progressive failure analysis of 2D woven composites at the meso-micro scale. Compos Struct 178:395–405. https://doi.org/10.1016/j.compstruct.2017.07.023
11. Wang YQ, Cheng P, Liu XS et al (2018) Study on a cracked single-layer 2D woven composite plate: fracture mode and damage analysis. Strength Mater 50:800–806. https://doi.org/10.1007/s11223-018-0025-3
12. Xue J, Kirane K (2021) Effect of the shape of the softening damage law on the predicted tensile fracturing and energy dissipation in textile composites

Investigation of Composite Material for a Remote Sensing CubeSat Structure: An Alternative to Traditional Aluminum Alloys

Adesh Phalphale, Nishita Sanghvi, Sunil V. Dingare, and Shubhamoy Sen

Abstract The CubeSats market is drastically growing leading to the development of smaller sensors that could be accommodated within smaller launch vehicles to comprehensively address most of the earth science needs. The positive spiral emanating from this development has resulted in lower costs, shorter mission development times, and an accelerated absorption of contemporary technologies, and commercial off-the-shelf components for space applications. This paper addresses the structural subsystem of a 3U CubeSat that is designed for glacier monitoring with a camera payload. The structure of a CubeSat must attain its objectives for the static and dynamic forces of the testing and the launch in the zero-gravity space thermal environment. Choosing the right material and structural design is essential in reducing the weight of CubeSat, as the structural elements usually make up a significant proportion of its overall mass. Typically, Aluminum alloys like Al 6061 or 7075 and respective series are widely used in CubeSat structure development. The use of composite material can play a vital role in reducing structural mass. As the 3U CubeSat in this paper is designed for the remote sensing application, reducing structural mass with the use of composite material can give more space for a better camera payload without compromising the structural integrity. The paper focuses on investigating the use of carbon fiber a widely used space-grade composite material by performing structural and thermal analysis on two 3U CubeSat structures with Aluminum 6061 and carbon fiber. A comparative analysis establishes that carbon fiber as the structural material can yield similar stiffness while reducing the structural mass substantially.

Keywords CubeSat structural analysis · Transient thermal analysis · Carbon fibre · Aluminum 6061

A. Phalphale (✉) · N. Sanghvi · S. V. Dingare
Department of Aerospace Engineering, MIT Art, Design and Technology University, Pune, Maharashtra, India
e-mail: adeshphalphale34@gmail.com

S. Sen
i4s Laboratory, Indian Institute of Technology Mandi, Mandi, Himachal Pradesh, India

R. K. Ray et al. (eds.), *Advances in Theoretical and Applied Mechanics*, Lecture Notes in Mechanical Engineering, https://doi.org/10.1007/978-981-97-0418-7_25

1 Introduction

CubeSats are miniature satellites measured in units of standardized dimensions (10 × 10 × 10 cm). Accordingly, a single unit is called 1U of mass up to 1.3 kg. These tiny satellites have several advantages from the aspects of lower cost, faster design, and manufacturing compared to their conventional counterparts [1]. Nonetheless, weight is one the major constraints in the design process of a CubeSat. The CubeSat Structure is built up of a lightweight material, aluminum alloys being a typical choice that deliver acceptable interfaces to all other subsystems to assure secure progression via all stages of the mission. The manufacturability and integration, lightweight, and sufficient area for the payload are the primary elements of the CubeSat structural subsystem. This subsystem also determines the capability of the structure to house considerable payload sensors simply integrated into the subsystem. Today composites are taking over aluminum as it is playing an inherent role in weight reduction and offers an increase in the payload [2].

The CubeSat structure should be able to withstand all the loads encountered during the launch and its entire mission period. Generally, the loads involved on the structure during its lifetime can be static and dynamic loads. The static part of load originates due to accelerations and launch forces that CubeSat experiences into the designated orbit. The dynamic loads result due to engine thrust, sonic forces, forces generated by air turbulence effect on the launcher, impulse thrust for orbital adjustment, and loads caused by transportation from the manufacturing site to the launching site. Selecting materials and designing space structural systems involves a delicate balance between mass, stiffness, and strength. Stiffness is necessary to ensure the durability of other sensitive subsystems. However, reducing the mass allows for increased payload capacity, enhancing the agility and effectiveness of the mission's objectives [3].

This research focuses on performing an analytical investigation on a composite material that can be used in designing a CubeSat for remote sensing applications. Using composites can benefit the platform, reduce its structural mass, and also give more space for the camera payload. The aim is to perform Modal, Static structural, and Transient thermal analysis on 3U CubeSat designed per requirements for the payload for glacier monitoring applications and obtain comparative results to check the feasibility of composite materials for building CubeSat.

CubeSats are embarked as secondary payloads with other primary satellites on commercial launching rockets, like Polar Satellite Launch Vehicle (PSLV). They are commonly anchored and deployed by a special deployer called P-Pod (Poly-Picosatellite Orbital Deployer). The thrust developed by these rockets during the launch stage varies from 800 to 4000 KN, and these small satellites have to endure this thrust as well as an assortment of additional forces. This implies that the CubeSats have to resist these forces and stay intact and operational. The satellite also has to handle the thermal shock when it is abruptly exposed to a very low temperature of space during the orbiting phase. To have these abilities built into the satellite specific tests are conducted on it, of which tests like Modal Analysis, Structural Analysis, and Thermal Analysis are of special importance [2].

In this paper these analyses are performed using Ansys Workbench 2021 R2 to simulate the 3U CubeSat model with two different materials, Carbon Fibre a widely used composite for space application and Aluminum 6061-T6. This analysis gives an insight into the benefits that composite material can offer if used in place of traditional aluminum alloys.

2 Mission Description

2.1 Mission Objectives

The CubeSat's core goal in this project is to make use of all available space-based resources and combine them to create a remote sensing CubeSat that can be utilized to better understand glacier phenomena and assist in the improvement of residents who live close to intense glacial activity. This paper is a part of the project that has the primary objective to identify the practical limits and factors that must be taken into account for the technical design of a CubeSat-based remote sensing system that will operate in Low Earth Orbit (LEO), with particular attention to the orbit, payload, and structural subsystems. While here the paper focuses specifically on the structural subsystem and takes the support of inputs from other subsystems for defining the boundary conditions for analysis performed.

2.2 Orbit Design

Although there are many different orbital configurations, it is crucial to select the one that is best for CubeSat's specific operation. For the remote sensing CubeSat, a sun-synchronous near-polar low-earth orbit has been chosen. With this orbital configuration, the CubeSat may pass over the Earth's poles during each round and see almost the entire planet as it revolves beneath it. These orbits are employed for weather forecasting and remote sensing/earth observation [4]. Due to the camera system payloads on the CubeSat, near-polar LEO is best suited with the goal of providing high-resolution imagery of the Himalayan terrain.

Table 1 lists significant orbital parameters that were determined using the analytical formulae. Kepler's equations were used to compute the orbital components. Keplerian elements serve as the foundation for the design and implementation of the orbit propagator and describe the size and form of the CubeSat's orbit [5].

These orbital parameters are essential in defining the boundary conditions for the transient thermal analysis that gives a temperature profile of a CubeSat in its solar phase. The solar and eclipse phases are nothing but the CubeSat in a sunlit state

Table 1 Orbital parameters

Orbital elements	Values
Semi-major axis, a	6,878,000 m
Eccentricity, e	0
Inclination, i	97°
Longitude of the ascending node, Ω	76°
Argument of periapsis, ω	0°
True anomaly, ν	150°

and vice versa. The orbit solar time is an essential boundary condition for transient thermal analysis and is calculated from orbital parameters [6]. The calculated solar time for this mission was around 2500 s.

3 Simulation Approach

3.1 CubeSat Configuration

To accommodate the needs of a remote sensing CubeSat mission with the least amount of modification, a standard stacked platform configuration is used. The standard platform places the utmost emphasis on the modularity and reconfigurability of each component. The key drivers for a standard CubeSat platform are the inclusion of recent developments in mission payloads; reuse in university-developed educational satellites; maximum reuse of hardware and software; optimization for more volume than conventional platforms; highly reliable interface; minimal assembly, integration, and testing; and minimal flight model production period [7, 8].

Furthermore, depending on the chosen camera sensor for this mission for the payload module, the weight of the CubeSat was estimated to vary from 2.2 to 3.3 kg. So, under most given conditions, the CubeSat would size up to 3U. This was taken as the upper limit and therefore the 3U CubeSat conceptual model was designed using SolidWorks CAD software and all further analyses are performed on this 3U CubeSat (Fig. 1).

3.2 Material Properties

The selection of either raw materials or the prefabricated structure depends on two several property constraints. Some of the commonly used materials in space architectures are Aluminum alloys (5005, 5052, 6061, and 7075), Kevlar, CFRP, and FML. The literature survey shows that the widely used metal and composite for the CubeSat structure were Aluminum 6061 and Carbon Fiber respectively [1–3]. Hence

Fig. 1 Conceptual design of the 3U CubeSat primary structure

Table 2 Material properties [9, 10]

Material	E (GPa)	ν	ρ (kg/m^3)	k (W/m K)	c_p (J/kg K)	Tensile strength (GPa)	Shear strength (GPa)	ε
Aluminum 6061	68	0.33	2700	202	895	0.29	0.21	0.6
AS4 carbon fiber	230	0.3	1790	6.83	1130	1.5	0.07	0.8

for the analysis, these two materials were used. The analysis results are essential to understand whether the structure with composite material performs optimally under a given set of conditions that the CubeSat would be required to meet.

A space-approved composite material HexTow AS4 Carbon fiber made by Hexcel was selected and assigned in form of quasi-isotropic composite panels in the present application. The 3U CubeSat structures one assigned with Aluminum 6061 standard properties and the other with AS4 Carbon fiber properties are subjected to further analysis. The properties of the material under analysis are presented in Table 2.

3.3 *Finite Element Model*

The finite element model of the geometry is created with a 9 mm element size consisting of 81,483 nodes and 32,656 tetrahedral elements in Ansys Workbench 2021 Student Version as is seen in Fig. 2. The plates between two levels contain hex mesh with elements created with quadratic formulation, and the other parts are

Fig. 2 Mesh model of a 3U CubeSat

meshed with tetrahedron elements. This is because the structure contains intricate patterns which make it difficult to produce pure hex mesh on all parts. As for the mesh quality, the aspect ratio is maintained under 25, with a maximum skewness of 0.4, which is standard practice for obtaining a good-quality mesh.

The mesh model is assigned to selected materials at a time which are Aluminum 6061 and Carbon fiber. All further analyses are done with these two material assignments.

3.4 Modal Analysis

In the modal analysis of a structural system, the initial and crucial step is to determine the natural frequencies of the structure based on the applied boundary conditions. The most crucial modal factor of a space structure, like the CubeSat, is the natural frequency threshold. The first natural frequency of the structure must be beyond a distinct value, which usually is defined by the launch vehicle [11].

According to the Polar Satellite Launch vehicle's specifications, the natural frequencies of the CubeSat should not be less than 35 Hz on the longitudinal axis and 20 Hz on the lateral axis. Also, as there will be high amplitude harmonic frequencies under the 100 Hz during the launching process, CubeSat natural frequencies should be more than 100 Hz to prevent resonance [12].

In the investigation, it is assumed that the CubeSat is positioned inside a Poly-Pico Satellite Orbital Deployer (P-POD). As a result, the side of the CubeSat that is in contact with the ground is completely fixed, simulating the bottom of the P-POD during launch. The P-POD utilizes a spring-loaded deployment technique that allows for minimal vertical movement during launch, while the side rails ensure that the CubeSat remains securely fixed in two directions. Therefore, to replicate the rigid sides of the P-POD and minimize any warping during takeoff, the side of the CubeSat facing the sky is constrained in all directions except for the vertical direction (+ y direction in our model).

To meet the CubeSat specifications, it is required that the structure has a first resonant frequency greater than 100 Hz. The majority of significant forcing amplitudes during launch occur below 100 Hz. The objective is to design a CubeSat that can withstand forces near or at resonant frequencies without being destroyed during the launch phase. The analysis is limited to a maximum of four modes. The frequency range is set from a minimum of 0 Hz to a maximum of 1000 Hz. These settings enable us to reduce the runtime by focusing on the modal frequencies we expect to encounter and consider [11].

3.5 Quasi-Static Analysis

During the launch, the CubeSat must endure the acceleration loads imposed by the launching sequence. To evaluate the stress levels on the CubeSat structure resulting from these acceleration loads, a quasi-static analysis is conducted.

For this analysis setup, the model is loaded with an acceleration of 75 m/s^2 in the + y direction and another 137.5 m/s^2 in the + z direction simultaneously. Lateral acceleration is in the + y direction because the structure is more prone to flex around the x-axis than the y-axis [12]. The values for analysis are not rounded concerning the factor of safety. The factors of safety for the already available CubeSat mission are typically 1.25, but as this 3U CubeSat is a conceptual model and not flight-ready it is assumed to tolerate a loading as the worst-case loading condition [3].

3.6 Transient Thermal Analysis

CubeSats must withstand the harsh space environment, which includes radiation from the Sun, albedo radiation reflected from the Earth, and infrared radiation emitted by the Earth. To assess the temperature distribution of the CubeSat structure, a transient thermal analysis is necessary, taking into account the worst-case scenario [6] (Fig. 3).

The CubeSat's external structure was also subjected to transient thermal analysis with boundary conditions and corresponding assumptions for the analysis namely the initial temperature of 25 °C, an ambient temperature of − 20 °C, heat flux of around 2108 W/m^2 (Albedo, Solar Flux, Earth IR) and the time 2500 s which was calculated in Sect. 2.2. As Ansys workbench cannot simulate the orbit profile, all faces were assumed to be exposed to the radiation. Also, the structure was assumed to have no thermal insulation. This gives the worst-case scenario.

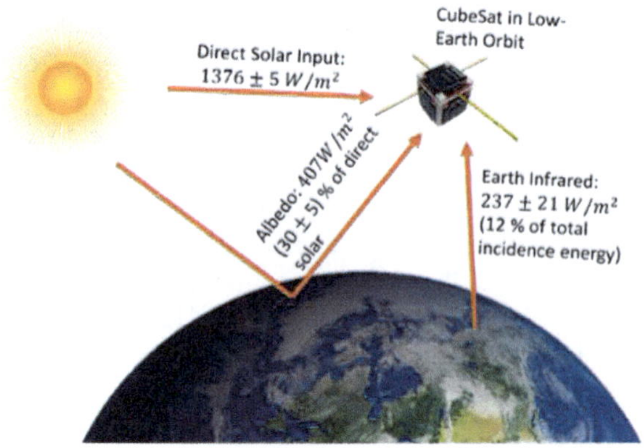

Fig. 3 Space thermal environment [6]

4 Results and Discussions

Comparison between Al 6061 and Carbon fiber structure gives 266.43 g of mass difference. These shows building a model using carbon fiber gives about a 33.3338% decrease in structural mass, which can then be utilized for other essential subsystems majorly the payload.

The first 4 eigenfrequencies and modes of simplified 3U CubeSat structure designed as per the mission requirements were obtained. The Carbon fiber structure gives a frequency of 848.59 Hz and the aluminum 6061-T6 structure gives a frequency of 705.18 Hz at mode 1. The lower limit for the first eigenfrequency for the specific mission is set above 90 Hz as per PSLV Datasheet. Thus, the structure of carbon fiber can be considered adequately robust with the additional advantage of mass reduction as compared to the Aluminum 6061 structure (Tables 3 and 4).

In both cases, the results of the static analysis demonstrated that the stresses and deformations are remarkably minimal. It should be noted that the input conditions for the conceptual model were not adjusted with a factor of safety, thus these results cannot be directly compared with those of a launch-ready CubeSat. When comparing materials, the composite material once again exhibited superior performance, with smaller maximum stress and deformation values compared to the aluminum alloy. According to these results, the accelerations specified in the PSLV can cause virtually no flex in the structure [3, 12] (Table 5).

The composite structure made from carbon fiber achieves a maximum temperature of around 30 °C whereas the aluminum structure achieves 177.27 °C. This shows a drastic reduction in temperature as it has the thermal conductivity of carbon fiber is 40 times less than aluminum alloy. This makes the carbon fiber structure most suitable for remote sensing applications increasing its thermal stability which subsequently increases its durability in the space environment.

Table 3 4 eigenfrequencies and mode shapes of the structure

1. Composite (carbon fiber) 3U CubeSat structure—532.85 g			
		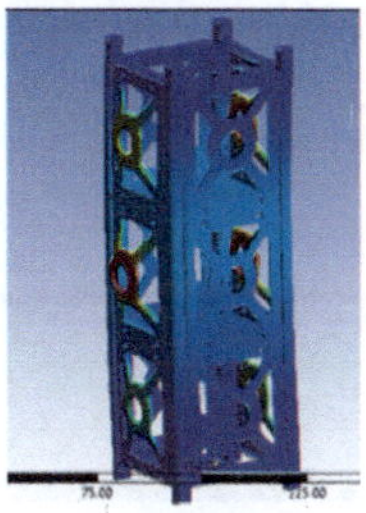	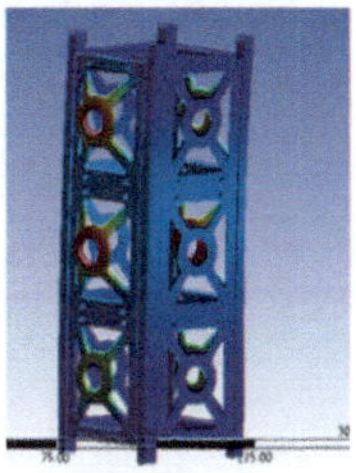
Mode 1 (848.59 Hz)	Mode 2 (863.51 Hz)	Mode 3 (930.67 Hz)	Mode 4 (936.37 Hz)
2. Aluminum (Al 6061) 3U CubeSat structure—799.28 g			
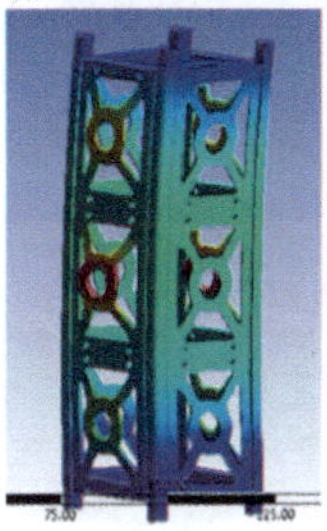	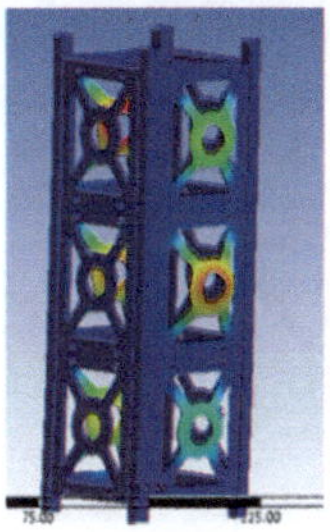	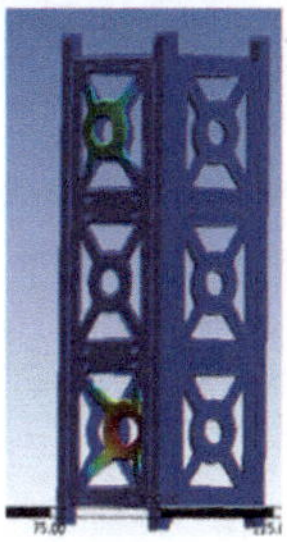	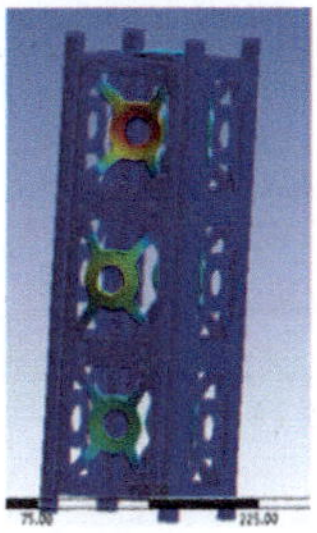
Mode 1 (705.18 Hz)	Mode 2 (729.17 Hz)	Mode 3 (762.8 Hz)	Mode 4 (770.32 Hz)

5 Conclusion and Future Prospects

This paper investigates the use of carbon fiber composite material in designing a 3U CubeSat for remote sensing applications. The study aims to evaluate the benefits of using a composite structure for CubeSats in remote sensing. The main finding of this work is the successful design achievement using composite material.

The composite structure exhibited a first eigenfrequency that exceeded the lower limit typically required for space structures. When comparing the Aluminum and Composite structures, an increase in the first eigenfrequency was observed, along with a significant reduction in the mass of the structure, amounting to approximately 35%. This means the structure with composite material can offer similar strength and stiffness as the aluminum with the advantage of weight reduction. The Quasi-Static Analysis showed that incorporating composite material can give better stress and deformation values. This also qualifies the robustness of the material. Thermally, the composite material can be more stable as compared to aluminum as a drastic temperature difference was observed. Hence the overall investigation implies that replacing composites could be a good option to obtain better structural performance

Table 4 Quasi-static analysis results

	Carbon fiber structure	Al 6061 structure
Equivalent stress		
	2.68 MPa	6.7948 MPa
Total deformation		
	0.01 mm	0.04 mm

and reduction in weight. This can benefit other subsystems in terms of mass and space.

Although this paper qualifies composites for the CubeSat structure it is essential to study the material in more detail manner. The future scope of this study will be to check the material feasibility in terms of the flight-ready model. The future prospects will be to investigate the composite material with vital considerations like its manufacturability, cost and development complexity. The practical model and testing can qualify the usage and effectiveness of composite material.

Table 5 Transient thermal results

For composite structure 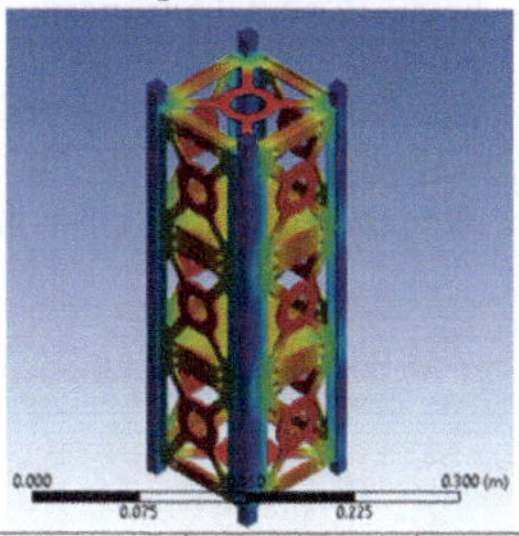				For aluminum structure 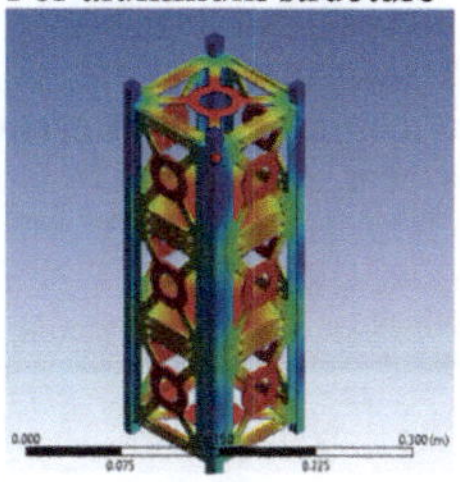			
T. [s]	Min. [°C]	Max. [°C]	Avg. [°C]	*T*. [s]	Min. [°C]	Max. [°C]	Avg. [°C]
25	25.001	25.082	25.05	25	41.716	48.066	44.224
33.333	25.004	25.104	25.065	45.326	55.476	63.672	58.576
41.667	25.008	25.123	25.079	65.652	68.63	77.225	71.813
66.667	25.023	25.181	25.121	126.63	100.87	108.32	103.61
141.67	25.075	25.346	25.242	224.96	134.69	139.63	136.52
1116.7	25.838	27.288	26.694	506.96	170.9	171.77	171.22
1366.7	26.062	27.789	27.055	575.41	173.39	173.94	173.6
1616.7	26.296	28.287	27.412	643.86	174.96	175.31	175.09
1866.7	26.538	28.782	27.768	712.31	175.95	176.17	176.04
2116.7	26.786	29.275	28.121	780.76	176.58	176.72	176.63
2366.7	27.039	29.763	28.473	849.21	176.97	177.06	177
2500	27.176	30.023	28.66	2500	177.21	177.27	177.63

References

1. Corpino S, Caldera M, Nichele F, Masoero M, Viola N (2015) Thermal design and analysis of a nanosatellite in low earth orbit. Acta Astronaut 115:247–261
2. Piedra S, Torres M, Ledesma S (2019) Thermal numerical analysis of the primary composite structure of a CubeSat. Aerospace 6:97
3. Ampatzoglou A, Baltopoulos A, Kotzakolios A, Kostopoulos V (2014) Qualification of composite structure for cubesat picosatellites as a demonstration for small satellite elements. Int J Aeronautics Aerosp Res 1(1):1–10
4. Ali E, Saeed B, Khan A, Akram F (2013) Conceptual design of an earth observation satellite
5. Zayan M, Eltohamy F (2008) Orbits design for remote sensing satellite: 1–9. https://doi.org/10.1109/AERO.2008.4526494
6. Phalphale A, Gaikwad V, Sharmale R, Ravuri N, Bihani V, Singh Y (2021) Optimal power generation and thermal analysis for ÆtherSat. IEEE Pune Sect Int Conf (PuneCon) 2021:1–6. https://doi.org/10.1109/PuneCon52575.2021.9686502
7. Nieto-Peroy C, Emami MR (2019) CubeSat mission: from design to operation. Appl Sci 9(15):3110. https://doi.org/10.3390/app9153110
8. Sanghvi N, Bihani V, Shaikh M, Gaikwad V, Sharmale R, Singh Y (2021) Design and analysis of the external structure and internal configuration of ÆtherSat. IEEE Pune Sect Int Conf (PuneCon) 2021:1–6. https://doi.org/10.1109/PuneCon52575.2021.9686509

9. Matweb material property data. Retrieved from http://www.matweb.com/search/DataSheet.aspx
10. HEXCEL Corporation (2016) HexTow AS4 carbon fibre datasheet. Hexcel Corporation, Stamford, CT, USA
11. Modal analysis of a satellite, Ansys innovation space. https://courses.ansys.com/index.php/courses/modal-analysis-of-a-satellite/
12. Guvenc CC, Topcu B, Tola C (2018) Mechanical design and finite element analysis of a 3 unit CubeSat structure

Stress Field Prediction in Particulate Polymer Composite Materials Using Paired Image-To-Image Translation Approach

Sristi Gupta, Vinod Kushvaha, and Divyesh Varade

Abstract Stress analysis is essential in understanding the mechanical behavior of designed composite structures. Conventionally, Finite Element Modeling (FEM) is employed for the computational analysis of newly fabricated materials. Ideally, for composite structures, the stress observations from the computational analysis using software such as ABAQUS must agree with the experimental yields of stresses. However, traditional computational techniques are known to have higher complexity and longer runtimes while further requiring intensive labor. Alternatively, in recent studies, to reduce the constraints of software such as ABAQUS, various machine learning approaches have been explored for examining the mechanical behavior of Particulate Polymer Composite (PPC). This study demonstrates the advantages and potential of using advanced deep learning architectures for the investigation of polymer stresses. In the proposed study, a paired image-to-image translation with Generative Adversarial Networks (GANs) is employed. The PPC stresses are developed using ABAQUS at six time steps and used for training the GAN. The results reveal that the implementation of the paired image-to-image translation with GANs aids an accurate and efficient investigation of stress fields in newly fabricated PPC materials. The performance of the approach is evaluated using the correlation between the reference data and the results. Further, the least mean square error (L2 norm) is investigated to analyze the performance of the network. The maximum correlation for the test results for epoxy, and polymer fibers are observed to be 0.9975, and 0.99975. The value of L2 Norm is observed to be less than 0.005 for stresses in x direction in both the materials used in the preparation of particulate polymer composite two-dimensional micromechanics model.

Keywords Neural networks · Mechanical behavior · Finite element modeling · Micromechanics

S. Gupta (✉) · V. Kushvaha · D. Varade
Department of Civil Engineering, Indian Institute of Technology Jammu, Jammu and Kashmir, India
e-mail: sristi.gupta@iitjammu.ac.in

R. K. Ray et al. (eds.), *Advances in Theoretical and Applied Mechanics*, Lecture Notes in Mechanical Engineering, https://doi.org/10.1007/978-981-97-0418-7_26

1 Introduction

Particulate Polymer Composites (PPC) have recently gained the keen attention of engineers for their exceptional mechanical behavior, custom-based applications, and numerous other benefits. Engineering sectors make use of polymer composites for their efficient and effective mechanical properties that include non-corrosive behavior, high thermal resistance, good rigidity, high impact resistance, and high strength-to-weight ratio [1, 2]. Polymer composites have wider applications in various engineering and medical domains [3–5]. Thus, precise information about the mechanical behavior of materials used in the fabrication of polymer composites is an essential process. Further, it is significant to understand the stress fields primarily in the design of material systems and for deciding the role of the specific polymer composites for various customized applications [6, 7].

Finite Element Method (FEM) has been the traditionally used numerical approach for stress analysis of polymer composite structures requiring solutions of complex Partial Differential Equations (PDE) [8, 9]. FEM-based software such as ABAQUS is popular for the computational analysis of stresses in composite systems. In contrast, with increasing complexity in the polymer composite structure, at times it is challenging to assess precisely the mechanical behavior of the materials quantitatively, thus constrained by high computational costs. This study focuses on reducing the cumbersome efforts for evaluating stresses, primarily the expensive computations and labor-intensive experimental procedures.

Over conventional FEM-based approaches, neural computing has gained significant popularity for the prediction of polymer composite stresses [10, 11]. Techniques such as convolutional neural networks (CNN) have been used successfully for predicting stress maps within the two-dimensional cross-section of fiber reinforced polymer composites [12, 13]. Advanced Deep Learning (DL) architectures have been recently employed in various studies for the assessment of 2D stress maps in polymer composites. DL-based Generative Adversarial Networks (GANs) are used for 3D microstructure generation where it is able to learn the properties of the material and produce unique 3-D microstructures with similar properties to the original data expeditiously with consistently high quality [14]. Furthermore, conditional GANs learn the mapping from images and random noise fed to the algorithm [15]. The conditional adversarial networks are potential approach for many image-to-image translation tasks, intended for receiving highly structured output. In a study, a controller is introduced to the pix2pix algorithm to improve the classification performance and training stability [16]. In order to enhance the predicted results, hybrid solution has also been studied for purely paired and unpaired training data such that both qualitatively and quantitatively improved results are obtained [17]. Both the selection and extraction of the salient features from the pixel space of the images can be either supervised or unsupervised. However, supervised learning provides more insights in the selection process while capturing the response for material structures [11].

This study focuses on predicting the stress fields in PPC using supervised learning. Here, the PPC is composed of short carbon-filled epoxy and is useful for its various customized industrial applications. The stress fields in the fiber reinforced polymer composite are predicted using pix2pix GANs for a set of images derived from the FEA-based software ABAQUS CAE/2020, Dassault Systémes. We have used a popular image-to-image translation algorithm to predict the stresses directly from the material microstructure based on ABAQUS simulations. The deep learning architecture with conditional Generative Adversarial Network (cGAN) with a CNN-based U-net architecture is utilized in Generator and discriminator for acquiring the predicted results.

2 Materials and Methods

2.1 Image Generation

In the present study, the PPC is prepared by reinforcing micron rod-shaped particulate carbon fibers in the thermosetting-based epoxy. Table 1 shows the elastic properties of epoxy and carbon fibers used for the preparation of the PPC. The average longitudinal dimension of the fibers is 60 μm, and the diameter is 8 μm. These casted samples of short carbon fibers mixed in the epoxy [18] are further machined into dog-bone shape specimens. These specimens are further tested against the tension at a static loading rate of 2 mm per minute. The experimentally validated computational results for two-dimensional representative area element of 600 × 600 μm^2 are analyzed through ABAQUS, an FEA-based software to build the image dataset for training the deep learning model. The images for the dataset are extracted from the output database (odb) files using an indigenously developed Python code for each of the 256 simulations run against the varied ranges of ± 30% values of Young's modulus of both the materials (epoxy and carbon fibers) in Abaqus.

Table 1 Elastic properties of carbon fibers and epoxy system

Material	Poisson's ratio θ	Elastic modulus (GPa)
Epoxy	0.3	4
Carbon fibers	0.2	225

2.2 Pix2pix Conditional Generative Adversarial Network (cGAN)

The primary implication of utilizing an AI framework is to reduce the computational runtime in deriving a comprehensive stress assessment on the individual components of a PPC. In previous studies, the CNN-based frameworks have remarkably improved the computational efforts in comparison to the finite element simulations as observed in previous studies [19, 20]. Recent studies show the application GANs for improved estimation of stresses in PPCs. Although generative modeling is defined as an unsupervised learning task, GANs formulate the learning problem in a supervised framework with two sub-models [15]. A generator, which is trained to generate like new samples, and a discriminator assesses the likelihood of real samples. In the conditional GANs (cGAN), the input images are conditioned to satisfy a given criterion to the generator network. An extension of cGAN is the pix2pix algorithm, where the satisfying condition to the generator is provided by means of an image itself. In the proposed approach, we utilize the pix2pix base cGAN deep learning framework. The architecture of the pix2pix cGAN, the generator consists of a U-net architecture with several encoding and decoding blocks and the discriminator comprises a patchGAN discriminator (an extension of the conventional CNN).

A simplified flowchart detailing the computational and analytical process involved in the assessment of stresses in PPC is illustrated in Fig. 1. In the proposed investigation, the predictions are given for negative stresses developed in both the materials which are epoxy and carbon fibers. In both cases, the input visually remains of same pattern as shown in Fig. 1. However, the input images are generated corresponding to each ground truth image of the stresses developed in each material. The greyscale images as shown to have three strips that comprise Young's modulus of each material (epoxy and carbon fibers) and the time steps referring to the frames in ABAQUS. Further illustrating that the input image comprises the parameters where first two grey-colored strips detail Young's modulus of carbon fibers and epoxy. However, the third parameter shows the time steps (frame numbers in ABAQUS) that show an increment of stresses in each frame. Additionally, the third strip refers to the third important label which is time steps. The reference data for the deep learning model is derived from simulations using the FEA software (ABAQUS), such that the first reference (epoxy) delineates the negative stresses focused around the carbon fibers in the PPC, while the second reference (carbon fibers) delineates the negative stresses focused within the carbon fibers. The FEA simulations from ABAQUS are used as input for training the pix2pix cGAN model with 1280 samples used for training from the 1536 samples generated from 6-time steps. The results are evaluated by statistically comparing the estimated stresses with the reference based on the correlation coefficient and the L2 norm.

Fig. 1 A simplified data-driven flowchart showing the steps followed in the development of the proposed methodology for the estimation of stresses in PPC

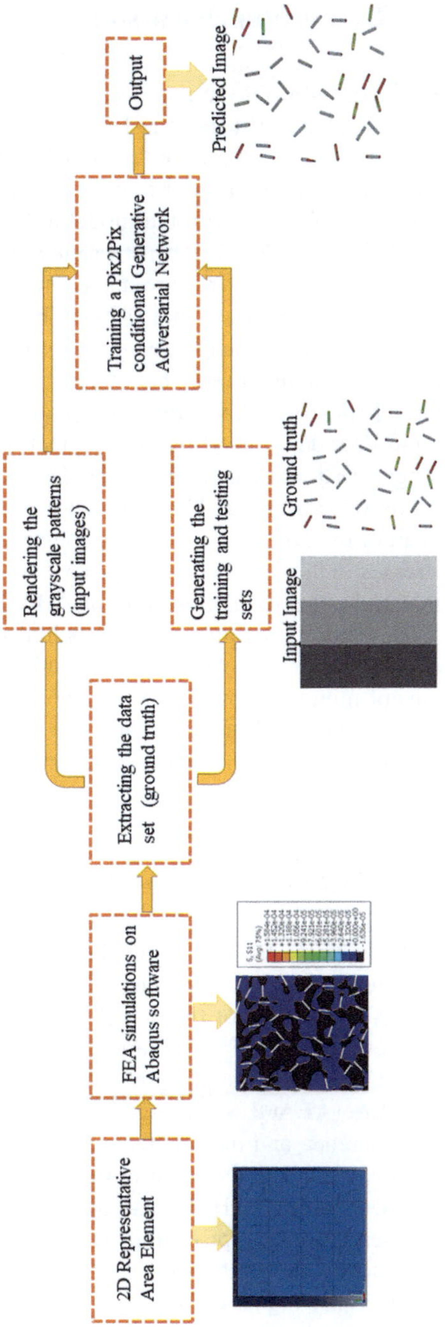

3 Results and Discussion

We have followed the approach discussed in Sect. 2, the training sample images' dataset is derived using the FEA software (ABAQUS) for pix2pix cGAN framework. As discussed earlier, the model performance is evaluated using the correlation coefficient and the L2 norm observed between the estimated stresses and the reference data. We have run the model at 2000 iterations and observed the correlation coefficient and the L2 norm at varying volumes ($\pm$ 30% values of Young's modulus) of the image datasets retrieved from the FEA software. The performance is evaluated in both cases of epoxy and carbon fibers. The results show a significant agreement between the input images, which are used to learn and subsequently imitate the information of the features in the reference data and predicted stresses. It is observed that the DL-based cGAN framework predicts the stresses depending on the quality of input data determined by various factors including the magnitude of stresses, the range of stresses, individual sample size (image size), and the volume of sample set used for training. All these factors contribute to improving the learning mechanism of the DL cGAN model accounting for the information outlining intensity of stresses in the fibers and the epoxy. Particularly, for analysis of polymer composite materials, the simulation time in ABAQUS is substantially more than that in the stress prediction from a trained paired image-to-image translation (pix2pix) with cGAN framework. Although simulations in ABAQUS require a runtime of a few tens of minutes. However, the proposed approach based on pix2pix cGAN usually takes about 15–20 min for running 256 simulations under 2000 iterations, which significantly reduces the run time for the investigation of stresses in the PPC.

3.1 Negative Stresses in the Epoxy

In the first case, the evaluation of the pix2pix cGAN model is conducted for the epoxy only reference data to predict stresses in the epoxy. The preliminary results based on 2000 iterations of the model run indicate significant performance of prediction of stresses in the epoxy, where it is observed that the distribution of the stress fields in the predicted results agrees strongly with the reference derived from the FEA software (ABAQUS). In Fig. 2, within the stress spectrum, the grey magnitudes in the reference and the predicted images correspond to positive stresses, which are, however, not focused upon in this DL algorithm. For the epoxy case, the negative stresses are observed to be marginal, and subsequently, the reference and the predicted stresses largely comprise almost null values corresponding to minor stresses. Overall, statistically, the predicted stress and the reference compare notably as observed by a significantly higher correlation coefficient and lower L2 norm values, as illustrated in Fig. 3. The least and highest agreement for the training dataset as determined by the correlation coefficient was observed to be 0.9775 and 0.9975, respectively, with the maximum least square error marginally above 0.02. The least and highest

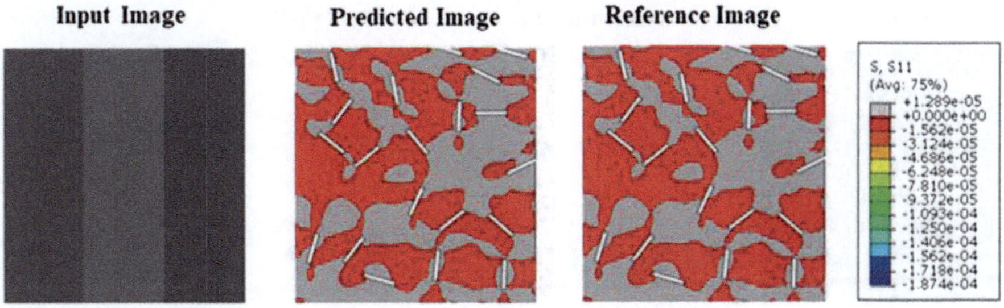

Fig. 2 Greyscale input images used for learning in the pix2pix model, the predicted image of stresses in the epoxy from the model, and the reference derived from the FEA software

agreement for the test dataset was observed to be marginally above 0.9775 and 0.9975, respectively, with the maximum least square error below 0.02.

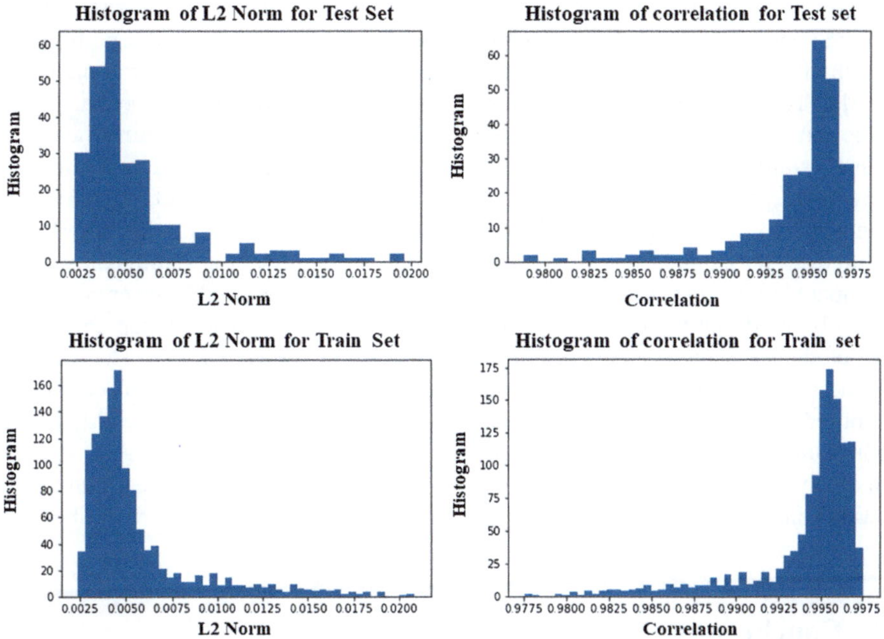

Fig. 3 Observations of agreement between the predicted stresses in epoxy from the model and the reference from FEA software based on the correlation coefficient and the L2 norm for the training and test datasets

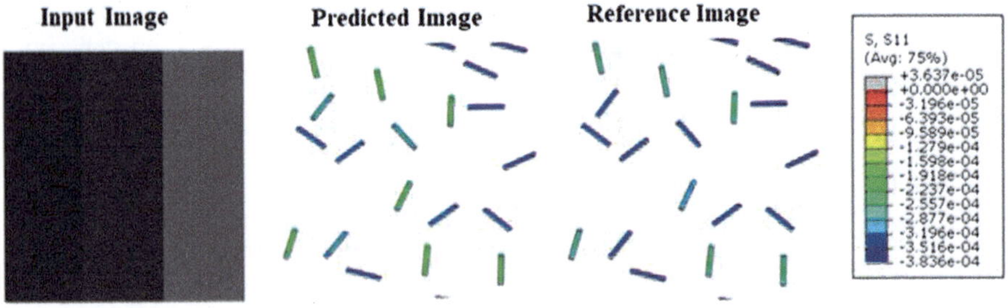

Fig. 4 Greyscale input images used for learning in the pix2pix model, the predicted image of stresses within the carbon fibers from the model, and the reference derived from the FEA software

3.2 Negative Stresses in the Carbon Fibers

In the second case, the evaluation of the pix2pix cGAN model is conducted for the carbon fibers only reference data to predict stresses in the carbon fibers. Similar to the approach followed for predicting the stresses in the epoxy, the preliminary results for the stresses in carbon fibers are determined from 2000 iterations of the model. The majority of the negative stresses observed in the carbon fibers are in the blue and the green bands as shown in Fig. 4. The input images for training the model remain similar to the case of the epoxy as greyscale images, with a different reference now corresponding to the stresses in the carbon fiber. The number of magnitudes of the negative stresses in carbon fibers, in this case, however, are significantly lower. Moreover, the number of pixels for the training of the model is significantly lower compared to the first case, thereby affecting the model accuracy. However, relatively, it is observed that the model still shows outstanding performance when comparing the predicted results with the reference statistically, as shown in Fig. 5. The least and highest agreement for the training dataset as determined by the correlation coefficient is observed to be 0.99775 and 0.99975, respectively, which is a significantly higher correlation value to that of the correlation obtained for the case of negative stresses in epoxy. Furthermore, minimum and maximum least square error was found to be 0.0005 and 0.0045, respectively, for the minor negative stresses in carbon fibers.

4 Conclusions

In this study, we have investigated the capabilities of the pix2pix cGAN as a deep learning model for predicting the stresses in the PPC. We focus on showing the predictions of the negative stresses in the epoxy and the carbon fibers of the PPC in this preliminary investigation. The DL approach is efficient in capturing stresses in both the negative and positive stress spectrum. The study aims to delineate the efficiency of deep learning methods over the conventional FEA-based software for simulating the stresses in PPC, primarily in terms of the computational run-time

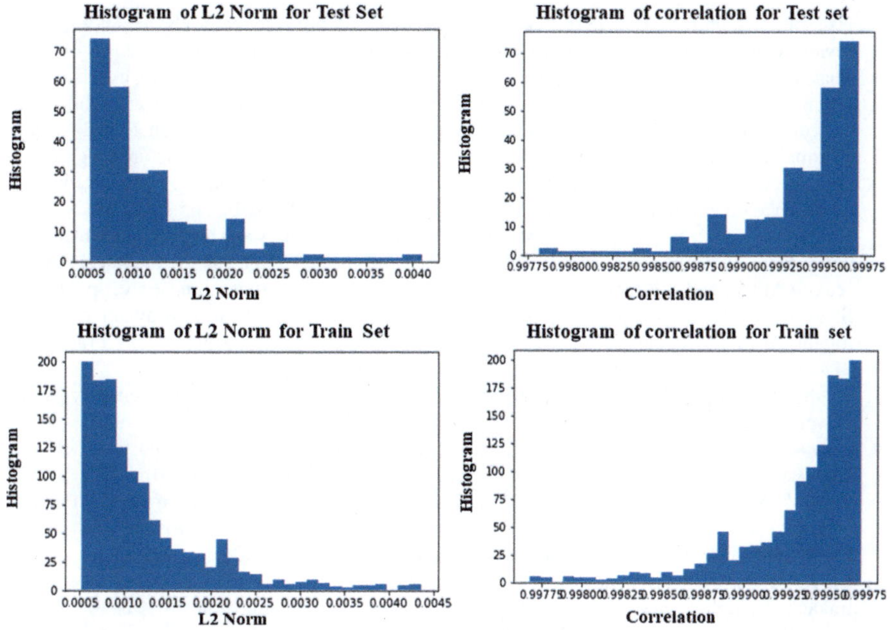

Fig. 5 Observations of agreement between the predicted stresses in carbon fibers from the model and the reference from FEA software based on the correlation coefficient and the L2 norm for the training and test datasets

over varied properties of material comprising both epoxy and the carbon fibers. The general observations for the pix2pix cGAN also hold in the present study, where the controlling factors are observed to be the number of iterations, the number of simulated images in the training dataset, and the number of parameters for the simulation of the stresses. However, we have observed that even in the constrained environment with 2000 iterations and 256 samples, the pix2pix cGAN model has shown considerable performance explained by the statistical agreement between the predicted stresses and the ABAQUS generated reference. The observed results also indicate the potential of the proposed approach based on the pix2pix cGAN for the prediction of positive stresses in the PPC.

References

1. Nagaraja KC, Rajanna S, Prakash GS, Rajeshkumar G (2021) Mechanical properties of polymer matrix composites: effect of hybridization. Mater Today Proc 34:536–538. https://doi.org/10.1016/j.matpr.2020.03.108
2. Gupta S, Sharma A, Varma RK, Kushvaha V (2022) A review on performance of near-surface mounted-carbon fiber-reinforced polymer laminates bonded into slits. Polymer Compos 1. https://doi.org/10.1002/pc.26795

3. Hsissou R, Seghiri R, Benzekri Z et al (2021) Polymer composite materials: a comprehensive review. Compos Struct 262:113640. https://doi.org/10.1016/j.compstruct.2021.113640
4. Sharma A, Kushvaha V (2020) Predictive modelling of fracture behaviour in silica-filled polymer composite subjected to impact with varying loading rates using artificial neural network. Eng Fract Mech 239:107328. https://doi.org/10.1016/j.engfracmech.2020.107328
5. Sharma A, Madhushri P, Kushvaha V, Kumar A (2020) Prediction of the fracture toughness of silicafilled epoxy composites using K-nearest neighbor (KNN) method. In: 2020 international conference on computational performance evaluation (ComPE). IEEE, pp 194–198
6. Kumar V, Isanaka BR, Gupta S, Kushvaha V (2021) Future trends and technologies in additive and substractive manufacturing. In: Mavinkere Rangappa S, Gupta MK, Siengchin S, Song Q (eds) Additive and subtractive manufacturing of composites. Springer, Singapore, pp 227–247
7. Sharma A, Anand Kumar S, Kushvaha V (2020) Effect of aspect ratio on dynamic fracture toughness of particulate polymer composite using artificial neural network. Eng Fract Mech 228:106907. https://doi.org/10.1016/j.engfracmech.2020.106907
8. Hughes TJR (2000) The finite element method: linear static and dynamic finite element analysis. Dover Publications, Mineola, N.Y.
9. Belytschko T (1989) The finite element method: linear static and dynamic finite element analysis: Thomas J. R. Hughes. Comput-Aided Civil Infrastr Eng 4:245–246
10. Yang Z, Yu C-H, Guo K, Buehler MJ (2021) End-to-end deep learning method to predict complete strain and stress tensors for complex hierarchical composite microstructures. J Mech Phys Solids 154:104506. https://doi.org/10.1016/j.jmps.2021.104506
11. Liu R, Yabansu YC, Agrawal A et al (2015) Machine learning approaches for elastic localization linkages in high-contrast composite materials. Integr Mater Manuf Innov 4:192–208. https://doi.org/10.1186/s40192-015-0042-z
12. Gupta S, Mukhopadhyay T, Kushvaha V (2022) Microstructural image based convolutional neural networks for efficient prediction of full-field stress maps in short fiber polymer composites. Defence Technol 262:619. https://doi.org/10.1016/j.dt.2022.09.008
13. Ajit A, Acharya K, Samanta A (2020) A review of convolutional neural networks. In: 2020 international conference on emerging trends in information technology and engineering (ic-ETITE), pp 1–5
14. Henkes A, Wessels H (2022) Three-dimensional microstructure generation using generative adversarial neural networks in the context of continuum micromechanics. Comput Methods Appl Mech Eng 400:115497. https://doi.org/10.1016/j.cma.2022.115497
15. Isola P, Zhu J-Y, Zhou T, Efros AA (2017) Image-to-image translation with conditional adversarial networks. In: 2017 IEEE conference on computer vision and pattern recognition (CVPR). IEEE, Honolulu, HI, pp 5967–5976
16. Wang X, Yan H, Huo C et al (2018) Enhancing Pix2Pix for remote sensing image classification. In: 2018 24th international conference on pattern recognition (ICPR), pp 2332–2336
17. Tripathy S, Kannala J, Rahtu E (2019) Learning image-to-image translation using paired and unpaired training samples. In: Jawahar CV, Li H, Mori G, Schindler K (eds) Computer vision—ACCV 2018. Springer International Publishing, Cham, pp 51–66
18. Kushvaha V (2016) Synthesis, processing and dynamic fracture behavior of particulate epoxy composites with conventional and hierarchical micro-/nano-fillers. Doctoral, Auburn University
19. Heidenreich JN, Gorji MB, Mohr D (2022) Modeling structure-property relationships with convolutional neural networks: yield surface prediction based on microstructure images. Int J Plast: 103506. https://doi.org/10.1016/j.ijplas.2022.103506
20. Minfei L, Yidong G, Ze C et al (2022) Microstructure-informed deep convolutional neural network for predicting short-term creep modulus of cement paste. Cem Concr Res 152:106681. https://doi.org/10.1016/j.cemconres.2021.106681

Influence of Alloying Elements on Mechanical Deformation of AlCoCrFeNi High-Entropy Alloy

Subrata Barman, Kritesh Kumar Gupta, and Sudip Dey

Abstract In the present study, a set of molecular dynamics (MD) simulations are carried out to investigate the effect of individual alloying elements on the mechanical deformation of AlCoCrFeNi high entropy alloy (HEA). Different HEA system configurations were considered by varying the atomic fraction (at.%) of the individual alloying elements. The shear strain-shear stress responses are recorded during the mechanical deformation of the individual HEA system. In addition to that, the microstructural deformation evolution of the HEAs is also investigated under the shear loading. This study proposes a bottom-up design approach for the synthesis of high-entropy alloys based on the desired characteristics.

Keywords AlCoCrFeNi high-entropy alloy · Molecular dynamics simulations · Shear deformation · Atomistic simulation · Deformation mechanism

1 Introduction

Over the past years, significant effort has been made by scientists and engineers to develop various advanced materials beyond the limitations of traditional materials. These advanced materials are widely used in multiple applications, which accomplish the requirements of the fast-growing industries and energy sectors. Among these advanced materials, high-entropy alloys (HEAs) are the recently developed multiple principle element alloys that include outstanding mechanical properties like strength, yield stress, toughness, corrosion resistance, and hardness [1–10]. HEAs are composed of many principal elements, resulting in a high amount of mixing entropy and leading to the solid solutions' high stability [9]. HEAs consist of at least five alloying elements, having an equimolar atomic fraction [10] or percentage of atomic

S. Barman (✉) · S. Dey
Department of Mechanical Engineering, National Institute of Technology Silchar, Silchar, India
e-mail: subrata21_rs@mech.nits.ac.in

K. K. Gupta
Amrita School of Artificial Intelligence, Amrita Vishwa Vidyapeetham, Coimbatore, India

R. K. Ray et al. (eds.), *Advances in Theoretical and Applied Mechanics*, Lecture Notes in Mechanical Engineering, https://doi.org/10.1007/978-981-97-0418-7_27

fraction between 5 and 30% of each element [11]. However, as per recent studies, the properties of the HEAs are not optimal with the equimolar ratio of alloying elements [12]. HEAs can be tailored for the anticipated applications by considering the non-equimolar compositions of the constituents [10, 13]. By tailoring the compositions and types of individual alloying elements, lightweight HEAs can be developed for their prospective applications in the domain of transportation, defense, and aerospace [9, 14]. Aluminum (Al) is frequently used as an alloying element in various alloys and HEAs that are lightweight, robust materials for structural applications [13–15]. Tung et al. [16] reported that adding Al content to the AlCoCrCuFeNi HEAs facilitates the formation of BCC phase and improves the hardness. Ji et al. [17] prepared a lightweight AlMgZnCuSi HEAs system and investigated mechanical and corrosion behavior by changing the proportion of alloying elements. With the increase in the molar ratio of Si–Mg, the content of the Al–Mg–Zn phase decreases, which enhances the mechanical and corrosion properties of the HEA system.

Among the HEAs, AlCoCrFeNi is a relatively low-density HEA and has been widely explored in the past [18–25]. Many studies reported that the mechanical properties of AlCoCrFeNi HEAs could be tuned for the desired application by tailoring the compositions of the alloying elements [20, 24]. Jiang et al. [25] conducted a molecular dynamics simulation to investigate the influence of Al concentration on the evolution of the microstructural deformation and mechanical properties of AlCoCrFeNi HEAs under the uniaxial tension. They have shown that the high concentration of Al reduces the tensile properties of AlCoCrFeNi HEAs.

Molecular dynamics (MD) simulations are widely utilized to investigate the properties of materials at the nanoscale [26–39]. MD simulations allow a better understanding of the mechanical properties at the atomistic level by controlling the atomic fraction of alloying elements. Many literatures reported the MD simulation of AlCoCrFeNi HEAs and investigated the mechanical properties and microscopic deformation evolution under the uniaxial tensile and compression loading conditions. However, the shear deformation-dependent structural evolution of the lightweight AlCoCrFeNi HEAs is still unexplored at the nanoscale level. Hence, it is crucial to understand the influence of individual alloying elements on the atomistic deformation and structural evolution of low-density AlCoCrFeNi HEAs. The present MD simulation is performed to investigate the effect of individual alloying elements (Al, Co, Cr, Fe, Ni) on the mechanical deformation and atomistic deformation evolution of the AlCoCrFeNi HEAs under shear loading.

2 Simulation Methods

In the present study, MD simulation of the HEA systems is conducted using the 'LAMMPS' package [40]. Structural visualization of the atomistic HEA systems is performed on the open-source 'OVITO' software [41]. An embedded-atom-method (EAM) potential function for the AlCoCrFeNi HEA systems is considered for the

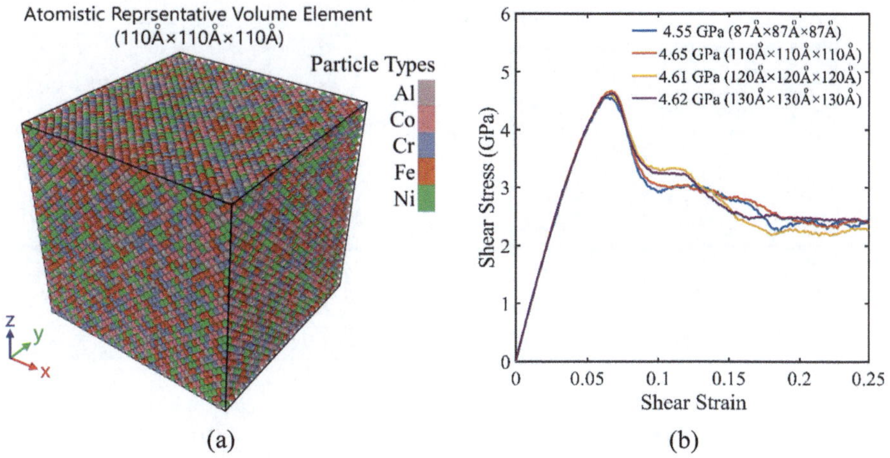

Fig. 1 **a** Atomistic representative volume element (RVE) of the AlCoCrFeNi HEAs. The inset color legend represents the different alloying element types of the AlCoCrFeNi HEAs; **b** shear stress and shear strain responses of AlCoCrFeNi HEAs with different sizes of atomistic RVE

present study [42]. The mechanical deformation is performed by applying the shear-loading on the atomistic representative volume elements (RVE) of size 110A° × 110A° × 110A° (refer to Fig. 1a). The size of the atomistic volume elements for the present study is settled after performing the size convergence study of the atomistic RVE, as shown in Fig. 1b. Figure 1b illustrates the RVE size-dependent shear stress and shear strain responses of AlCoCrFeNi HEAs. It is evident that with the variation in the size of atomistic RVE of the HEAs, no significant change is observed in the shear stress-shear strain response. Hence, in further investigation, the size configuration of atomistic RVE is considered as 110A° × 110A° × 110A°.

3 Shear Deformations

After the energy minimization and equilibration of the atomistic RVE model of the HEAs, the shear loading is applied by varying the tilt factor in the *xy* plane, as shown in Fig. 2. Shear strain is calculated as the ratio of the displacement along the *x*-axis (δ_x) and the length of the simulation box perpendicular to the direction of the shear loading (L_y). . The shear stress is calculated from the virial stress response of the system. Different configurations of the HEA systems are designed by considering (a) HEA-A ('zero' atomic fraction (at.%) of one element, while keeping the equal atomic fraction (at.%) for the remaining four elements); (b) HEA-B (different atomic fraction (at.%) of Al).

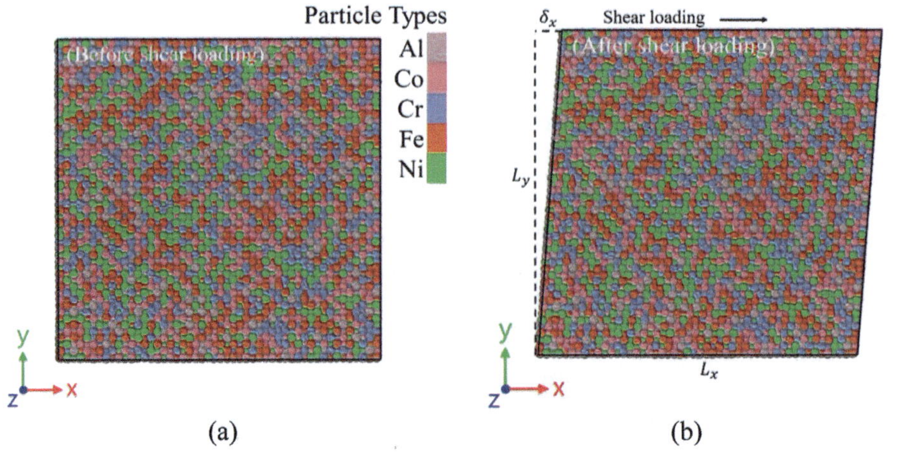

Fig. 2 Atomistic RVE of AlCoCrFeNi HEAs under shear loading **a** before shear loading; **b** after shear loading

4 Result and Discussion

Figure 3a shows the influence of the alloying elements on the shear stress and shear strain responses of HEA-A systems. Here, we have considered 0 at.% of an individual element while keeping the other elements at an equal atomic fraction. It is evident from the figure that 0 at.% of Al concentration provides the highest shear stress, i.e., 10.958 GPa at a strain of 0.107. The shear stress reduced to 5.552 GPa at a strain of 0.077, with 0 at.% concentration of Co. The 0 at.% of Cr, Fe, and Ni shows the inadequate shear strength of the AlCoCrFeNi HEAs and starts to yield at the low strain of 0.059, 0.049, and 0.056 (represented by the yellow, violate, and green color, respectively, in Fig. 3a). The low concentration of the Al provides the high yield stress, indicating that yielding starts at a high strain of 0.107. However, it is noteworthy to mention that the absence of Al results in a relatively high density (8.3360 g/cm^3) of the HEA-A system and increases the HEA system's weight. The density of each HEA-A system and corresponding shear stresses are shown in Fig. 3b. The HEA-A system with 0 at.% of Co provides adequate shear strength (5.552 GPa) with reasonable low-density (6.427 g/cm^3) among other HEA-A systems.

The microstructural evolution of the HAEs under shear deformation is investigated with the help of common neighbor analysis (CNA) [43] and visualized in OVITO software. Snapshots of the microstructural evolution of HEA-A systems at a yield strain of 0.107 are shown in Fig. 4a–e. Figure 4a depicts that at the verge of yielding, most of the structure remains in the face-centered cubic (FCC) phase with 0 at.% of the Al. The face-centered cubic (FCC) transformed into hexagonal close-packed (HCP) and body-centered cubic (BCC) phase for the HEA-A system with 0 at.% of Co, indicating that the system begins to deform before the strain of 0.107 (refer to Fig. 4b).

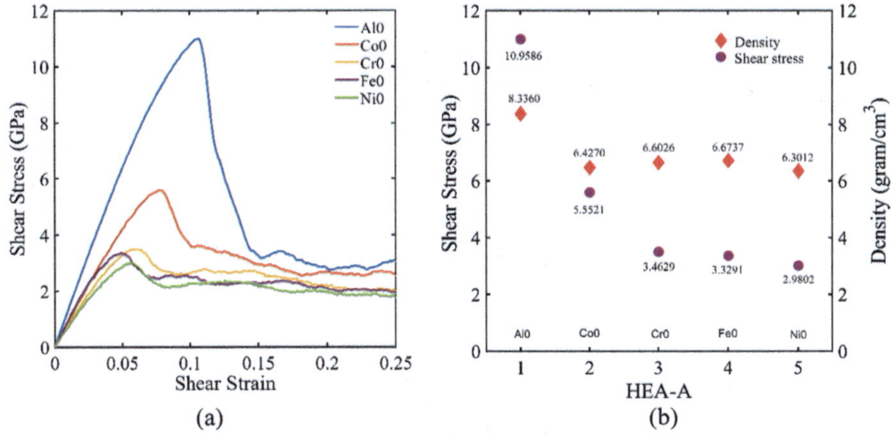

Fig. 3 **a** Shear stress and shear strain responses of HEA-A systems under shear loading; **b** shear stress and density responses of HEA-A systems. *x*-axis represents the HEA-A systems with 0 at.% of individual alloying elements in AlCoCrFeNi HEAs

The same can be observed in Fig. 4c–e, which shows the microstructure deformation evolution of the HEA-A system with 0 at.% of Cr, Fe, and Ni, respectively.

Figure 4a reveals that the absence of the Al in the AlCoCrFeNi HEAs provides a stable microstructure compared to other HEA-A systems (0 at.% of Co, Cr, Fe, and Ni), even at a high strain of 0.107. However, among all alloying elements, the low concentration of Al shows a significant increase in the density of AlCoCrFeNi HEAs.

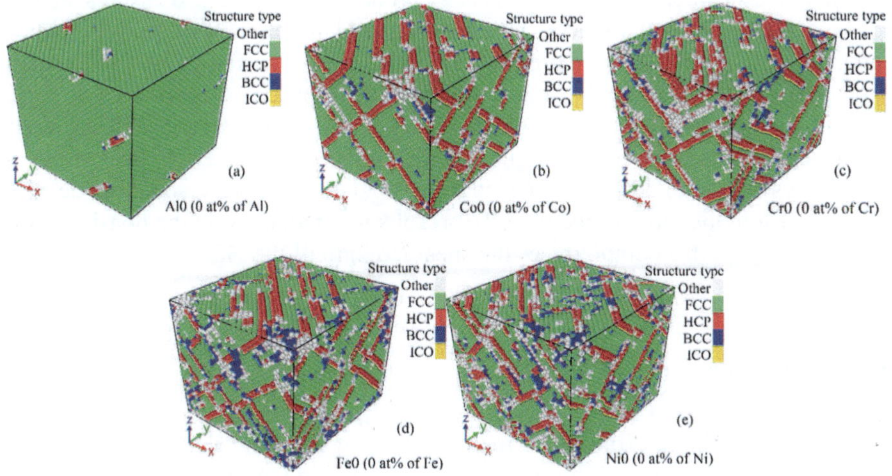

Fig. 4 Microstructure of the HEA-A systems with 0 at.% of individual alloying elements (Al, Co, Cr, Fe, and Ni) at a strain of 0.107. **a** 0 at.% of Al; **b** 0 at.% of Co; **c** 0 at.% of Cr; **d** 0 at.% of Fe; **e** 0 at.% of Ni; The inset color legend shows the structure types

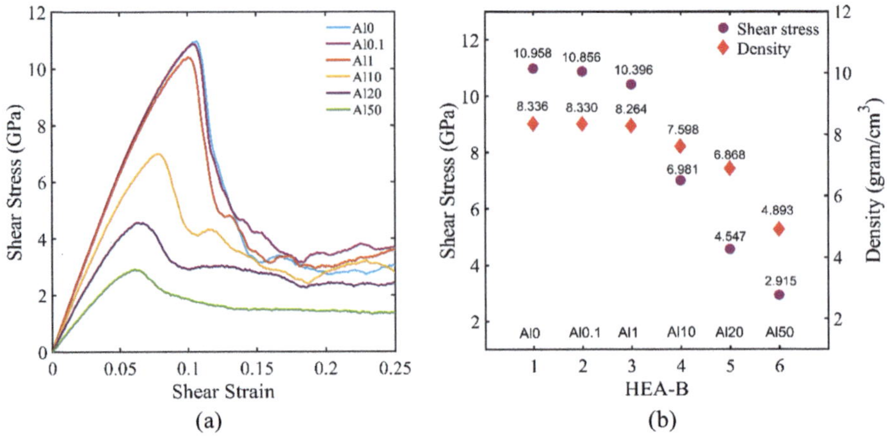

Fig. 5 **a** Shear stress and shear strain responses of HEA-B systems with different atomic fractions of Al under shear loading; **b** shear stress and density responses of HEA-B system. *x*-axis represents the HEA-B systems with different atomic fractions of Al

So, in view of that, further study is carried out to show the influence of different Al concentrations (at.%) on the shear strength and density of the AlCoCrFeNi HEAs. It is clear from Fig. 5a that, with the decrease in at.% of Al on the AlCoCrFeNi HEAs, the yield stress increases, indicating that the deformation of the HEAs under shear loading decreases. This is because as the aluminum (Al) concentration reduces, ductility also decreases, contributing to the higher shear strength of the system. At the low concentration of the Al, the phase transformation from face-centered cubic (FCC) to body-cantered cubic (BCC) is less prone to occur and also ensures the low active shear bands of the HEAs. This makes HEAs more resistant to deformation and increases their mechanical properties [25, 44]. The same can be evident from the explanation of Fig. 4a. As the atomic fraction of Al increases, the degree of drops in yield stress reduces. These observations are found in agreement with the study conducted by Jiang et al. [25]. Figure 5b shows the magnitudes of shear stress of HEA-B systems with different concentration of Al and their corresponding density. The increase in the atomic fraction of Al results in a reduction in the density of the HEAs. However, this compromises the shear strength of the HEAs.

5 Conclusions

This work presents the molecular dynamics study of mechanical deformation of the AlCoCrFeNi HEAs with different atomic fractions (at.%) of Al, Co, Cr, Fe, and Ni under the shear loading. The low concentration of Al in AlCoCrFeNi HEAs significantly improves the strength against the deformation under shear loading and impedes the phase transformation at high yield stress. However, low concentration Al also leads to the undesired increment in the density of HEAs. Among all AlCoCrFeNi HEAs studied here, HEA with 0 at.% of Co provides adequate shear strength (5.552 GPa) with reasonably low density (6.427 g/cm^3).

Acknowledgements Subrata Barman would like to acknowledge the financial support from the Ministry of Education (MoE), Govt. of India, during this research work.

References

1. Youssef KM, Zaddach AJ, Niu C, Irving DL, Koch CC (2015) A novel low-density, high-hardness, high-entropy alloy with close-packed single-phase nanocrystalline structures. Mater Res Lett 3(2):95–99
2. Liu S, Gao MC, Liaw PK, Zhang Y (2015) Microstructures and mechanical properties of AlxCrFeNiTi0. 25 alloys. J Alloy Compd 619:610–615
3. Stepanov ND, Shaysultanov DG, Salishchev GA, Tikhonovsky MA (2015) Structure and mechanical properties of a light-weight AlNbTiV high entropy alloy. Mater Lett 142:153–155
4. Stepanov ND, Yurchenko NY, Skibin DV, Tikhonovsky MA, Salishchev GA (2015) Structure and mechanical properties of the AlCrxNbTiV (x= 0, 0.5, 1, 1.5) high entropy alloys. J Alloy Compd 652:266–280
5. Senkov ON, Woodward C, Miracle DB (2014) Microstructure and properties of aluminum-containing refractory high-entropy alloys. Jom 66(10):2030–2042
6. Zou Y, Maiti S, Steurer W, Spolenak R (2014) Size-dependent plasticity in an Nb25Mo25Ta25W25 refractory high-entropy alloy. Acta Mater 65:85–97
7. Wang Y, Yang Y, Yang H, Zhang M, Ma S, Qiao J (2018) Microstructure and wear properties of nitrided AlCoCrFeNi high-entropy alloy. Mater Chem Phys 210:233–239
8. Yao MJ, Pradeep KG, Tasan CC, Raabe D (2014) A novel, single phase, non-equiatomic FeMnNiCoCr high-entropy alloy with exceptional phase stability and tensile ductility. Scr Mater 72(73):5–8
9. Kumar A, Gupta M (2016) An insight into evolution of light weight high entropy alloys: a review. Metals 6(9):199
10. Conway PL, Klaver TPC, Steggo J, Ghassemali E (2022) High entropy alloys towards industrial applications: high-throughput screening and experimental investigation. Mater Sci Eng A 830:142297
11. Tokarewicz M, Grądzka-Dahlke M (2021) Review of recent research on AlCoCrFeNi high-entropy alloy. Metals 11(8):1302
12. Chen KT, Wei TJ, Li GC, Chen MY, Chen YS, Chang SW, Yen HW, Chen CS (2021) Mechanical properties and deformation mechanisms in CoCrFeMnNi high entropy alloys: a molecular dynamics study. Mater Chem Phys 271:124912
13. Abd El-Hameed AM, Abdel-Aziz YA (2021) Aluminium alloys in space applications: a short report. J Adv Res Appl Sci Eng Technol 22(1):1–7
14. Feng R, Gao MC, Lee C, Mathes M, Zuo T, Chen S, Hawk JA, Zhang Y, Liaw PK (2016) Design of light-weight high-entropy alloys. Entropy 18(9):333
15. Menou E, Tancret F, Toda-Caraballo I, Ramstein G, Castany P, Bertrand E, Gautier N, Díaz-Del PEJR (2018) Computational design of light and strong high entropy alloys (HEA): obtainment of an extremely high specific solid solution hardening. Scripta Mater 156:120–123
16. Tung CC, Yeh JW, Shun T, Chen S-K, Huang Y-S, Chen H-C (2007) On the elemental effect of AlCoCrCuFeNi high-entropy alloy system. Mater Lett 61(1):1–5
17. Ji C, Ma A, Jiang J (2022) Mechanical properties and corrosion behavior of novel Al–Mg–Zn–Cu–Si lightweight high entropy alloys. J Alloy Compd 900:163508

18. Kao YF, Chen TJ, Chen SK, Yeh JW (2009) Microstructure and mechanical property of as-cast,-homogenized, and-deformed AlxCoCrFeNi ($0 \leq x \leq 2$) high-entropy alloys. J Alloy Compd 488(1):57–64
19. Li C, Li JC, Zhao M, Jiang Q (2010) Effect of aluminum contents on microstructure and properties of AlxCoCrFeNi alloys. J Alloy Compd 504(1):5
20. Wang WR, Wang WL, Yeh JW (2014) Phases, microstructure and mechanical properties of AlxCoCrFeNi high-entropy alloys at elevated temperatures. J Alloy Compd 589:143–152
21. Sharma A, Balasubramanian G (2017) Dislocation dynamics in Al0.1CoCrFeNi high-entropy alloy under tensile loading. Intermetallics 91:31–34
22. Shun TT, Du YC (2009) Microstructure and tensile behaviors of FCC Al0.3CoCrFeNi high entropy alloy. J Alloys Compounds 479(1–2):157–160
23. Lin CM, Tsai HL (2011) Evolution of microstructure, hardness, and corrosion properties of high-entropy Al0.5CoCrFeNi alloy. Intermetallics 19(3):288–294
24. Yang T, Xia S, Liu S, Wang C, Liu S, Zhang Y, Xue J, Yan S, Wang Y (2015) Effects of Al addition on microstructure and mechanical properties of AlxCoCrFeNi high-entropy alloy. Mater Sci Eng A 648:15–22
25. Jiang J, Chen P, Qiu J, Sun W, Saikov I, Shcherbakov V, Alymov M (2021) Microstructural evolution and mechanical properties of AlxCoCrFeNi high-entropy alloys under uniaxial tension: a molecular dynamics simulations study. Mater Today Commun 28:102525
26. Huang H, Van Swygenhoven H (2009) Atomistic simulations of mechanics of nanostructures. MRS Bull 34(3):160–166
27. Gupta KK, Mukhopadhyay T, Roy L, Dey S (2022) Hybrid machine-learning-assisted quantification of the compound internal and external uncertainties of graphene: towards inclusive analysis and design. Mater Adv 3(2):1160–1181
28. Gupta KK, Mukhopadhyay T, Roy L, Dey S (2022) High-velocity ballistics of twisted bilayer graphene under stochastic disorder. Adv Nano Res 12(5):529–547
29. Gupta KK, Dey S (2019) Effect of temperature on the fracture strength of perfect and defective monolayered graphene. In: Advances in computational methods in manufacturing. Springer, Singapore, pp 793–804
30. Gupta KK, Roy A, Dey S (2020) Comparative study of various defects in monolayer graphene using molecular dynamics simulation. In: Advances in applied mechanical engineering. Springer, Singapore, pp 539–546
31. Gupta KK, Roy L, Dey S (2022) Hybrid machine-learning-assisted stochastic nano-indentation behaviour of twisted bilayer graphene. J Phys Chem Solids 167:110711
32. Roy A, Gupta KK, Dey S (2022) Probabilistic investigation of temperature-dependent vibrational behavior of hetero-nanotubes. Appl Nanosci 12(7):2077–2089
33. Gupta KK, Roy A, Mukhopadhyay T, Roy L, Dey S (2022) Probing the stochastic fracture behavior of twisted bilayer graphene: efficient ANN based molecular dynamics simulations for complete probabilistic characterization. Mater Today Commun 32:103932
34. Gupta KK, Mukhopadhyay T, Dey S (2023) Probing the molecular-level energy absorption mechanism and strategic sequencing of graphene/Al composite laminates under high-velocity ballistic impact of nano-projectiles. Appl Surf Sci 629:156502
35. Gupta KK, Mukhopadhyay T, Roy A, Roy L, Dey S (2021) Sparse machine learning assisted deep computational insights on the mechanical properties of graphene with intrinsic defects and doping. J Phys Chem Solids 155:110111
36. Gupta KK, Mukhopadhyay T, Roy A, Dey S (2020) Probing the compound effect of spatially varying intrinsic defects and doping on mechanical properties of hybrid graphene monolayers. J Mater Sci Technol 50:44–58
37. Gupta KK, Roy L, Dey S (2022) Machine learning-based molecular dynamics simulations of monolayered graphene. In: Recent advances in computational and experimental mechanics, vol. II: select proceedings of ICRACEM 2020. Springer Nature, Singapore, pp 251–263
38. Gupta KK, Roy L, Dey S (2022) Ballistic performance of Bi-layer graphene: artificial neural network based molecular dynamics simulations. In: Machine learning applied to composite materials. Springer Nature, Singapore, pp 135–148

39. Roy A, Gupta KK, Naskar S, Mukhopadhyay T, Dey S (2021) Compound influence of topological defects and heteroatomic inclusions on the mechanical properties of SWCNTs. Mater Today Commun 26:102021
40. Thompson AP, Aktulga HM, Berger R, Bolintineanu DS, Brown WM, Crozier PS, in 't Veld PJ, Kohlmeyer A, Moore SG, Nguyen TD, Shan R (2022) LAMMPS-a flexible simulation tool for particle-based materials modeling at the atomic, meso, and continuum scales. Comput Phys Commun 271:108171
41. Stukowski A (2009) Visualization and analysis of atomistic simulation data with OVITO–the open visualization tool. Modell Simul Mater Sci Eng 18(1):015012
42. Farkas D, Caro A (2020) Model interatomic potentials for Fe–Ni–Cr–Co–Al high-entropy alloys. J Mater Res 35(22):3031–3040
43. Honeycutt JD, Andersen HC (1987) Molecular dynamics study of melting and freezing of small Lennard-Jones clusters. J Phys Chem 91(19):4950–4963
44. Chen S, Xie X, Chen B, Qiao J, Zhang Y, Ren Y, Dahmen KA, Liaw PK (2015) Effects of temperature on serrated flows of Al0. 5CoCrCuFeNi high-entropy alloy. JOM 67(10):2314–2320

Importance of Separation Gaps for Asymmetric Tall Buildings Under Earthquakes

Tiyas Maity and Manish Bharti

Abstract The poor performance of structures under strong ground motion can be attributed to structural asymmetry. Asymmetry strongly contributes to translational-torsional coupling in the seismic response, which can result in higher lateral deflections, increased member forces, and eventually collapse. During large earthquakes, structures with asymmetry cause structural damages or the collapse of the entire structure. This paper aims to investigate the seismic response of horizontal asymmetric buildings with and without separation gaps. Three asymmetric building plans ("*T*" shape, "+" shape, and "*L*" shape) have been chosen in order to investigate the beneficial effects of separation gaps. Comparative study has been carried out in terms of developed horizontal and vertical forces; moments and reinforcement required in columns. It is observed from the analysis results that the amount of reinforcement is quite less in case the separation gaps at strategic locations are introduced.

Keywords Asymmetric building · Earthquake analysis · Separation gap · Seismic resistant design

1 Introduction and Objective

The frequent earthquakes around the globe have brought our country to the urgent need for a strategic program to ensure aseismic design of all new buildings. The recent disastrous earthquakes and the resulting horrific loss of life and property demonstrate that there is an urgent need to reassess the status of multistory building design in our country, particularly in seismic-prone locations. One of the positive outcomes of the recent earthquakes is a growing public and government awareness of the need for seismically resistant building design.

T. Maity (✉)
Department of Architecture, Jadavpur University, Kolkata, India
e-mail: tiyasmaity@gmail.com

M. Bharti
Cortex Construction Solutions Pvt. Ltd., New Delhi, India

R. K. Ray et al. (eds.), *Advances in Theoretical and Applied Mechanics*, Lecture Notes in Mechanical Engineering, https://doi.org/10.1007/978-981-97-0418-7_28

Asymmetric structures are being used increasingly nowadays because of architectural and functional demands. In addition, the architects have to make irregular plan of the buildings because of irregular shape of the available land and the pressure from the developers for the maximum utilization of the available space. A few common asymmetric shapes of buildings in plan are *Y*, *C*, *L*, *U*, *E*, *F*, *T* etc. The discontinuities may exist because of the presence of large openings, cutouts, re-entrant corners, etc. During earthquakes, structures that possess an uneven distribution of stiffness and strength in their floor plan experience interconnected lateral and torsional movements [5]. Such abrupt changes result in torsion and stress concentration [4]. Damage to the structure typically begins at the location of structurally weak planes in the building system. The lack of structural symmetry can significantly contribute to the inadequate performance of buildings under intense seismic forces [6]. The vulnerability generated further damage during earthquakes [7], resulting in structural collapse. However, IS 456:2000 [2] mentions the introduction of expansion joints at places where there is an abrupt change in plan. However, architects and clients generally do not prefer such expansion joints in buildings because of the presence of susceptible points for leakage.

Earthquake forces are generated by the dynamic response of the building due to seismic-induced ground motion. This makes earthquake actions fundamentally different from any other imposed loads. Through the careful selection of appropriate, well-distributed lateral load resisting systems, and by ensuring the building is reasonably regular in both plan and elevation, the influence of many second-order effects, such as torsional effects, can be minimized and significant simplifications can be made to model the dynamic building response. An efficient seismic design of buildings [8] is one where the designer takes control of the buildings by dictating how the building is to respond. This can be attained by choosing of the desired response mode and suppressing the increase of detrimental response modes that could lead to building collapse. The random, three-dimensional motion generated by earthquakes may introduce twisting and torsional effects into the response, either directly as a function of the input ground motion, or because of variations in the spatial distribution of the structure being irregular in plan.

Thus, it is clear that the separation gaps are crucial for enhancing the structural resilience and safety of such buildings in earthquake-prone areas. Seismic events generate powerful forces that can lead to destructive vibrations and ground shaking, posing significant risks to buildings. However, incorporating separation gaps in the design of asymmetric tall buildings allows them to effectively absorb and dissipate seismic energy. In summary, separation gaps in asymmetric tall buildings under earthquakes are essential for enhancing structural resilience, safeguarding occupants, and ensuring long-term durability.

The objective of the present study is to investigate the behavioral response of tall buildings having plan in irregularity under seismic forces. The three-dimensional analysis has been carried out by the seismic coefficient method as prescribed in IS1893 (Part 1):2016 [1] using commercial software STAAD.Pro. The percentage of reductions in moments, shear forces, axial forces and steel reinforcements of the various building components will be compared with the addition of separation gaps at some strategic locations. The parametric study shows that it is possible to relieve the additional stresses in the asymmetric buildings by inserting appropriate separation gaps.

2 Building Modeling, and Analysis

2.1 *Finite Element Formulation*

The building frame, comprising beams and columns, is designed as a three-dimensional structure. In this model, both the beams and columns are represented as frame elements. The stiffness matrix, denoted as $[K]$, is employed to describe the behavior of a two-dimensional element, and it can be expressed as follows:

$$[K] = t \iint_A [B]^T [D][B] \mathrm{d}x\mathrm{d}y. \tag{1}$$

Here, t is the thickness of element, $[D]$ is the constitutive matrix and $[B]$ is the strain displacement relationship matrix which may be expressed as:

$$[B] = \begin{bmatrix} \frac{\partial N}{\partial x} & 0 \\ 0 & \frac{\partial N}{\partial y} \\ \frac{\partial N}{\partial y} & \frac{\partial N}{\partial x} \end{bmatrix}. \tag{2}$$

The structure undergoes analysis utilizing two-dimensionaframe elements, where each node is equipped with three degrees of freedom. The associated stiffness matrix for the element can be expressed as follows:

$$[K] = \begin{bmatrix} H_1 & 0 & 0 & -H_1 & 0 & 0 \\ 0 & 12H_2 & 6H_2L & 0 & -12H_2 & 6H_2L \\ 0 & 6H_2L & 4H_2L^2 & 0 & -6H_2L & 2H_2L^2 \\ -H_1 & 0 & 0 & H_1 & 0 & 0 \\ 0 & -12H_2 & -6H_2L & 0 & 12H_2 & -6H_2L \\ 0 & 6H_2L & 2H_2L^2 & 0 & -6H_2L & 4H_2L^2 \end{bmatrix} \tag{3}$$

Here, $H_1 = AE/L$, $H_2 = EI_z/L^3$, I_z represents the second moment of area about the z-axis, L denotes the length of the element, A signifies the cross-sectional area, and E represents the modulus of elasticity.

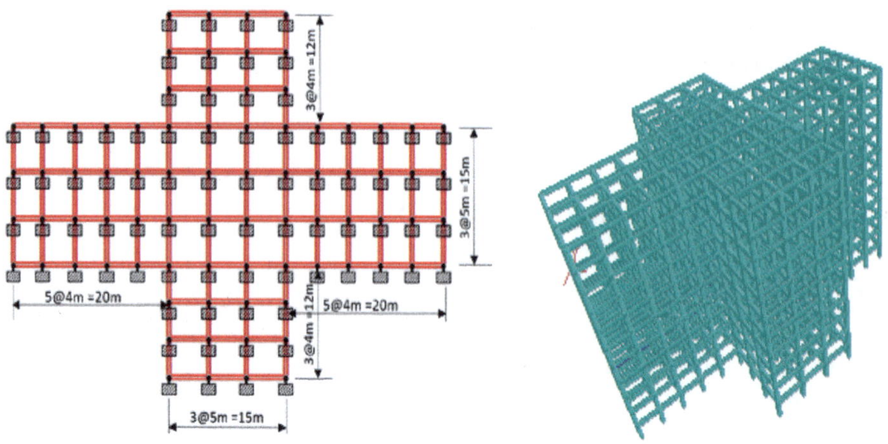

(a) Plan (b) 3D rendered view

Fig. 1 Plus (+) shaped $G + 10$ storied building model

2.2 *Building Configuration*

Three asymmetric building plans ("+" shape, "*T*" shape, and "*L*" shape) have been chosen in order to investigate the beneficial effects of separation gaps at strategic locations. The buildings are of $G + 10$ storied reinforced concrete bare framed and are assumed to be situated at seismic Zone-5. The building supports have been considered as fixed. The story height of all the three buildings is of assumed 3.0 m and the height of ground floor is assumed as 2.5 m. A three-dimensional model of the building is generated in the software STAAD.Pro. Centerline dimensions are followed for the numerical modeling of the buildings. The plan layout of three different shaped buildings, their dimensions, and 3D rendered view are indicated in Figs. 1, 2 and 3. Column to column distance and their positions are also shown in these figures, i.e., Figs. 1, 2 and 3.

2.3 *Loads Considered in Analysis and Design*

2.3.1 Dead Load and Live Load

Dead loads on the structure include all permanent loads attached to the structure, such as the weight of beams, columns, walls, partitions, floors, roofs, stairways, etc. The unit weights of various building materials are taken as per IS-875 (Part-I) [3]. External masonry walls of 250 mm thick with 12 mm plaster on the internal side and 20 mm thick plaster on the external side are considered. Internal walls are considered to be 125 mm thick with 12 mm plaster on both sides. The masonry walls are assumed

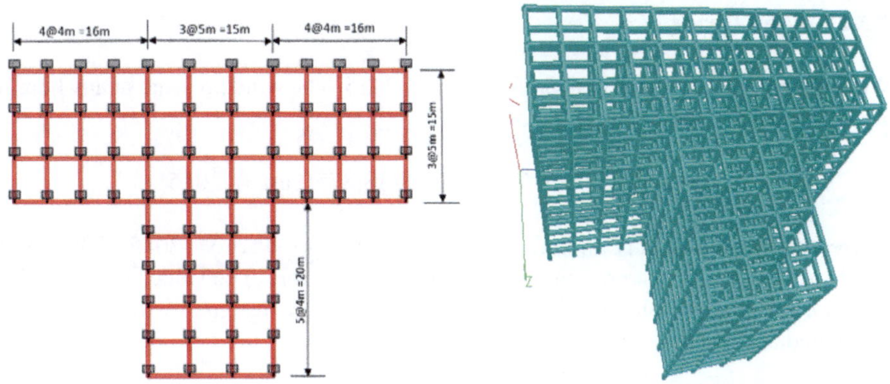

(a) Plan (b) 3D rendered view

Fig. 2 Tee (T) shaped $G + 10$ storied building model

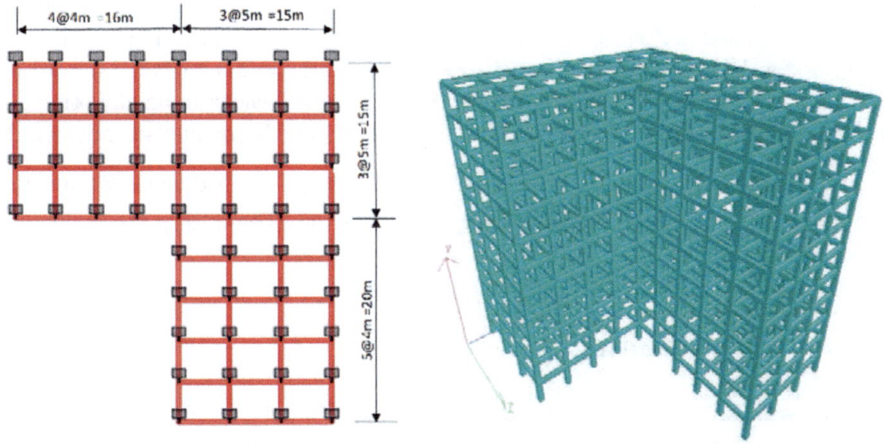

(a) Plan (b) 3D rendered view

Fig. 3 L shaped $G + 10$ storied building model

to be non-structural members. In order to account for the opening due to doors and windows, the dead load due to the walls is reduced by 10%. Floor slabs and roofs are considered to be 125 mm thick. Live loads on the floors and roof comprise all loads other than the dead loads. The live loads on different occupancies are considered as per IS: 875 (Part 2).

2.3.2 Wind Loads

Wind load is considered as per IS-875 (Part-3)-2015. The wind pressure at any height above mean ground level is calculated as,

$$p_z = 0.6V_z^2 \quad \text{[Ref. Clause 7.2 of IS:875(Part 3): 2015]},$$

Here, $V_Z = k_1k_2k_3k_4V_b$ [Ref. *Clause* 5.3 of IS:875 (Part 3): 2015]. The basic wind speed, V_b is considered in the present case as 50 m/s. The values of k_1, k_2, k_3, and k_4 are assumed following IS 875 (Part 3): 2015. The designed wind pressure is calculated and applied on the structural model.

2.3.3 Earthquake Loads

Earthquake loads on the structure are determined as per IS 1893 (Part I): 2016. Seismic loads are considered in the horizontal direction only. The building is considered to be situated in Zone V. The design lateral forces are considered in two orthogonal horizontal directions of the structures and the structure is analyzed according to *Clause* 6.3.2 of IS 1893 (Part 1): 2016. The design base shear along any principal direction of the building is computed as $V_B = \mathrm{A_h}W$. The horizontal seismic coefficient (A_h) is calculated as, $A_h = \frac{\left(\frac{Z}{2}\right)\left(\frac{Sa}{g}\right)}{\left(\frac{R}{I}\right)}$.. The building is considered to be situated in medium soil. The Zone factor (Z), Importance factor (I), and Response reduction factor (R) are considered as 0.36, 1.2, and 5, respectively.

2.3.4 Load Combinations

The following abbreviations are used in specifying different load combinations along the two principal axes X and Z. The following load combinations are considered for the analysis of the superstructure.

1. 1.5 × (DL + LL)	10. 1.2 × (DL + LL + SLX)
2. 1.5 × (DL + SLX)	11. 1.2 × (DL + LL − SLX)
3. 1.5 × (DL − SLX)	12. 1.2 × (DL + LL + SLZ)
4. 1.5 × (DL + SLZ)	13. 1.2 × (DL + LL − SLZ)
5. 1.5 × (DL − SLZ)	14. 1.2 × (DL + LL + WLX)
6. 1.5 × (DL + WLX)	15. 1.2 × (DL + LL − WLX)
7. 1.5 × (DL − WLX)	16. 1.2 × (DL + LL + WLZ)
8. 1.5 × (DL + WLZ)	17. 1.2 × (DL + LL − WLZ)
9. 1.5 × (DL − WLZ)	

Here, DL = dead load, LL = live load, SLX = seismic load in *X* direction, SLZ = seismic load in *Z* direction, WLX = wind load in *X* direction, WLZ = wind load in *Z* direction.

2.4 Introduction of Separation Gaps

The beam-column layouts of the three different asymmetric building structures are shown in Figs. 1, 2 and 3. To make those plans symmetric, a separation gap of 200 mm has been introduced at suitable places so that plans of those buildings become symmetric. The line of separations has been marked in the figures in red color (Fig. 4).

3 Results and Discussions

An effort has been made to estimate the forces and moments of the above-mentioned three RC buildings under gravity as well as wind and earthquake forces. The results presented in this paper are based on linear analyses, performed on asymmetrical bare framed buildings. Finite element analyses of these building frames are performed in software STAAD.Pro.

3.1 Comparison of Forces and Moments

Maximum moments, horizontal, and vertical forces under various loading combinations are calculated for all the three structures with and without separation gaps and are compared in graphical and tabular forms to understand the beneficial effects of introduction of separation gaps. The node numbers for a particular plan are referred in Fig. 4 and a few typical supports have been considered for comparison where the differences may occur. From Fig. 5, it is observed that the major differences in horizontal forces are in the inner columns where chances of torsion are developed in the intact structure. Forces in vertical direction are greater for intact building and less for buildings with separation gap for a few columns as can be seen from Fig. 6. A few inner columns such as columns associated with node numbers 6 and 20 have relatively larger vertical forces while separation gap is not provided. Similarly, moments on inner columns are reduced significantly (Ref. Fig. 7) while separation gaps are introduced in the asymmetric plan. Similar results for maximum horizontal force, vertical force, and resultant moment are seen for "*T*", and "*L*" shaped plan as seen for "+" shaped plan above.

Fig. 4 Proposed separation gap on different asymmetric buildings

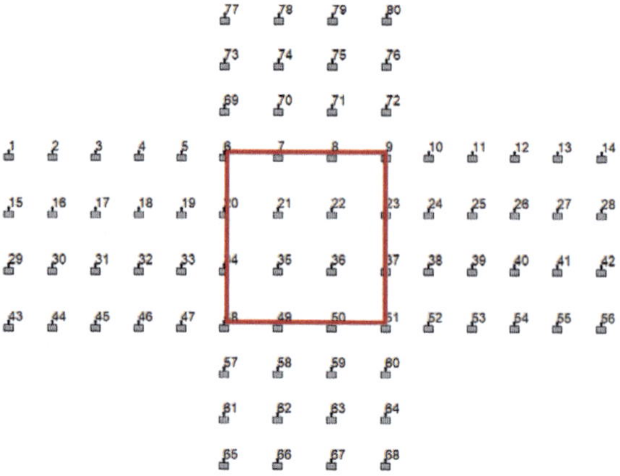

(a) Separation gap on plus shaped building

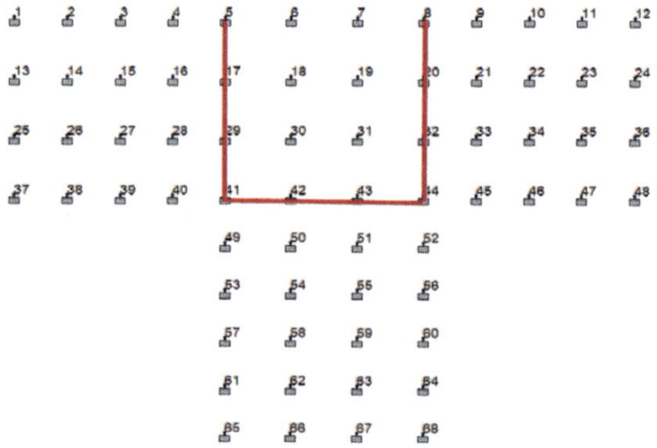

(b) Separation gap on T shaped building

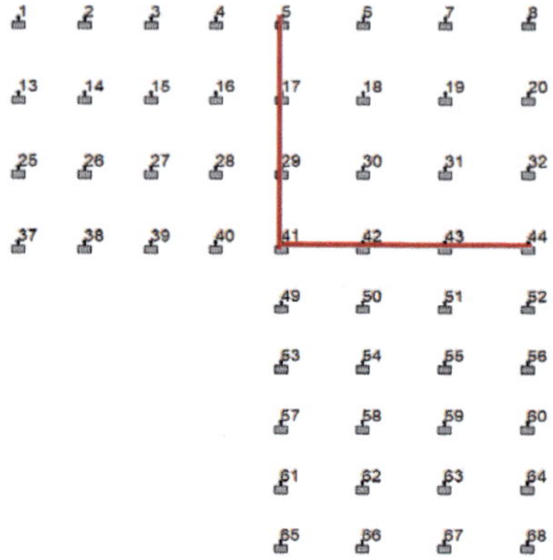

(c) Separation gap on L shaped building

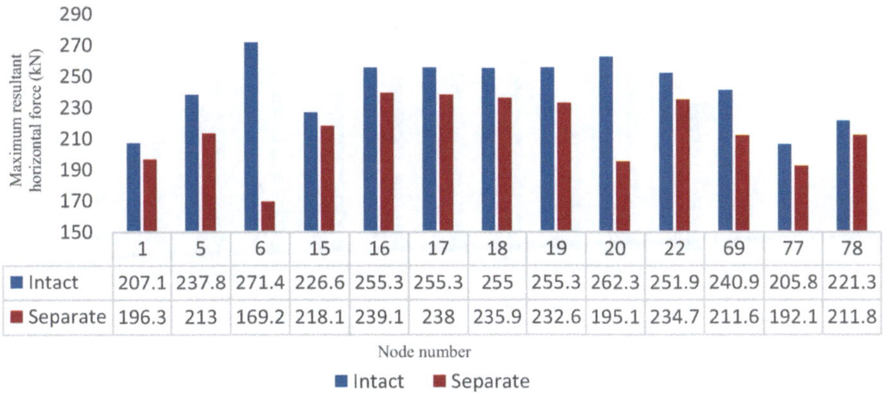

	1	5	6	15	16	17	18	19	20	22	69	77	78
Intact	207.1	237.8	271.4	226.6	255.3	255.3	255	255.3	262.3	251.9	240.9	205.8	221.3
Separate	196.3	213	169.2	218.1	239.1	238	235.9	232.6	195.1	234.7	211.6	192.1	211.8

Fig. 5 Maximum resultant horizontal forces in columns for plus-shaped plan

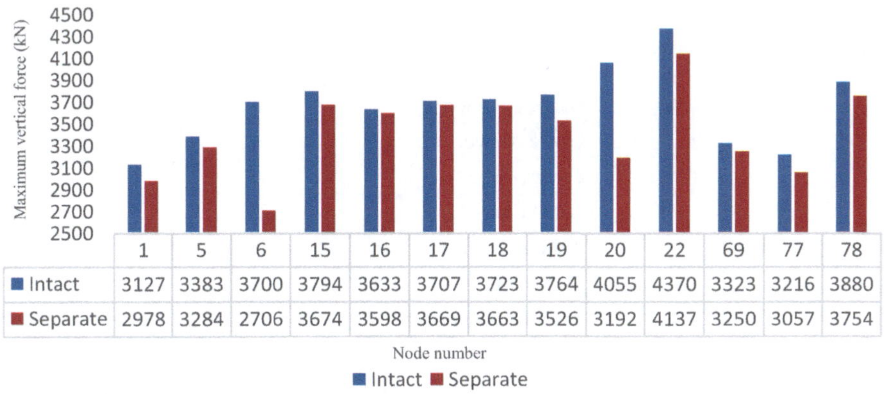

	1	5	6	15	16	17	18	19	20	22	69	77	78
Intact	3127	3383	3700	3794	3633	3707	3723	3764	4055	4370	3323	3216	3880
Separate	2978	3284	2706	3674	3598	3669	3663	3526	3192	4137	3250	3057	3754

Fig. 6 Maximum vertical forces in columns for plus-shaped plan

3.2 Comparison of Steel Reinforcement in Columns

The design of the above-mentioned buildings is carried out based on the design code IS 456:2000. The parameters as shown in Table 1 are considered for the design of these RC buildings. The required reinforcement in various columns as obtained from the STAAD.Pro analysis and design is compared considering with and without separation gaps and is reported in tabular form (Tables 1 and 2) for all the three cases. The results indicate that the reduction in reinforcement can be achieved if appropriate separation gaps are provided making unsymmetrical structures to a symmetrical one. Moreover, it is noticed that relatively large amount of reinforcement is necessary on columns near notch areas of the structure. Similar results have been observed for "*T*" and "*L*" shaped plan.

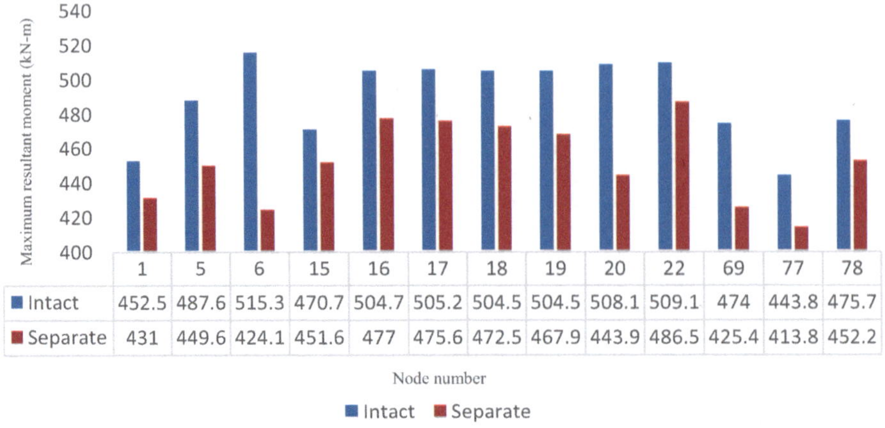

Fig. 7 Maximum resultant moments in columns for plus-shaped plan

Table 1 Design parameters

Parameters	Value
Grade of concrete	M25
Grade of steel	Fe500
Clear covers	
Beam (mm)	25
Column (mm)	40
Slab (mm)	20
Size of column	
Upto $G + 5$	550 mm × 550 mm
From $G + 6$ to $G + 10$	450 mm × 450 mm
Size of all beams	350 mm × 500 mm
Thickness of slab	125 mm

4 Conclusions

An effort has been made to estimate the forces and moments of three reinforced concrete buildings having asymmetric plan under wind and earthquake forces. The comparative study reveals the importance of insertion of separation gaps for irregular shaped structures to make them wind and seismic resistant. The importance of achieving a regular building plan layout, with a well-distributed lateral load-resisting system is highlighted in this paper. It is observed that the amount of reinforcement can be drastically reduced and thus can be made cost-effective if the unsymmetrical building can be made symmetric by simply introducing appropriate separation gaps. Thus, it can be stated that the modern buildings can be designed to be safe under extreme earthquake and wind loadings with collapse being avoided even in case

Table 2 Reduction in reinforcement having separation gaps in "+" shape plan

Column attached to node no	Reinforcement required in mm^2		% of reduction in reinforcement
	Building without separation gap	Building with separation gap	
1	4263	3942	8.1
2	5082	4825	5.3
5	4840	4477	8.1
6	5566	2894	92.3
7	6292	4444	41.6
15	5808	4840	20.0
16	4966	4583	8.4
17	4921	4641	6.0
18	4916	4631	6.2
19	5019	4262	17.8
20	5808	3664	58.5
70	5222	4042	29.2
74	5206	4351	19.7
77	4846	3681	31.6
78	6292	5082	23.8

of asymmetric buildings. Further investigations can be done considering non-linear analysis for the same.

References

1. IS: 1893-2016 (2016) Criteria for earthquake resistant design of structures. Bureau of Indian Standards, New Delhi
2. IS: 456-2000 (2000) Plain and reinforced concrete—code of practice. Bureau of Indian Standards, New Delhi
3. IS 875 (Part I to Part V): 1987 (1987) Design loads (other than earthquake) for buildings and structures—dead loads
4. Khan MJJ, Seshadri ST, Bellam SKP (2017) Seismic performance of symmetrical and asymmetrical buildings with heavy loads. Int J Appl Eng Res 12:12008–12015. ISSN 0973-4562
5. Ladjinovic DZ, Folic RJ (2008) Seismic analysis of asymmetric in plan buildings. In: The 14th world conference on earthquake engineering, 12–17 Oct 2008, Beijing, China
6. Pawar AS, Deshmukh RR, Patil PR, Lad NM, Jadhav VR (2022) Seismic analysis of symmetric and unsymmetric buildings in plan. Int J Res Publ Rev 3(6):4430–4434
7. Priestley MJN (1993) Myths and fallacies in earthquake engineering—conflicts between design and reality. New Zealand Nat Soc Earthq Eng Bull Wellington 26:329–335
8. Raghuprasad BK, Vinay S, Amarnath K (2016) Seismic analysis of buildings symmetric & asymmetric in plan. SSRG Int J Civil Eng 3(5)

Programming Shape Based on Mechanics of Origami

A. Sharma and T. Mukhopadhyay

Abstract Patterns created with Miura rigid origami are finding more and more applications in fields like engineering and technology. However, the techniques to do geometric analysis that is employed in this research study are, in most cases, unplanned. In this paper, in order to acquire the various geometric forms, kinematic analysis and geometric analysis are both utilized. Kinematic analysis is performed using the D-H matrix approach in order to acquire the dihedral angle, and then these angles are incorporated into the geometric modelling in order to generate a wide range of different geometrical tessellations. In addition to this, it is demonstrated how the geometric correlations have the potential to be combined into programmable geometric tessellations.

Keywords Origami · Miura-ori pattern · Kinematic analysis · Rigid foldable geometry

1 Introduction and Objective

Origami, the art of folding paper into a variety of three-dimensional geometries, has created a new discipline of science and technologies that focuses on the fabrication of three-dimensional structures from initially two-dimensional form structures using folding processes. Origami folding patterns are attracting a greater amount of attention due to its mechanics for achieving programming features. Researchers [1–4] have been persistently trying to use origami to create programming shapes. The development of a unique material meta-structure of origami/kirigami enables programmable deformation-dependent stiffness and shape modulation-dependent distant actuation properties [5, 6]. Through proper design changes on the original pattern, one may employ the Miura-ori, a globally planar pattern in its partly folded states, to systematically build either "globally planar" or "globally curved" designs [7]. In order to create new patterns while maintaining the flat-foldability constraint at each node,

A. Sharma (✉) · T. Mukhopadhyay
Department of Aerospace Engineering, Indian Institute of Technology Kanpur, Kanpur, India
e-mail: aditis20@iitk.ac.in

R. K. Ray et al. (eds.), *Advances in Theoretical and Applied Mechanics*, Lecture Notes in Mechanical Engineering, https://doi.org/10.1007/978-981-97-0418-7_29

one decreases the symmetry of the Miura-ori. Researchers get to the conclusion that, by appropriately designing variants on the original pattern, the Miura-ori may be used to methodically construct a range of innovative designs [8]. As a result of the 4-degree vertex origami tessellation, which has the capacity to self-Thus, there is a significant motivation to create distinctive shapes based on the fundamentals of Miura origami in this field. The work depicts the construction of a webbed-foot-inspired gripper created by meticulously combining many basic Miura-ori and waterbomb-ori modules. The suggested webbed gripper's gripping behaviour may be altered by altering the two-dimensional crease arrangement. This may be accomplished by utilizing the form and kinematic automation that is provided for each basic module [9]. The purpose of this paper is to introduce kinematics analysis of linkages in order to achieve various programming forms of the Miura origami design.

In this paper, motion behaviour of Miura origami is analysed by treating it as rigid origami. Rigid origami allows continuous mobility between the folded and unfolded states along the prescribed creases without stretching or bending the facets. The material characteristics of origami must be considered in future research for it to be generally applicable to engineering applications. To effectively depict the considerable deformation, rigid-flexible connection, and other folding features during transformation, the dynamic and kinematic models of origami structures also need to be enhanced [6, 10–12]. The kinematic relationships of Miura origami are then established by treating creases as links and facets as joints. Kinematic relationships obtained using kinematic analysis are then incorporated to do the geometric analysis of Miura origami (Fig. 1).

These geometric relationships are used to obtain programming shapes of Miura origami. In addition to thorough computational evaluations, we have created physical prototypes of a few combinations to show how programmed shape changes work. This research focuses on deployable structures, robotics, and metamaterials, all of which are closely connected to space technology and the advancement of aerospace sectors.

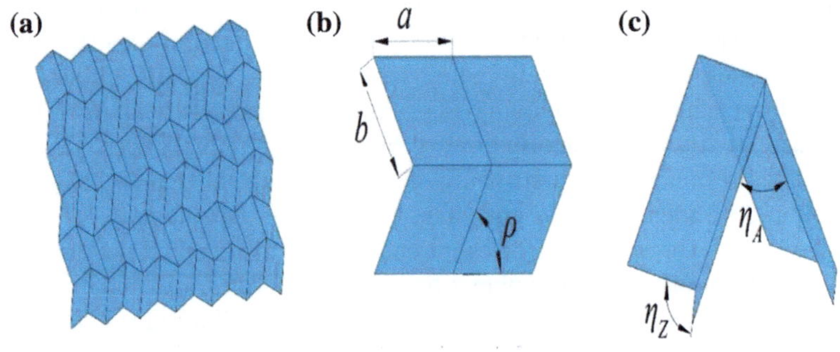

Fig. 1 Miura-ori tessellation and parameterization of pattern. **a** Unfolded and developed Miura-ori tessellation, **b** parameters of unit cell crease pattern of Miura and **c** partially folded geometry of unit cell with configuration variables

2 Mathematical Formulation

In order to do a kinematic analysis of the Miura pattern, the pattern is represented as a spherical linkage, with the facets serving as joints and the creases as links. The kinematic analysis is done by *D-H* matrix method where closing equation is given as,

$$Q_{21} Q_{32} \ldots Q_{1K} = I \tag{1}$$

where, Q is the transformation matrix of spherical linkage.

$$Q_{(i+1)i} = \begin{bmatrix} \cos\theta_i & -\cos\rho_{i(i+1)}\sin\theta_i & \sin\rho_{i(i+1)}\sin\theta_i \\ \sin\theta_i & \cos\rho_{i(i+1)}\cos\theta_i & -\sin\rho_{i(i+1)}\cos\theta_i \\ 0 & \sin\rho_{i(i+1)} & \cos\rho_{i(i+1)(i+1)} \end{bmatrix} \tag{2}$$

And the inverse transformation matrix of spherical linkage,

$$Q_{i(i+1)} = \begin{bmatrix} \cos\theta_i & \sin\theta_i & 0 \\ -\cos\rho_{i(i+1)}\sin\theta_i & \cos\rho_{i(i+1)}\cos\theta_i & \sin\rho_{i(i+1)} \\ \sin\rho_{i(i+1)}\sin\theta_i & -\sin\rho_{i(i+1)}\cos\theta_i & \cos\rho_{i(i+1)} \end{bmatrix} \tag{3}$$

In the current analysis, the Miura-ori pattern is treated as a spherical linkage with a 4-degree vertex, as shown in Fig. 2. Miura unit pattern linkage has three Mountain creases (in blue) and one valley crease.

The closing equation for the Miura-ori can be expressed as:

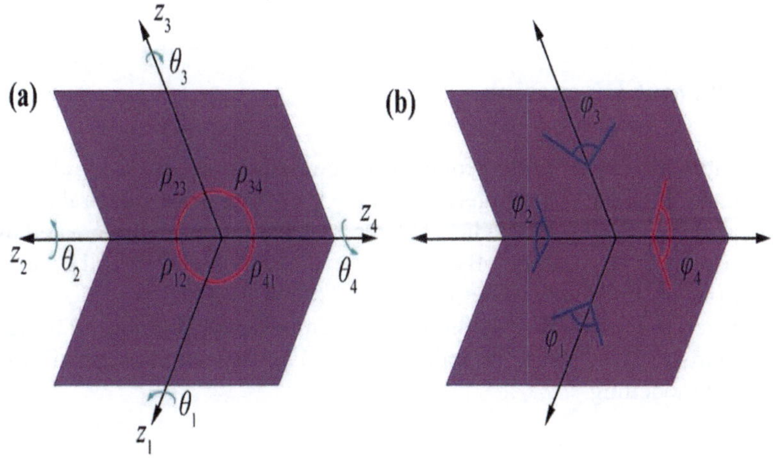

Fig. 2 Miura-ori Pattern Kinematic Analysis. a Kinematic variable (θ) along z axes and ρ as angles at the vertices, **b** unit linkage showing dihedral angle (φ) shown on mountain crease in blue and valley crease in red

$$Q_{21}Q_{32}Q_{43} = Q_{41} \tag{4}$$

By putting Eqs. (2) and (3) in Eq. (4), equating the matrix elements, and putting kinematic variable (θ) in form of dihedral angle (φ) as $\theta = \Pi - \varphi$ for mountain crease and $\theta = \varphi - \Pi$ for valley crease, eventually we get relationship between dihedral angles($\varphi_{1,}$ φ_2) and sector angle (ρ).

Now using these dihedral angles, edge angles η_{A} and η_{z} can be found out as

$$\eta_A = \sin^2 \rho \cos \varphi_1 - \cos^2 \rho \tag{5}$$

$$\eta_Z = \sin^2 \rho \cos \varphi_2 + \cos^2 \rho \tag{6}$$

For a pattern shown in Fig. 1, n denotes the number of straight lines and l denotes the number of zigzag lines. The vertex of intersection is denoted by $U_{p,q}$ where p varies from 1, 2, …, n and q varies from 1, 2, …, l. The 3D coordinates (x, y, z) given as:

$$x = \begin{cases} (p-1)a \sin \frac{\eta_A}{2} & \text{for odd } q \\ (p-1)a \sin \frac{\eta_A}{2} + b \cos \frac{\eta_Z}{2} & \text{for even } q \end{cases} \tag{7}$$

$$y = (q-1)b \sin \frac{\eta_Z}{2} \tag{8}$$

$$z = \begin{cases} 0 & \text{for odd } p \\ a \cos \frac{\eta_A}{2} & \text{for even } p \end{cases} \tag{9}$$

By finding the (x, y, z) coordinates and varying the dihedral angle (φ_2), we can achieve different folding shapes as shown in Fig. 3.

3 Results and Discussion

Programmable variation of rigid foldable shapes can be achieved by altering the dihedral angle. In order to accurately depict the folding process, dihedral angles are typically preferred in origami studies. Dihedral angles are the angles created by two adjacent faces in an origami fold. By varying the dihedral angle, we will get different configurations of Miura assemblies (as referred to in Fig. 3).

As depicted in Fig. 4b–d, programmable assemblies can be achieved by altering one characteristic of the Miura-ori unit pattern and incorporating kinematic analysis in geometric modelling.

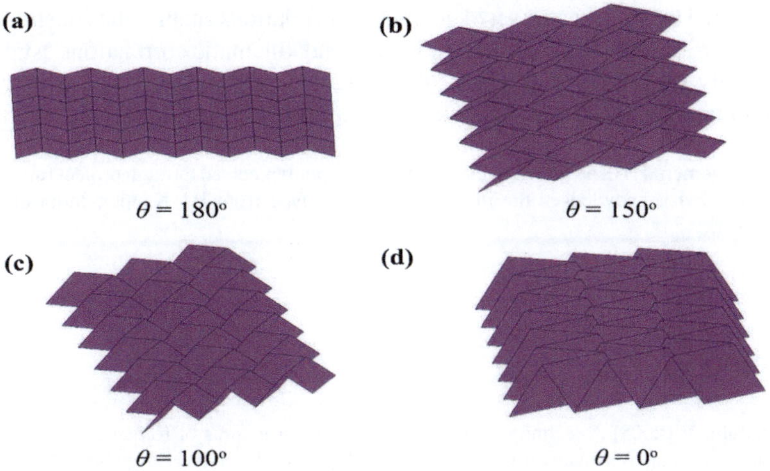

Fig. 3 Programmable varying Miura-ori assemblies. **a** Unfolded flat foldable configuration of at dihedral angle $(\theta) = 180^{o}$, **b** half-foldable configuration at dihedral angle $(\theta) = 150^{o}$, **c** half-foldable configuration at dihedral angle $(\theta) = 100°$, **d** fully foldable configuration at a dihedral angle $(\theta) = 0^{o}$

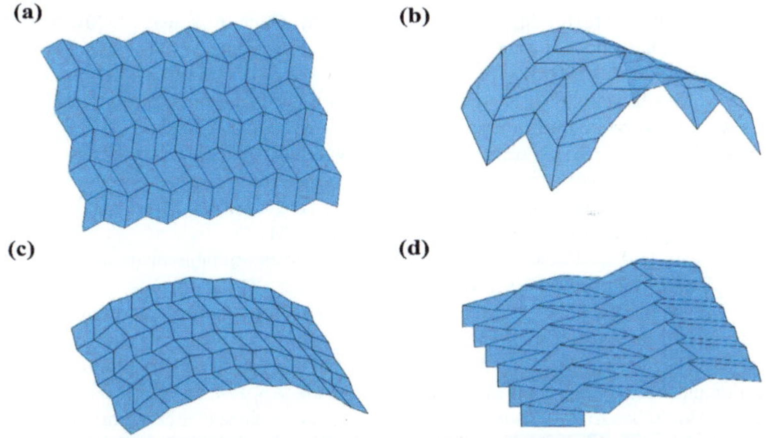

Fig. 4 Programmable shapes based on Miura origami. **a** Miura tessellation, **b** arc pattern tessellation, **c** arc Miura tessellation, **d** non-developable Miura tessellation

4 Conclusion

To conclude, this paper has presented the geometric tessellations of rigidly foldable assemblies generated from the Miura-ori pattern. The geometric coordinates of the Miura-ori base pattern were obtained by geometric and kinematic analysis. The geometric tessellations permit programmable folding action. Nevertheless, it was demonstrated that the created set of geometric correlations enables the construction of new geometrical tessellations. The geometric and kinematic correlations established in this study should serve as a basis for designers and engineers to deploy rigid—Miura origami designs for a variety of applications.

Acknowledgements AS acknowledges the financial support received through project funding from IIT Kanpur. TM acknowledges the initiation grant received from IIT Kanpur, India during the research work.

References

1. Khaliulin V (2005) A technique for synthesizing the structures of folded cores of sandwich panels. Russian Aeronautics C/C of Izvestiia Vysshie Uchebnye Zavedeniia Aviatsionnaia Tekhnika 48(1)
2. Gattas JM, Wu W, You Z (2013) Miura-base rigid origami: parameterizations of first-level derivative and piecewise geometries. J Mech Des 135(11)
3. Chen Y, Yan J, Feng J (2019) Geometric and kinematic analyses, and novel characteristics of origami inspired structures. Symmetry 11:1101. https://doi.org/10.3390/sym11091101
4. Tachi T (2009) Generalization of rigid-foldable quadrilateral-mesh origami. J Int Assoc Shell Spatial Struct 50(3):173–179
5. Mukhopadhyay T, Ma J, Feng H, Hou D, Gattas JM, Chen Y, You Z (2020) Programmable stiffness and shape modulation in origami materials: Emergence of a distant actuation feature. Appl Mater Today 19:100537
6. Sinhaa A, Mukhopadhyay T. Kirigami-inspired modular materials for programming constitutive laws: Mixed-mode multi-directional auxeticity and contact-induced stiffness modulation
7. Sareh P, Guest SD (2015) A framework for the symmetric generalisation of the Miura-ori. Int J Space Struct 30(2):141–152
8. Sareh P, Guest SD (2015) Designing symmetric derivatives of the Miura-ori. In: Advances in architectural geometry 2014. Springer, Cham, pp 233–241
9. Wang C, Guo H, Liu R, Yang H, Deng Z (2021) A programmable origami-inspired webbed gripper. Smart Mater Struct 30(5):055010
10. Dudte LH, Vouga E, Tachi T, Mahadevan L (2016) Programming curvature using origami tessellations. Nat Mater 15(5):583–588
11. Wang H, Zhao D, Jin Y, Wang M, Mukhopadhyay T, You Z (2020) Modulation of multi-directional auxeticity in hybrid origami metamaterials. Appl Mater Today 20:100715
12. Fang H, Chu SCA, Xia Y, Wang KW (2018) Programmable self-locking origami mechanical metamaterials. Adv Mater 30(15):1706311

Nonlinear Dynamics of Bistable Piezoelectric Harvester with Symmetric Elastic Constraints

Shubhanshu Maheshwari, Aravindan Muralidharan, and Shaikh Faruque Ali

Abstract A nonlinear vibro-impact piezoelectric energy harvesting system is studied in the present study. The harvesting system comprises a bistable inverted cantilever beam with a tip mass impacting a rigid frame. The governing equations of the non-smooth system are developed using Lagrange's method. The dynamics and harvesting performance are studied using numerical simulations. The results show that the proposed vibro-impact model for energy harvesters has the potential to harvest from ambient vibrations more efficiently as compared to conventional linear piezoelectric energy harvesters.

1 Introduction

The ever-improving living standards have led to a significant increase in the energy demand for small-scale electronic devices. Due to the need for replacing or recharging batteries and environmental concerns while disposing of the same, alternative energy sources have become more prevalent in the last decade. Technology breakthroughs in energy conversion from ambient sources, including solar [1], wind [2], thermal [3], and ambient vibration [4], to mention a few, have given rise to recent studies into powering small-scale electronic devices. Since ambient vibration sources are omnipresent, it is one of the suitable sources of power for small electronic devices.

The energy harvesters that transform mechanical vibrational energy from the environment into electrical power are termed vibration energy harvesters. It is possible to successfully power sensors and other monitoring equipment in structures and machinery without the use of external power sources by making use of the generally unwanted vibrations, especially in remote situations. Vibrational energy harvesters based on lead zirconate titanate (PZT) are the most widely used. Jeon et al. [5] published the first studies on MEMS harvesters based on PZT thin films. The most

S. Maheshwari (✉) · A. Muralidharan · S. F. Ali
Indian Institute of Technology Madras, 600036 Chennai, India
e-mail: shubhanshu2201@gmail.com

S. F. Ali
e-mail: sfali@iitm.ac.in

R. K. Ray et al. (eds.), *Advances in Theoretical and Applied Mechanics*, Lecture Notes in Mechanical Engineering, https://doi.org/10.1007/978-981-97-0418-7_30

popular vibrational energy harvesters are based on lead zirconate titanate (PZT). The proposed d_{33} mode PZT cantilever was designed with a proof mass, and the power generated was calculated at the resonant frequency. Lee et al. [6] proposed piezoelectric MEMS generators based on cantilevers operating in the d_{33} and d_{31} modes and taking advantage of PZT films. The effective operating range to harvest power is reduced by the fact that typical linear vibrational energy harvesters only function in close proximity to their resonance. Depending on the device's design, amplitude, and frequency of the source vibration, the total amount of power that can be salvaged is also limited. This suggests that the device needs to be made exactly because even the smallest deviation could result in large power reductions, even if the source of the ambient vibration has a well-established history and has a known frequency.

The efficiency of the vibrational energy harvester can be enhanced by using techniques like resonance tuning techniques, power amplification, and the addition of nonlinear oscillations [7]. In order to boost the power produced by the vibrational piezoelectric energy harvester and broaden its working bandwidth, many publications reported on alternate harvester architectures and methodologies. A PZT-based MEMS array with parallel connections was described by Liu et al. [8] as having a significant increase in bandwidth. A device where a low-frequency cantilever strikes high-frequency piezoelectric cantilever results in a wideband operation frequency range was created by Liu et al. [8] and presented as a MEMS energy harvester based on frequency-up conversion. Recent studies have shown that nonlinear piezoelectric structures have advantages over broadband energy harvesting [9, 10]. The system's geometrical design or the addition of an external nonlinear force can both generate nonlinearity. To this end, Karličić et al. [11] proposed the nonlinearity resulting from harvesters using Duffing-type oscillators, causing the bandwidth to widen as a result of the negative feedback brought on by the Duffing stiffening. Nonlinear behaviour can also be obtained by multi-stable (bistable and tristable) structures, demonstrating two mechanically stable positions or more. The frequency response's bandwidth can be greatly widened, and its transitions between these positions happen rapidly with the right external excitations, boosting the velocity of displacement even at low frequencies and enhancing the voltage [12].

An alternative way of harvesting broadband energy out of a linear device is through impact. In this study, a vertical cantilever beam with tip mass is considered and its deformation is restricted using a stopper around it. Along these lines, this paper studies a new harvesting system undergoing vibro-impacts, considering its capacity to capture a significant amount of energy. The harvester configuration consists of an inverted cantilever beam attached with a piezoelectric patch and tip mass, impacting a nearly rigid stopper. The base excitation given to the beam leads to an impact on the frame on either side (Fig. 1). The system shows a rich dynamical behaviour as the result of the vibro-impact of a nonlinear beam.

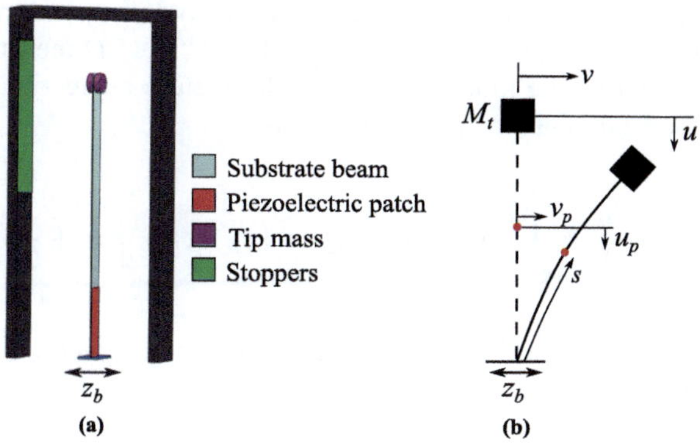

Fig. 1 **a** Schematic representation of the harvesting system with inverted cantilever beam and rigid stoppers on either side, undergoing base excitation z_b. **b** Inverted beam harvester schematic in a simplified form. M_t stands for the elastic beam's attached tip mass, while u and v stand for the vertical and horizontal displacements of tip mass

2 Description of System

In this study, the system includes an inverted elastic cantilever beam of length L with a tip mass M_t, attached at a length L_t from the fixed end, impacting rigid stoppers on either side when the base is excited harmonically as shown in Fig. 1(a). The beam has thickness h and width w with a piezoelectric MFC patch [13] of length L_{mfc}, thickness h_p and width w_p, attached to it. The clearance between the equilibrium position of the beam and the nearly rigid stopper is assumed to be δ. The beam has a cross-sectional area of A, a mass density of ρ, Young's modulus of Y, and a second moment of inertia of I. The system is given a base excitation of amplitude z_b, which results in the beam impacting the nearly-rigid stoppers on either side.

Since there is a time-dependent contact between the unimorph and the stoppers during vibro-impacts, the system is idealized as piecewise oscillators. It is assumed that the beam's thickness is smaller than the length; hence, the effects of shear deformation and rotary inertia of the beam are neglected. The tip mass (M_t) is maintained symmetric with respect to the centre line, and the parameters of the beam are specified such that the first torsional resonance frequency is significantly greater than the excitation frequency. As a result, the analysis ignores the beam's torsional modes and the vibration is solely planar. This motivates the use of Euler–Bernoulli's beam theory in the modelling.

2.1 *Mechanical Model of Beam*

The mechanical model of the unimorph is derived using the Lagrangian method, assuming Euler–Bernoulli's beam theory [14]. Figure 1b shows a schematic of the inverted cantilever beam with a tip mass. At the tip mass, the elastic displacements in the horizontal and vertical directions are v and u, respectively. Consider an arbitrary point on the beam, P, at a distance s, measured along the neutral axis from the base. At point P, the elastic displacements in the horizontal and vertical directions are v_p

and u_p, respectively. Let $\phi_p(s, t)$ denote the rotation of the beam at s, and hence, the rotation at the tip mass is $\phi(t) = \phi_p(L_t, t)$ measured at the centre of gravity of tip mass. The kinetic energy of the system neglecting the effect of rotary inertia of the beam mass as given by [15] is

$$T = \frac{1}{2}\rho A \int_0^L \left[\left(\dot{v}_p(s,t) + \dot{z}\right)^2 + \left(\dot{u}_p(s,t)\right)^2\right] ds + \frac{1}{2}M_t\left[(\dot{v} + \dot{z}_b)^2 + \dot{u}^2\right] + \frac{1}{2}I_t\dot{\phi}^2, \tag{1}$$

where the dot denotes differentiation with respect to time, t. The potential energy of the system is

$$V = \frac{1}{2}EI \int_0^L \left[v''_p(s,t)\left(1 + \frac{1}{2}\left(v'_p(s,t)\right)^2\right)\right]^2 ds - \rho A g \int_0^L u_p(s,t)\, ds - M_t g u, \tag{2}$$

where prime denotes differentiation with respect to s and g is the gravitational constant. Any point along the beam can be represented by the displacement as the product of the tip mass displacement and a beam deformation function. Considering position of tip mass at a length L_t,

$$v_p(s,t) = v_p(L_t, t)\,\psi(s) = v(t)\,\psi(s). \tag{3}$$

Approximating the function $\psi(s)$, such that it satisfies the boundary conditions, is given by [14]

$$\psi(s) = \lambda_t\left\{1 - \cos\left(\frac{\pi s}{2L}\right)\right\}, \tag{4}$$

where λ_t is a constant such that its value at the position of tip mass is unity, i.e., $\psi(L_t) = 1$, giving

$$\lambda_t = \frac{1}{\left\{1 - \cos\left(\frac{\pi L_t}{2L}\right)\right\}}. \tag{5}$$

Using Lagrange's equations, the system's equation of motion is determined in terms of the displacement of the tip mass as

$$m\ddot{v} + m_{nl}\left(v^2\ddot{v} + v\dot{v}^2\right) + c\dot{v} + kv + k_{nl}v^3 = F(t), \tag{6}$$

where m, c, k and θ are the equivalent mass, damping, stiffness, and electromechanical coupling coefficients, and the subscript nl represents the corresponding coefficients for nonlinear terms.

2.2 Coupled Electromechanical Model

The addition of a piezoelectric layer on the substrate layer results in the generation of voltage $V(t)$ due to the moment about the beam's neutral axis. The relation between moment $\bar{\chi}$ and voltage generated as given in [16]

$$\bar{\chi}(s,t) = \alpha_p V(t)\,. \tag{7}$$

The electromechanical constant α_p is influenced by the geometry, configuration, and electromechanical characteristics of the piezoelectric material being used. For unimorph with d_{31} poling configuration,

$$\alpha_p = Y d_{31} w \left(h + h_p - \bar{h}\right), \tag{8}$$

where h_p is the thickness of piezoelectric patch and $\bar{h}$ is the effective neutral axis. The work done by piezoelectric patches, having active length L_a in extracting electric charge,

$$W_e = \int_0^{L_a} \bar{\chi}(s,t) v_p'' \left(1 + \frac{1}{2} {v'_p}^2\right) ds. \tag{9}$$

Using displacement approximation for v, given by Eq. (3), the work done is

$$W = \left(\theta v + \frac{1}{3}\theta_{nl} v^3\right) V, \tag{10}$$

where

$$\theta = \alpha_p \int_0^{L_a} \psi''(s)\, ds = \alpha_p \psi'(L_a) \tag{11}$$

$$\theta_{nl} = 3\alpha_p \int_0^{L_a} \frac{1}{2}\psi''(s) \left(\psi'(s)\right)^2 ds = \frac{1}{2}\alpha_p \left(\psi'(L_a)\right)^3. \tag{12}$$

Equation (6) can now be written with additional terms defined by Eq. (10) as

$$m\ddot{v} + m_{nl}\left(v^2\ddot{v} + v\dot{v}^2\right) + c\dot{v} + kv + k_{nl}v^3 - \theta V - \theta_{nl} v^2 V = F(t)\,. \tag{13}$$

Considering piezoelectric patches as capacitors, the charge produced can be given in Eq. (10). Considering an electrical circuit with resistance connected across the piezoelectric layer, the electrical equation becomes [16]

$$C_p\dot{V} + \frac{V}{R_l} + \theta\dot{v} + \theta_{nl}v^2\dot{v} = 0. \tag{14}$$

2.3 Modelling the Impact

Figure 2 shows a lumped mass model of the vibro-impact system. The lumped mass (m) of the primary beam attached with spring of equivalent stiffness k_{nl}, when given base excitation of magnitude z_b, displaces by v. The contact between the beam and stoppers is such that the system behaves as a piecewise oscillator. The impact force is modelled as a spring force with spring stiffness k_c and is given by

$$f_i = \begin{cases} k_c\,(\delta - v) \text{ if } v < \delta \\ 0 \qquad\qquad \text{otherwise.} \end{cases} \tag{15}$$

The intermittent contact occurring between the beam and stoppers induces non-linearity in the system. The Signum function incorporates the impact force given in Eq. (14). From the above discussions, Eqs. (13) and (14) can be written as

$$m\ddot{v} + m_{nl}\left(v^2\ddot{v} + v\dot{v}^2\right) + c\dot{v} + kv + k_{nl}v^3 - \theta V - \theta_{nl}v^2V = F\,(t) + f_i$$
$$C_p\dot{V} + \frac{V}{R_l} + \theta\dot{v} + \theta_{nl}v^2\dot{v} = 0.$$

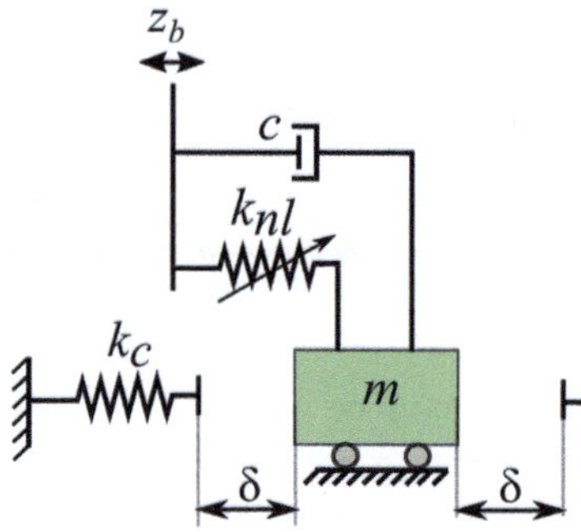

Fig. 2 Simplified lumped mass model of the vibro-impact system

3 Numerical Simulations

The clearance between the equilibrium position of the beam and stoppers is considered to be 12 mm for the analysis. Proportional damping is considered in the beam with a damping coefficient (c) to be 1%. The parameters for the numerical simulations are given in Table 1.

The equilibrium position of the tip mass is depicted in Fig. 3a, and it should be noticed that there are two equilibrium points for the post-buckling response. Figure 3 shows the linearized system's pre- and post-buckled natural frequencies as a function of tip mass. With increasing tip mass, it is seen that the inverted cantilever beam's natural frequency decreases until it reaches zero at the Euler buckling load, which corresponds to a tip mass of 5.5 g. As tip mass increases, the beam also buckles, and tip mass also causes an increase in the natural frequencies near stable equilibrium. The system's dynamics is studied for monostable and bistable cases in view of the harvested power. Therefore, two cases are considered for analysis:

1. Tip mass is set to a value that makes the system monostable, i.e., $M_t = 4.5$ g.
2. Tip mass is set to a value that makes the system bistable, i.e., $M_t = 20$ g.

Table 1 Parameters and properties of substrate layer used for numerical simulation

Parameter	Beam	MFC patch [13]
Length, L (mm)	150	37
Width, w (mm)	10	10
Thickness, h (mm)	0.50	0.30
Density, ρ (kg/m^3)	7880	5440
Young's modulus, Y (GPa)	210	30.37
Coupling coefficient, d_{31} (m/V)	–	-1.70×10^{-10}
Capacitance, C_p (Farads)	–	2.15×10^{-8}

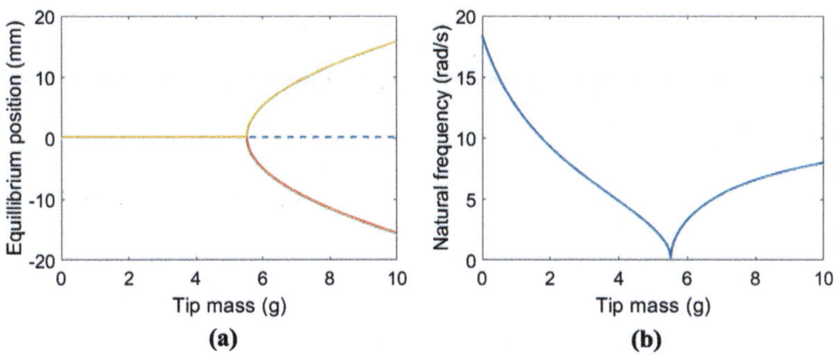

Fig. 3 Effect of the tip mass on **a** the equilibrium position and **b** associated natural frequencies for stable equilibrium positions. The dashed line shows the positions of unstable equilibrium

4 Results and Discussions

4.1 Analysis of Monostable System

The system is first analysed for a value of tip mass such that it is monostable, i.e., $M_t = 4.5$ g, with zero initial conditions. The system is excited near the resonant frequency with an input power of 20 mW. Numerical simulation is done without a rigid stopper and in the presence of a stopper to study the effect of impact.

Comparing Fig. 4a, b, when the system undergoes impact, there are no high-velocity impacts and the displacement is also constrained due to the stoppers. This can be well understood from the displacement time histories at the steady state as shown in Fig. 5. The stoppers restrict the movement of the beam without any significant change in the velocity. The dynamics is changed due to the stopper, but it does not have any appreciable difference in the magnitude of displacement and velocity of the beam.

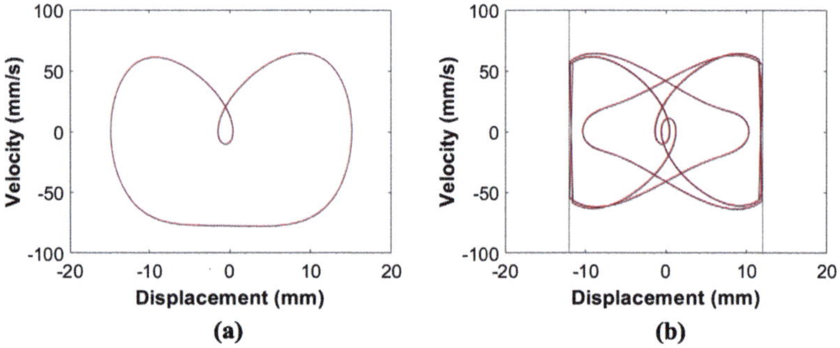

Fig. 4 Phase portrait of a monostable system **a** without stopper, **b** with stopper, when the system is excited with an excitation frequency near to the resonance frequency of the system

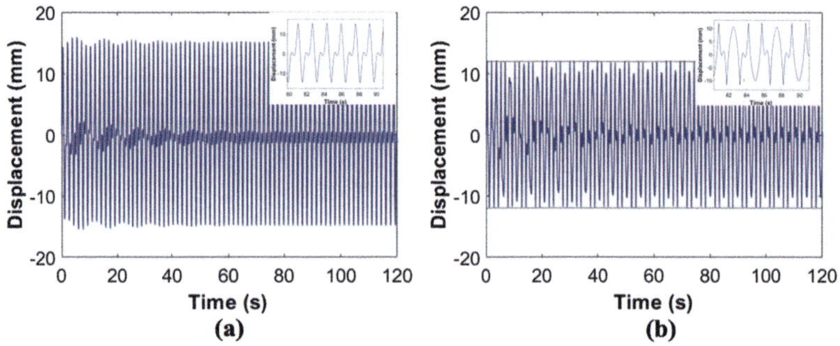

Fig. 5 Displacement time histories of a monostable system **a** without stopper, **b** with stopper, when the system is excited with an excitation frequency near to the resonance frequency of the system

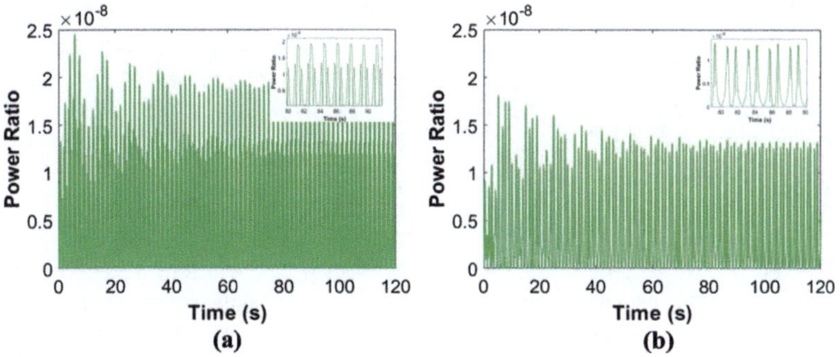

Fig. 6 Ratio of electrical power generated to the input power of a monostable system **a** without stopper, **b** with stopper, when the system is excited with an excitation frequency near to the resonance frequency of the system

The power generated is simulated by introducing a load resistor of 10 MΩ. The ratio of electrical power generated to the input power, with and without impact, is shown in Fig. 6. It is observed that there is a slight reduction in electrical power generated due to the constrained motion of the vertical beam and low-velocity impacts. The impacts change the periodic motion since the stoppers hinder the actual movement. This study shows that though the dynamics of the system is changing due to impact, there is no significant change in power generated due to low-velocity impacts.

4.2 Analysis of Bistable System

The system is analysed for the value of tip mass such that it is bistable, i.e., $M_t = 20$ g, with zero initial conditions. The system is excited near the resonant frequency with an input power of 20 mW. The numerical simulation is done without a rigid stopper as well as in the presence of a stopper to study the effect of impact.

Figure 7 shows the phase portraits for both cases. As can be observed in Fig. 7b, when the system undergoes impact, there are very high-velocity impacts, and these impacts on the stoppers aid the inter-well movement of the beam. Contrary to the monostable case, the stoppers are leading to a more uniform inter-well movement, leading to a steady state, as can be seen in Fig. 8.

The power generated is simulated by introducing a load resistor of 10 MΩ. The ratio of electrical power generated to the input power, with and without impact, is shown in Fig. 9. Comparing Figs. 8 and 9, it can be observed that, though there is not much appreciable change in the magnitude of the displacement in the presence of stoppers, there is a huge difference in the magnitude of power generated with and without impact. The magnitude of power has become comparatively stable due to the restriction provided by the stoppers.

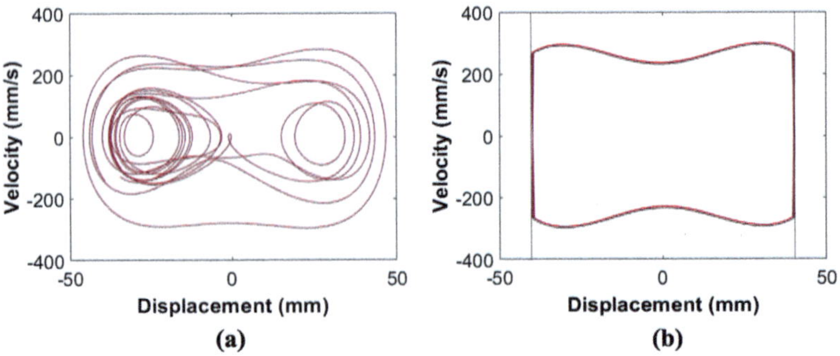

Fig. 7 Phase portrait of a bistable system **a** without stopper, **b** with stopper, when the system is excited with an excitation frequency near to the resonance frequency of the system

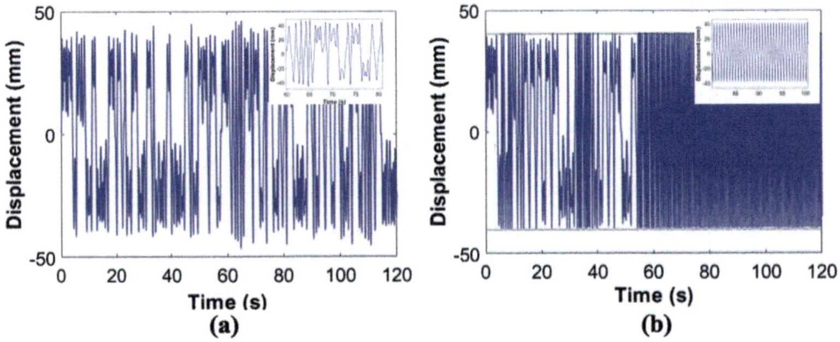

Fig. 8 Displacement time histories of a bistable system **a** without stopper, **b** with stopper, when the system is excited with an excitation frequency near to the resonance frequency of the system

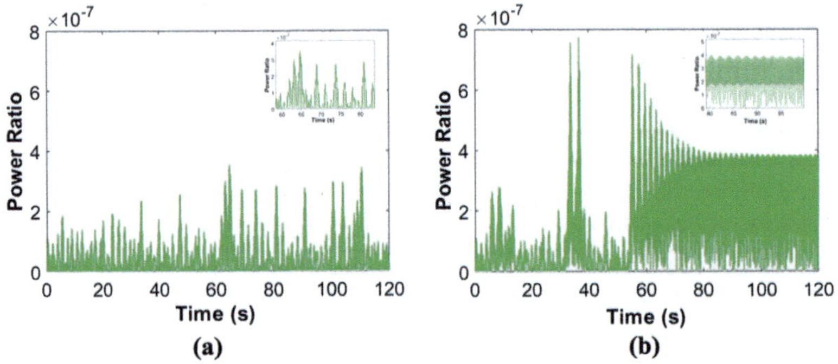

Fig. 9 Ratio of electrical power generated to the input power of a bistable system **a** without stopper, **b** with stopper, when the system is excited with an excitation frequency near to the resonance frequency of the system

This study shows that compared to oscillations between the potential wells, the electrical power production is lower when the tip mass oscillates near one of the equilibrium positions (monostable). But when the beam undergoes inter-well oscillations, the stoppers increase electrical power output significantly.

5 Conclusion

A vibro-impact-based energy harvester is modelled and governed by non-smooth equations of motion. The system comprises an inverted cantilever with a tip mass, impacting nearly rigid stoppers on either side. The critical mass is calculated, beyond which the system shows bistable behaviour. The system's dynamics is understood with and without impacts for the monostable and bistable cases. It has been observed from Fig. 6 that the stoppers are restricting the movement of the monostable beam, reducing the output power. On the contrary, for the bistable beam, Fig. 9 shows that using stoppers helps in maintaining inter-well vibrations, hence increasing the power generated. Therefore, using a stopper is nonbeneficial for monostable beams but beneficial for bistable beams. The bistable system is simulated for a frequency sweep with the ratio of excitation frequency to the natural frequency of the beam from 1 to 10 for an input power of 20 mW.

Figure 10 shows the ratio of the amplitude of average electrical power generated to the input power of the bistable beam for varying ratios of the excitation frequency. The figure (Fig. 10) shows that the vibro-impacts increase the range of excitation frequencies for which electrical power can be harvested, as well as an increase in the magnitude of harvested power.

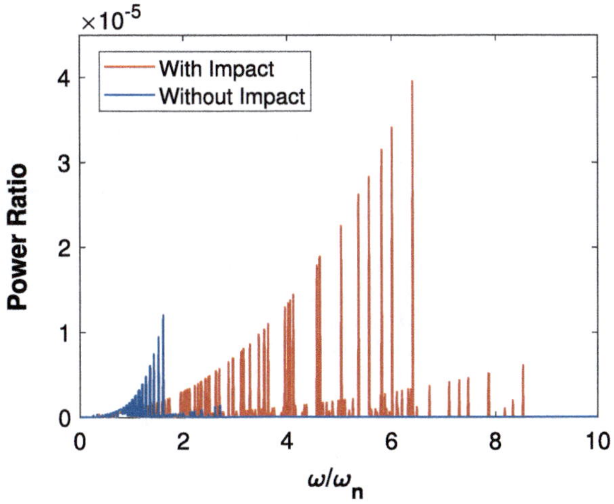

Fig. 10 Ratio of the amplitude of average electrical power generated to the input power of bistable beam under frequency sweep

The field of energy harvesting is emerging. The conventional linear vibrational energy harvesters are based on the principle of resonance, giving power output at the resonant frequency. The bandwidth of excitation frequency can be effectively increased by introducing nonlinearity in the system. This increase in bandwidth comes at the cost of the amplitude of generated power since the response of the beam gets distributed among a range of excitation frequencies. The magnitude of generated power can be increased by introducing other transduction methods, placing permanent magnets as tip mass, or using multifunctional materials that can effectively increase the magnitude of generated power. The present study focuses on characterizing the performance of vibro-impact energy harvester through numerical simulation. The results of the work shall aid in the optimization of harvester configuration for enhanced power generation.

References

1. Grätzel C, Zakeeruddin SM (2013) Recent trends in mesoscopic solar cells based on molecular and nanopigment light harvesters. Mater Today 16(1–2):11–18
2. Wen Q, He X, Lu Z, Streiter R, Otto T (2021) A comprehensive review of miniatured wind energy harvesters. Nano Mater Sci 3(2):170–185
3. Ramkumar A, Ramakrishnan M (2022) A comprehensive review on small-scale thermal energy harvesters: advancements and applications. Mater Today Proc 66:1552–1562
4. Todaro MT, Guido F, Mastronardi V, Desmaele D, Epifani G, Algieri L, De Vittorio M (2017) Piezoelectric MEMS vibrational energy harvesters: advances and outlook. Microelectron Eng 183–184:23–36
5. Jeon YB, Sood R, Jeong JH, Kim SG (2005) MEMS power generator with transverse mode thin film PZT. Sensors Actuators A Phys 122(1):16–22
6. Lee BS, Lin SC, Wu WJ, Wang XY, Chang PZ, Lee CK (2009) Piezoelectric MEMS generators fabricated with an aerosol deposition PZT thin film. J Micromech Microeng 19(6)
7. Tran N, Ghayesh MH, Arjomandi M (2018) Ambient vibration energy harvesters: a review on nonlinear techniques for performance enhancement. Int J Eng Sci 127:162–185
8. Liu H, Lee C, Kobayashi T, Tay CJ, Quan C (2012) Piezoelectric MEMS-based wideband energy harvesting systems using a frequency-up-conversion cantilever stopper. Sens Actuators A Phys 186:242–248
9. Wang Z, Li T, Du Y, Yan Z, Tan T (2021) Nonlinear broadband piezoelectric vibration energy harvesting enhanced by inter-well modulation, Energy Conversion Manage 246
10. Shim HK, Sun S, Kim HS, Lee DG, Lee YJ, Jang JS, Cho KH, Baik JM, Kang CY, Leng Y, Hur S, Song HC. On a nonlinear broadband piezoelectric energy harvester with a coupled beam array. Appl Energy: 120129
11. Karličić D, Cajić M, Paunović S, Adhikari S (2020) Nonlinear energy harvester with coupled Duffing oscillators. Commun Nonlin Sci Numer Simul 91
12. Zhou S, Lallart M, Erturk A (2022) Multistable vibration energy harvesters: principle, progress, and perspectives. J Sound Vibr 528
13. S. M. corporation, MFC types specifications. https://www.smart-material.com/MFC-product-P2V2.html. Accessed 15 Sept 2022
14. Friswell MI, Ali SF, Bilgen O, Adhikari S, Lees AW, Litak G (2012) Non-linear piezoelectric vibration energy harvesting from a vertical cantilever beam with tip mass. J Intell Mater Syst Struct 23(13):1505–1521

15. Esmailzadeh E, Nakhaie-Jazar G (1998) Periodic behavior of a cantilever beam with end mass subjected to harmonic base excitation. Int J Non-Linear Mech 33(4):567–577
16. Crawley EF, De Luis J (1987) Use of piezoelectric actuators as elements of intelligent structures. AIAA J 25(10):1373–1385

Analysis of Thermal Autofrettage of Disks Using von Mises Yield Criteria

Mohit Rajput, S. M. Kamal, and R. U. Patil

Abstract This work presents analytical solutions of elastic–plastic stresses, strains and residual stresses in the thermal autofrettage of a hollow disk based on von Mises yield criterion. During thermal autofrettage, the material is assumed to exhibit elastic-perfectly plastic behavior under plane stress condition. The numerical evaluation of the stresses and strains in a typical mild steel disk undergoing thermal autofrettage is carried out, and the corresponding solution is compared with the existing analytical solution employing Tresca's yield criterion.

Keywords Thermal autofrettage · Residual stress · Von Mises yield criterion

1 Introduction

Disks, which are in the form of very short thick-walled cylinders, are commonly used in various industrial applications designed to work against high tensile stresses at its inner side. For example, fastener holes, outer wall of a butterfly valve used in high-pressure pipeline and rotating disks used in turbomachinery. All these components are subjected to high tensile stresses at its inner side in service. In order to safeguard them against very high tensile stresses that may lead to catastrophic failure, strengthening mechanisms play a crucial role. Usually, the disks used in high-pressure/temperature applications or the disks that are engineered to rotate at extremely high speed in service, cold expansion is employed as a strengthening mechanism. In this work, autofrettage is considered to be employed for strengthening such disks.

Autofrettage is a process commonly used in thick-walled cylinders or spherical vessels to increase their load carrying capacity during use. This involves applying a load that plastically deforms the cylindrical or spherical component, inducing

M. Rajput (✉) · R. U. Patil
Department of Mechanical Engineering, Indian Institute of Technology Jammu, Jagti, India
e-mail: 2020rme2068@iitjammu.ac.in

S. M. Kamal
Department of Mechanical Engineering, Tezpur University, Tezpur, Assam 784028, India

R. K. Ray et al. (eds.), *Advances in Theoretical and Applied Mechanics*, Lecture Notes in Mechanical Engineering, https://doi.org/10.1007/978-981-97-0418-7_31

plastic deformation on the inner side. Typically, the material surrounding the outer wall undergoes elastic deformation. Once the desired level of plastic deformation is achieved, the plastically deforming load is unloaded, generating compressive residual stresses on the inside and surrounding area of the component [1, 2]. These stresses counteract the high tensile stresses experienced during use, allowing the component to carry more service load. This process not only increases the load carrying capacity but also enhances the component's resistance to stress corrosion cracking and improves its fatigue life.

Various autofrettage methods are available, including hydraulic, swage, explosive, thermal, and rotational autofrettage. Despite some drawbacks, hydraulic and swage autofrettage have been widely used in industries. However, hydraulic autofrettage involves the use of dangerously high hydraulic pressure [3], while swage autofrettage requires a large hydraulic pressure to push an oversized mandrel through the bore, resulting in enormous friction [4]. Explosive autofrettage, which applies an explosive charge to the inside of a thick-walled cylinder, is not commonly used due to operational difficulties [5]. To address these challenges, Kamal and Dixit [6, 7] proposed a novel thermal autofrettage method, which induces plastic deformation at the inner side and its neighborhood by subjecting the thick-walled cylinder to a radial temperature difference across its wall thickness. This is achieved by simultaneously heating the outer wall and cooling the inner wall, after which the inner wall spreads, generating beneficial compressive residual stresses of significant magnitude. Additionally, Zare and Darijani [8] conceived the idea of new rotational autofrettage, where a cylinder is assumed to be rotating at a sufficiently high angular velocity and then brought to a complete rest, resulting in the generation of compressive residual stresses within the cylinder.

All autofrettage processes primarily meant for strengthening long cylindrical vessels, *e.g.*, gun barrel, pressure vessels, and high-pressure pipelines. Nevertheless, few literatures analyze the applicability of autofrettage for strengthening very short cylinders or hollow disks. For example, Rees [9] analyzed the swaging of disks and lugs using both Tresca and von Mises yield criteria. Kamal and Dixit [7] carried out a extensively analysis of elastic–plastic loading and post-unloading residual stresses in thermal autofrettage of disks using Tresca yield criteria and its associated flow rule. The strengthening of disks used in rotating machinery, disks used in high-pressure/high-temperature environment inducing residual stresses by means of rotational autofrettage was exemplified by Kamal [10]. He used Tresca yield criteria and its associated flow rule incorporating material strain hardening for the analysis. Some recent works [11–13] on the analysis of rotational autofrettage of disks reveal that autofrettage could be a potential strengthening mechanism to reduce the overall stress level in disks.

It is worth noting that the literature on the analysis of thermal autofrettage of disks is limited. Only, Ref. [6] considered the strengthening aspect of hollow disks by means of thermal autofrettage using Tresca yield criteria. The process needs further analysis to assess its suitability for industrial application. It is recognized that the von Mises yield criteria predict the yielding behavior of many materials close to the reality. In view of this, in the current work, the thermal autofrettage process is considered to be analyzed using von Mises yield criteria. Elastic-perfectly plastic material model is used for the analysis.

2 Problem Definition

The schematic illustration of the thermal autofrettage process is shown in Fig. 1. A thick-walled hollow disk is assumed having inside radius a and outside radius b. The desired temperature difference can be achieved by heating the outer wall using an external source (heating element) and by circulating cold fluid through the bore simultaneously to keep the inner wall at a lower temperature. It is considered that external surface of the disk reaches to a desired high temperature T_b while maintaining the inner wall at sufficiently lower temperature T_a creating a temperature difference of $(T_b - T_a)$. This temperature difference is responsible for the plastic deformation of the inner wall of the disk. The temperature distribution under radial temperature difference in the disk wall at steady state is given by [14]

$$T = T_a + (T_b - T_a)\frac{\ln\left(\frac{r}{a}\right)}{\ln\left(\frac{b}{a}\right)} \tag{1}$$

It is to be noted that below a certain threshold $(T_b - T_a)$, the entire disk wall remains in the elastic condition. In Ref [6], it was shown that when $(T_b - T_a)$ reaches a certain first threshold, yielding initiates place first at the inside of the wall. Upon exceeding the first threshold $(T_b - T_a)$, a plastic region spreads from inner radius to a certain intermediate radius c, i.e., $a \le r \le c$. The outer region remains $c \le r \le b$ in the elastic state. When $(T_b - T_a)$ is increased further, a situation arises where the outer wall initiates yielding at a certain second threshold temperature difference. Upon exceeding this second threshold, simultaneous plastic deformation of inner and outer wall takes place resulting two plastic regions spreading from inner and outer radius to certain intermediate radii. At this stage, the intermediate wall

Fig. 1 The thermal autofrettage process

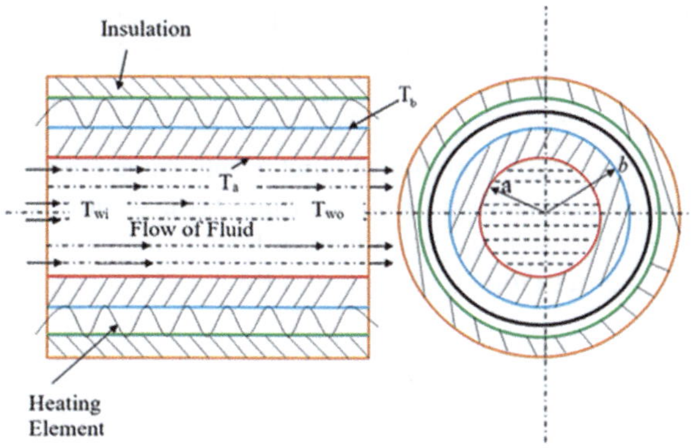

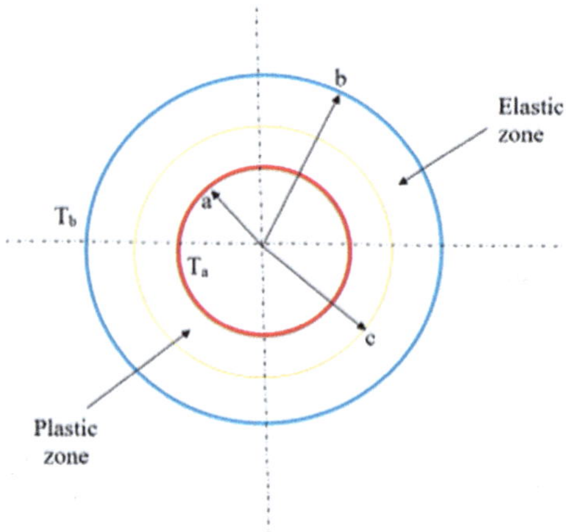

Fig. 2 The elastic and plastic regions in the disk wall

thickness is subjected to elastic deformation provided $(T_b - T_a)$ does not reach to the plastic collapse temperature difference.

The current analysis is based on the consideration of evolving only an inner plastic region and an outer elastic region as depicted in Fig. 2. This is the advantageous situation from the point of view of generation of lesser tensile residual stress at the external surface of the disk [6]. For a very short cylinder, the plane stress (axial stress, $\sigma_z = 0$) is justified. The disk is presumed to be axisymmetric for the analysis. The novelty of this work lies in the mathematical treatment of the problem, i.e., analysis of thermal autofrettage incorporating von Mises yield criterion.

2.1 *Initiation of Yielding*

When the disk wall is fully elastic, the thermo-elastic solution is well-known and available in any standard text, e.g., Timoshenko and Goodier [15]. It is assumed here that yielding occurs in the disk according to von Mises yield criteria, which is given by the following expression for plane stress condition:

$$\sigma_r^2 - \sigma_r\sigma_\theta + \sigma_\theta^2 = \sigma_y^2 \tag{2}$$

As yielding commences first at the inside radius, $r = a$ of the disk under radial temperature difference, the thermo-elastic solutions for σ_r and σ_θ [15] are evaluated at $r = a$ and then substituted in Eq. (2) to obtain the following expression for the yield onset radial temperature difference:

$$(T_b - T_a)|_{\text{initial yielding}} = \frac{2\sigma_y}{E\alpha\left\{\frac{2b^2}{b^2-a^2} - \frac{1}{\ln\frac{b}{a}}\right\}} \tag{3}$$

2.2 Stress Analysis During the Loading Stage of Thermal Autofrettage

To achieve thermal autofrettage in the disk, the wall of the disk is subjected to a radial temperature difference $(T_b - T_a)$ more than the temperature difference given by Eq. (3). If it is such that $(T_b - T_a)$ is not sufficient to initiate the yielding of the outer wall, then the disk wall is divided into an inner plastic region propagating from inside radius a to certain intermediate radius c and an outer elastic region propagating from c to the outside radius b as depicted in Fig. 1b. The stress analysis in the plastic region $(a \leq r \leq c)$ and in the elastic region $(c \leq r \leq b)$ is carried out in Sects. 2.2.1 and 2.2.2, respectively.

2.2.1 The Plastic Region $(a \leq r \leq c)$

In the plastic region, the yielding propagates according to the von Mises yield criteria given by Eq. (2). The equilibrium equation remains valid irrespective of the nature of deformation, i.e., elastic or plastic. Thus, the axisymmetric equilibrium equation for the plastic region can be stated as

$$\sigma_\theta = \sigma_r + r\frac{d\sigma_r}{dr} \tag{4}$$

Incorporating Eq. (4) in Eq. (2) and solving the resulting quadratic equation, the following expression is obtained:

$$r\frac{d\sigma_r}{dr} = \frac{-\sigma_r \pm \sqrt{4\sigma_y^2 - 3\sigma_r^2}}{2} \tag{5}$$

The term under square root in Eq. (5) is always positive. As the radial stress distribution as a function of radius at the inner side in the plastic region is tensile and an increasing function for $T_b > T_a$, its derivative will be positive. This is feasible only when a positive sign before the square root term is considered. Thus, Eq. (5) is treated with the positive sign before the square root term.

Now, using numerical Taylor's series expansion method (considering up fourth order term) with the initial boundary condition $\sigma_r(a) = 0$, the solution for Eq. (10) is obtained as

$$\sigma_r(r) = (r-a)\frac{\sigma_y}{a} - \frac{(r-a)^2}{2}\frac{3\sigma_y}{2a^2} + \frac{(r-a)^3}{6}\frac{3\sigma_y}{a^3} - \frac{(r-a)^4}{24}\frac{57\sigma_y}{8a^4} + \cdots \tag{6}$$

With the help of equilibrium Eqs. (4) and (6), the solution for hoop stress as a function of radial position in the plastic region is given as

$$\sigma_\theta = \sigma_r + r\left\{\frac{\sigma_y}{a} - (r-a)\frac{3\sigma_y}{2a^2} + \frac{(r-a)^2}{2}\frac{3\sigma_y}{a^3} - \frac{(r-a)^3}{6}\frac{57\sigma_y}{8a^4} + \cdots\right\} \tag{7}$$

where σ_r is given by Eq. (6).

Using the additive decomposition of elastic and plastic parts, the total radial and hoop strains along with the strain–displacement relations can be expressed as

$$\varepsilon_r = \frac{\mathrm{d}u}{\mathrm{d}r} = \varepsilon_r^e + \varepsilon_r^p \ , \quad \varepsilon_\theta = \frac{u}{r} = \varepsilon_\theta^e + \varepsilon_\theta^p \ , \tag{8}$$

where u is the radial displacement. Using plastic incompressibility and strain–displacement relations along with generalized Hooke's law provides

$$\frac{\mathrm{d}u}{\mathrm{d}r} + \frac{u}{r} = \frac{1-v}{E}(\sigma_r + \sigma_\theta) + 2\alpha T - \varepsilon_z^P \tag{9}$$

Solving Eq. (9), one obtains

$$u = \frac{\lambda - \int \varepsilon_z^p r\mathrm{d}r + C}{r} \tag{10}$$

where C is a constant of integration and

$$\lambda = \int\left\{\left(\frac{1-v}{E}\right)(\sigma_r + \sigma_\theta) + 2\alpha T\right\} r\mathrm{d}r \tag{11}$$

Using the second relation of Eq. (8), the plastic part of the hoop strain ε_θ^P can be computed as

$$\varepsilon_\theta^p = \frac{u}{r} - \varepsilon_\theta^e = \frac{u}{r} - \left\{\frac{1}{E}(\sigma_\theta - v\sigma_r) + \alpha T\right\} \tag{12}$$

Using Eq. (10) in Eq. (12), one obtains

$$\varepsilon_\theta^p = R(r) + \frac{C}{r^2} - \frac{\int \varepsilon_z^p r\mathrm{d}r}{r^2} \tag{13}$$

where,

$$R(r) = \frac{\lambda}{r^2} - \frac{1}{E}[\sigma_\theta - v\sigma_r] - \alpha T \tag{14}$$

The constant C in Eq. (13) can be evaluated using the boundary condition of vanishing plastic strains at the radii of elastic–plastic interface $r = c$. This provides $C = -c^2 R(c)$. Substituting the value of C in Eq. (13), the solution for plastic part of the hoop strain is expressed as

$$\varepsilon_\theta^p = R - \frac{c^2 R(c)}{r^2} - \frac{\int \varepsilon_z^p r \mathrm{d}r}{r^2} \tag{15}$$

The plastic constitutive relations for Mises material for plane stress case without strain hardening are given by [16]

$$\varepsilon_r^p = \frac{\varepsilon_{\mathrm{eq}}^p(r)}{\sigma_y}\left(\sigma_r - \frac{\sigma_\theta}{2}\right), \varepsilon_\theta^p = \frac{\varepsilon_{\mathrm{eq}}^p(r)}{\sigma_y}\left(\sigma_\theta - \frac{\sigma_r}{2}\right), \varepsilon_z^p = \frac{\varepsilon_{\mathrm{eq}}^p(r)}{\sigma_y}\left(-\frac{(\sigma_r + \sigma_\theta)}{2}\right) \tag{16}$$

Substituting Eq. (15) in the second relation of Eq. (16), the expression for the equivalent plastic strain $\varepsilon_{\mathrm{eq}}^p(r)$ is derived as

$$\varepsilon_{\mathrm{eq}}^p(r) = \frac{\sigma_y}{\left(\sigma_\theta - \frac{\sigma_r}{2}\right)}\left[R - \frac{c^2 R(c)}{r^2} + \frac{\int \left\{\frac{r\varepsilon_{\mathrm{eq}}^p(r)}{2\sigma_y}(\sigma_r + \sigma_\theta)\right\} \mathrm{d}r}{r^2}\right]. \tag{17}$$

2.2.2 The Elastic Region ($c \le r \le b$)

In the elastic region ($c < r < b$), the solution for thermal stresses is obtained by using generalized Hooke's law with the incorporation of equilibrium equation along with the strain compatibility and strain–displacement relations. The resulting solution involves two integration constants, which can be obtained by using the boundary condition of vanishing radial stress at the outer radius and at $r = c$, $\left\{(\sigma_z - \sigma_r)^2 + (\sigma_r - \sigma_\theta)^2 + (\sigma_\theta - \sigma_z)^2 = 2\sigma_y^2\right\}$. After substituting the elastic constants, the final solution for stresses in the elastic region is obtained as

$$\sigma_r(r) = \frac{E\alpha(T_b - T_a)}{2\ln\left(\frac{b}{a}\right)}\left\{\frac{1}{2} - \ln\frac{r}{a}\right\} + \left\{\frac{-B_P + \sqrt{B_P^2 - 4A_P C_P}}{2A_P}\right\} - \frac{\left\{\frac{E\alpha(T_b - T_a)}{2\ln\left(\frac{b}{a}\right)}\left\{\frac{1}{2} - \ln\frac{b}{a}\right\} + \left(\frac{-B_P + \sqrt{B_P^2 - 4A_P C_P}}{2A_P}\right)\right\} b^2}{r^2} \tag{18}$$

$$\sigma_\theta(r) = -\frac{E\alpha(T_b - T_a)}{2\ln\left(\frac{b}{a}\right)}\left\{\frac{1}{2} + \ln\frac{r}{a}\right\} + \left\{\frac{-B_P + \sqrt{B_P^2 - 4A_P C_P}}{2A_P}\right\} + \frac{\left\{\frac{E\alpha(T_b - T_a)}{2\ln\left(\frac{b}{a}\right)}\left\{\frac{1}{2} - \ln\frac{b}{a}\right\} + \left(\frac{-B_P + \sqrt{B_P^2 - 4A_P C_P}}{2A_P}\right)\right\} b^2}{r^2} \tag{19}$$

where

$$A_p = \left(1 + \frac{3b^4}{c^4}\right),\ B_P = \frac{6Pb^4}{c^4}\left(\frac{1}{2} - \ln\frac{b}{a}\right) - \frac{3Pb^2}{c^2} - 2P\ln\frac{c}{a},\ P = \left\{\frac{E\alpha(T_b - T_a)}{2\ln\left(\frac{b}{a}\right)}\right\},$$
$$C_p = P^2\left[\frac{3}{4} + \left\{\ln\left(\frac{c}{a}\right)\right\}^2\right] - \sigma_y^2 + \frac{3b^2}{c^2}P^2\left(\frac{1}{2} - \ln\frac{b}{a}\right)\left\{\frac{b^2}{c^2}\left(\frac{1}{2} - \ln\frac{b}{a}\right) - 1\right\} \tag{20}$$

3 Estimation of Thermal Residual Stresses

Residual thermal stresses are produced in a hollow disk when it is heated and subsequently cooled until its temperature equalizes. To determine the thermal residual radial and hoop stress in each zone, the elastic stresses [15] are subtracted from the corresponding elastic–plastic loading stresses in the respective zones, assuming that the unloading process is entirely elastic.

4 Solution Methodology

In order to numerically evaluate the unknown radii of elastic–plastic interface c and correspondingly the stresses and strains during loading and residual stresses after unloading, the following solution methodology is adopted:

a. To find out the value of radius c, the following boundary condition is used and then the resulting equation for c is solved using FZERO function in MATLAB.

$$\sigma_r^{(\text{Elastic zone})} = \sigma_r^{(\text{Plastic zone})} \quad \text{at} \quad r = c$$

b. Taking the initial guess for $\varepsilon_{\text{eq}}^P$ as zero on the right-hand side of Eq. (17) at all radii r, the updated values of plastic strain field $\varepsilon_{\text{eq}}^P(r)$ are obtained at all radial positions from left-hand side of Eq. (17) for the fixed value of c obtained in (a).
c. The process of updating the values of $\varepsilon_{\text{eq}}^P(r)$ as stated in (b) is repeated till the convergence in $\varepsilon_{\text{eq}}^P$ is achieved within a specified tolerance.

Once the value of c is known, the elastic–plastic stresses induced in the disk due to autofrettage can be evaluated by using the equations developed in Sect. 2 and residual stresses following the procedure as discussed in Sect. 3.

5 Numerical Simulation

A numerical evaluation of the stresses and strains in a typical mild steel disk induced by thermal autofrettage is carried out taking $a = 10$ mm, $b = 30$ mm. The material properties considered for mild steel are as follows: young's modulus of elasticity, $E = 200$ GPa; yield stress, $\sigma_y = 324$ MPa; and coefficient of thermal expansion, $\alpha = 13 \times 10^{-6}$/°C. The yield onset temperature difference for the present disk is 186.027 °C dictated by Eq. (3). Thus, thermal autofrettage in the disk is considered to be achieved by employing a temperature difference of 310 °C, which will cause yielding at the inner wall only forming an inner plastic region. As discussed in Sect. 4, the elastic–plastic interface radius, c is evaluated as 12.078 mm. Now, the elastic–plastic stress distribution in the disk are plotted using the equations developed in Sect. 2 along the positive radial direction and is shown in Fig. 3a along with the stress distribution due to Tresca obtained using the model presented in Ref. [6]. It is to be noted that the value of the elastic–plastic interface predicted by the present model is slightly underestimated by 2.91% when compared with the model predicted by Tresca [6]. The elastic–plastic stress distribution due to the present model shows that the hoop stress increases from the inner radius and becomes the maximum at the radius of elastic–plastic interface. However, due to Tresca, the hoop stress distribution is parallel to abscissa. Upon vanishing the induced temperature difference, the residual stresses generated in the disk wall are evaluated following the procedure described in Sect. 3 and is shown in Fig. 3b along the positive radial direction. It is noticed that significantly larger compressive residual stresses are induced at the inner radius of the disk and its neighborhood. At the inner radius, the magnitude of the compressive residual stress is 216 MPa.

The residual hoop stress at the inner wall remains compressive up to a radial depth of 1.85 mm from inner wall. The residual hoop stresses beyond this radius remain tensile with significantly smaller magnitude following a decreasing trend up to the

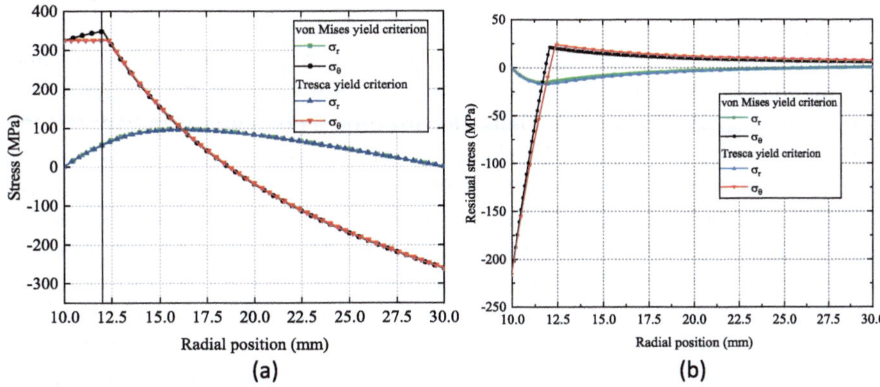

Fig. 3 **a** Comparison of elastic–plastic stress using Tresca [6] and von Mises yield criterion, **b** residual stress in mild steel disk

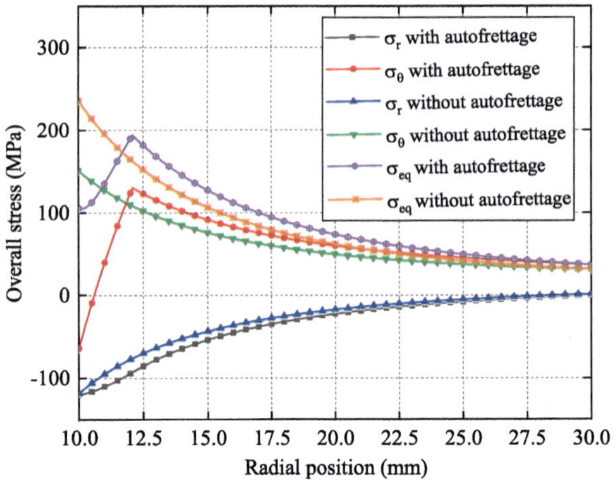

Fig. 4 Overall stresses in mild steel disk

outer radius. This generation of small tensile residual hoop stress at the outer wall is advantageous as it lowers the susceptibility to stress corrosion cracking. It is observed that the residual radial stress is wholly compressive vanishing at the inner and outer radii. The disk at this stage with compressive residual hoop stress in the vicinity of the inner wall is said to be thermally autofrettaged. The thermally autofrettaged mild steel disk is now considered to be subjected to working internal pressure of 120 MPa to assess the reduction in the maximum stress level in the pressurized disk. The resulting overall stress in the autofrettaged disk during pressurization is obtained by superposing the residual stresses to the Lamé stresses due to internal pressure [15]. In the non-autofrettaged counterpart, the stress distribution is given only by Lamé's solution. The overall stress distribution in the present autofrettaged disk as a function of radius along with the corresponding stress distribution in its non-autofrettaged counterpart is presented in Fig. 4.

In the case of the thermally autofrettaged disk, the magnitude of the maximum equivalent von-Mises stress (σ_{eq}) is found to be 17.85% lesser as compared to the corresponding non-autofrettaged disk. Through numerical evaluation, it is obtained that the present thermally autofrettaged disk can bear the maximum internal pressure of 218.04 MPa. On the other hand, the maximum pressure capacity of the non-autofrettaged disk is 165.93 MPa. Thus, the enhancement in the pressure bearing capacity of the disk due to thermal autofrettage is 31.40% as compared to the corresponding non-autofrettaged disk. Table 1 shows that the autofrettaged disk based on the von Mises criterion can able to handle the max pressure capacity of 218.04 MPa before yielding, whereas the autofrettaged disk using Tresca criterion can able to handle the max pressure capacity of 195 MPa before yielding. It is observed that maximum pressure capacity of the autofrettage disk based on von Mises model is 11.79% higher as compared to Tresca model. When compared to non-autofrettage disk, the maximum pressure capacity increase based on von Mises model is 31.4%, whereas the maximum pressure capacity increase based on Tresca model is 35.41%.

Table 1 Comparison of pressure load carrying capacity of the mild steel disk under the thermal autofrettage based on the von Mises model and Tresca model

Yield criterion	Maximum pressure capacity increase in service (in MPa)	Maximum pressure capacity for non-autofrettage disk (in MPa)
von Mises	218.04	165.93
Tresca	195	144

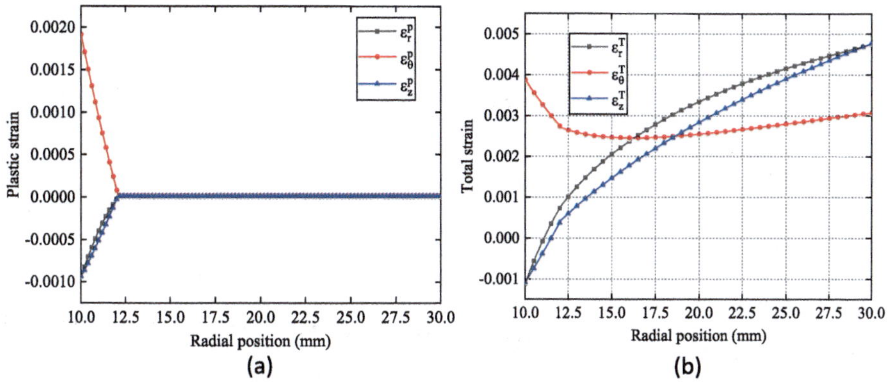

Fig. 5 **a** Plastic strain, **b** total strain in mild steel disk

The corresponding Tresca model overestimates this enhancement in pressure bearing capacity of the disk.

The plastic strain distribution in the disk during loading stage of thermal autofrettage is evaluated following the methodology described in Sect. 4 and is plotted in Fig. 5a along the positive radial direction. It is noticed that the plastic hoop strain induced in the plastic region is tensile and is the maximum at the inner radius. The plastic radial and axial strains are compressive in nature. In Fig. 5b, the total strains induced in the disk during elastic–plastic deformation are shown. It is observed that the total radial as well as the total axial strain changes from compressive to tensile along the positive radial direction. However, the total hoop strain remains tensile throughout. The maximum plastic deformation occurs in all directions (radial, hoop, and axial direction) at inner wall of the disk.

6 Conclusion

In this work, an analytical model for the disks subjected to thermal autofrettage based on von Mises yield criteria using elastic-perfectly plastic material model and plane stress condition. The present model is solved numerically to obtain the elastic–plastic stresses, strains as well as the residual stress distribution induced by thermal autofrettage in a typical mild steel disk. The corresponding solutions are compared with the model due to Tresca yield criteria. An enhancement of 31.40% in the pressure bearing capacity of the present thermally autofrettaged disk is achieved as compared to the corresponding non-autofrettaged disk. The maximum pressure capacity of the autofrettage disk based on von Mises model is 11.79% higher as compared to Tresca model. The corresponding Tresca model overestimates this enhancement in pressure bearing capacity of the disk.

Acknowledgements Funding from the Armament Research Board, DRDO (Govt. of India) through grant ARMREB/ADMB/2021/234, project Use of Thermal Autofrettage for Defence Application is gratefully acknowledged.

References:

1. Dixit US, Kamal SM, Shufen R (2019) Autofrettage processes
2. Shufen R, Dixit US (2018) A review of theoretical and experimental research on various autofrettage processes. J Press Vessel Technol Trans ASME 140(5):1–15
3. Avitzur B (1994) Autofrettage-stress distribution under load and retained stresses after depressurization. Int J Press Vessel Pip 57(3):271–287
4. Davidson TE, Barton CS, Reiner AN, Kendall DP (1962) New approach to the autofrettage of high-strength cylinders. Exp Mech 2(2):33–40
5. Howell G, Glick HS, Kaplan MA, Em E (1973) Éam E/Em 32 (19)
6. Kamal SM, Dixit US (2015) Feasibility study of thermal autofrettage process. Top Mining Metall Mater Eng: 81–107
7. Kamal SM, Dixit US (2015) Feasibility study of thermal autofrettage of thick-walled cylinders. J Press Vessel Technol Trans ASME 137(6)
8. Zare HR, Darijani H (2016) A novel autofrettage method for strengthening and design of thick-walled cylinders. Mater Des 105:366–374
9. Rees DWA (2011) A theory for swaging of discs and lugs. Meccanica 46(6):1213–1237
10. Kamal SM (2018) Analysis of residual stress in the rotational autofrettage of thick-walled disks. J Press Vessel Technol Trans ASME 140(6):1–10
11. Kamal SM, Kulsum R (2019) Parametric study of axisymmetric circular disk subjected to rotational autofreattge
12. Kamal SM, Dixit US (2021) Design of a disk-mandrel assembly for achieving rotational autofrettage in the disk. Proc Inst Mech Eng Part C J Mech Eng Sci 235(13):2452–2467
13. Kamal SM (2022) Estimation of optimum rotational speed for rotational autofrettage of disks incorporating Bauschinger effect. Mech Based Des Struct Mach 50(7):2535–2554
14. Noda N (1996) Thermal stresses
15. Timoshenko S, Goodier JN (1951) Theory of elasticity. McGraw Hill Education
16. Eraslan AN, Akis T (2003) On the elastic-plastic deformation of a rotating disk subjected to a radial temperature gradient. Mech Based Des Struct Mach 31(4):529–561

Transient Analysis of Graphene Reinforced FG-Porous Sandwich Plates Subjected to Underwater Blast

Mohammed Shakir and Mohammad Talha

Abstract The present study performs the transient analysis of sandwich plates under the influence of underwater blast loading. The sandwich plate consists of a functionally graded (FG) porous core reinforced with graphene nanoplatelets (GnP), and two isotropic face sheets. To estimate the effective properties of the reinforced core, such as Young's modulus and mass density, the Halpin–Tsai model and rule of mixture are respectively utilized. The distribution of porosity and GnP employs continuous functions to accomplish gradation along the thickness direction. The governing equations based on the higher-order plate kinematic model are developed using a variational approach. The transient response in terms of nondimensional center displacement has been accessed by the finite element method along with Newmark's time integration approach and the influence of various parameters such as porosity index, GnP's weight fraction, type of porosity distribution, etc. is also explored.

Keywords Finite element analysis · Functionally graded porous core · Graphene reinforcement · Sandwich plate · Transient analysis · Underwater blast

1 Introduction

The protective design of navy vehicles as well as ferries against underwater blast loading demand high specific rigidity, flexural strength, and energy dissipation capabilities, which can be possible with sandwich structures with a porous core [1]. However, the porosity causes a reduction in the overall stiffness of the structures, which can be enhanced by adding graphene nanoplatelets (GnP) [2]. Moreover, the sandwich structures mainly face delamination problems due to stress concentration at the interfaces, which can be sorted out using functionally graded porous (FG-porous) materials. In these materials, the density and size of the pores vary in a graded fashion

M. Shakir (✉) · M. Talha
Computational Design Lab, School of Mechanical and Materials Engineering, Indian Institute of Technology Mandi, Mandi, Himachal Pradesh 175075, India
e-mail: mshakirme@gmail.com

R. K. Ray et al. (eds.), *Advances in Theoretical and Applied Mechanics*, Lecture Notes in Mechanical Engineering, https://doi.org/10.1007/978-981-97-0418-7_32

such that the variation in the materials' properties can be attained continuously and smoothly [3].

A few researchers investigated the transient/dynamic response of sandwich structures with various core configurations. Yang et al. [2] presented buckling and vibration characteristics of FG-porous GnP-reinforced plates with uniform and nonuniform porosity distributions using Chebyshev-Ritz technique. Mohseni and Shakouri [4] employed classical thin plate theory to investigate sandwich plates with viscoelastic core and GnP-reinforced face sheets and found free and forced vibration responses. Gao et al. [5] characterized the dynamic response of sandwich plates with square-celled core along with viscoelastic material employing the first-order shear deformation theory (FSDT) model. Quan et al. [6] obtained a nonlinear vibration response using an analytical approach for FG-porous sandwich plates, which comprise a homogenous core and two FG material sheets. Karakoti et al. [7] analyzed the transient response of porous P and S-FGM sandwich plates subjected to blast and thermal loading conditions. Van Long et al. [8] considered FG-sandwich shallow shells and found a nonlinear dynamic response for the underwater blast loading in the FSDT framework.

Based on the literature, this study aims to conduct a transient analysis of sandwich plates, reinforced with GnP and FG-porous core, when exposed to underwater blast loading. The effective properties like Young's modulus and mass density of the core are obtained using the Halpin–Tsai model and rule of mixture, respectively, whereas the face sheets are considered isotropic. Using a variational approach, the governing equations for a higher-order kinematic-based plate model are formulated. The transient response in terms of nondimensional center displacement has been accessed by the finite element method along with Newmark's time integration approach and the effect of various parameters such as porosity index, GnP's weight fraction, etc. is also explored.

2 Theoretical Background

In the present study, a sandwich plate has been considered, which is composed of two thin face layers and an FG-porous graphene nanoplatelets (GnP)-reinforced core as shown in Fig. 1. The sandwich plate possesses length, width, and total thickness as *a*, *b*, and *h*, respectively, whereas the core thickness is denoted by t_c.

3 Modeling of Graphene Reinforced FG-Porous Material

The core of the sandwich plate is considered porous which follows functionally graded distributions, i.e., porosity type 1 (symmetric type) and type 2 (asymmetric type) along the thickness direction as demonstrated in Fig. 1. Porosity distributions in the core cause variation in the material properties, which are expressed as [2],

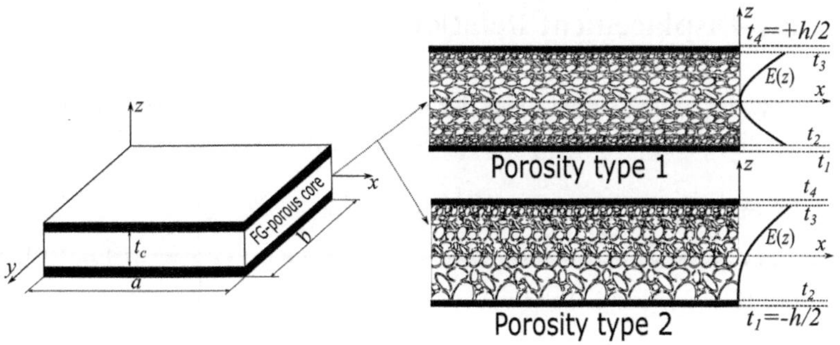

Fig. 1 Sandwich plate with FG-porous GnP-reinforced core

$$\begin{cases} E(z) = E_1[1 - k_0\,\varsigma(z)] \\ \rho(z) = \rho_1[1 - k_m\,\varsigma(z)] \end{cases}; \quad \varsigma(z) = \begin{cases} \cos(\pi z/t_c) & \text{Porosity type 1} \\ \cos(\pi z/2t_c + \pi/4) & \text{Porosity type 2} \end{cases} \tag{1}$$

In the above equation, k_0 and k_m are the porosity and mass density indices. The notations E and ρ denote Young's modulus and mass density whereas subscript '1' indicates the homogenized values of E and ρ as the FG-porous core is also reinforced with GnP. Hence, the homogenized Young's modulus (E_1) can be determined using the Halpin–Tsai model as [2],

$$E_1 = \frac{E_M}{8}\left[3\left(\frac{1 + \Upsilon_1^l\,\Upsilon_2^l\,V_{\mathrm{GnP}}}{1 - \Upsilon_2^l\,V_{\mathrm{GnP}}}\right) + 5\left(\frac{1 + \Upsilon_1^w\,\Upsilon_2^w\,V_{\mathrm{GnP}}}{1 - \Upsilon_2^w\,V_{\mathrm{GnP}}}\right)\right] \tag{2}$$

where

$$\Upsilon_1^l = 2l_{\mathrm{GnP}}/t_{\mathrm{GnP}};\ \Upsilon_1^w = 2w_{GnP}/t_{\mathrm{GnP}};\ \Upsilon_2^l = \frac{(E_{\mathrm{GnP}}/E_M) - 1}{(E_{\mathrm{GnP}}/E_M) + \Upsilon_1^l};\ \Upsilon_2^w = \frac{(E_{\mathrm{GnP}}/E_M) - 1}{(E_{\mathrm{GnP}}/E_M) + \Upsilon_1^w}$$

The subscripts 'M' and 'GnP' are employed for the metal matrix and graphene nanoplatelets, respectively.

Further, the rule of mixture is utilized to obtain homogenized mass density (ρ_1) and Poisson's ratio (ν_1) as [2],

$$\begin{aligned} \rho_1 &= \rho_{\mathrm{GnP}} V_{\mathrm{GnP}} + \rho_M(1 - V_{\mathrm{GnP}}) \\ \nu_1 &= \nu_{\mathrm{GnP}} V_{\mathrm{GnP}} + \nu_M(1 - V_{\mathrm{GnP}}) \end{aligned} \tag{3}$$

where V_{GnP} is the volume fraction of graphene nanoplatelets, which is accessed as discussed in our earlier report [3].

4 Strain–Displacement Relations

In this study, the higher-order shear deformable plate theory introduced by Pandya and Kant [9] is utilized and can be expressed as follows:

$$\left.\begin{aligned} u(x,y,z,t) &= u_0(x,y,t) + z\,\alpha_x(x,y,t) + z^3\,\beta_x(x,y,t) \\ v(x,y,z,t) &= v_0(x,y,t) + z\,\alpha_y(x,y,t) + z^3\,\beta_y(x,y,t) \\ w(x,y,z,t) &= w_0(x,y,t) \end{aligned}\right\} \Rightarrow \{u\} = [\Re]\{\Lambda\} \tag{4}$$

where $\{\Lambda\} = \left\{u_0\ v_0\ w_0\ \alpha_x\ \alpha_y\ \beta_x\ \beta_y\right\}^T$, i.e., a vector representing degrees of freedom and $[\Re]$ is the matrix of thickness coordinate, i.e., z.

Based on the above-mentioned displacement field, the strain–displacement relationship can be established as,

$$\{\in\} = \begin{Bmatrix} \in_x \\ \in_y \\ \in_{xy} \\ \in_{yz} \\ \in_{xz} \end{Bmatrix} = \begin{Bmatrix} \partial u/\partial x \\ \partial v/\partial y \\ \partial u/\partial y + \partial v/\partial x \\ \partial v/\partial z + \partial w/\partial y \\ \partial u/\partial z + \partial w/\partial x \end{Bmatrix} = \begin{Bmatrix} \in_x^0 \\ \in_y^0 \\ \in_{xy}^0 \\ \in_{yz}^0 \\ \in_{xz}^0 \end{Bmatrix} + z \begin{Bmatrix} \in_x^1 \\ \in_y^1 \\ \in_{xy}^1 \\ 0 \\ 0 \end{Bmatrix}$$
$$+ z^2 \begin{Bmatrix} 0 \\ 0 \\ 0 \\ \in_{yz}^2 \\ \in_{xz}^2 \end{Bmatrix} + z^3 \begin{Bmatrix} \in_x^3 \\ \in_y^3 \\ \in_{xy}^3 \\ 0 \\ 0 \end{Bmatrix} \Rightarrow \{\in\} = [\mathrm{H}]\{\bar{\in}\} \tag{5}$$

where $\left\{\bar{\in}\right\} = \left\{\in_x^0\ \in_y^0\ \in_{xy}^0\ \in_{yz}^0\ \in_{xz}^0\ \in_x^1\ \in_y^1\ \in_{xy}^1\ \in_{yz}^2\ \in_{xz}^2\ \in_x^3\ \in_y^3\ \in_{xy}^3\right\}^T$ which is defined in our earlier report [3].

Further, $\left\{\bar{\in}\right\}$ can also be written as,

$$\left\{\bar{\in}\right\} = [\chi]\{\Lambda\} \tag{6}$$

5 Stress–Strain Relations

The material used in this study follows Hook's law, the stress–strain relationship for the linear elastic material can be obtained as [3],

$$\{\sigma\} = [Q]\{\in\}$$

$$[Q] = \begin{bmatrix} E(z)/\left(1 - v_1^2\right) & v_1 E(z)/\left(1 - v_1^2\right) & 0 & 0 & 0 \\ v_1 E(z)/\left(1 - v_1^2\right) & E(z)/\left(1 - v_1^2\right) & 0 & 0 & 0 \\ 0 & 0 & G(z) & 0 & 0 \\ 0 & 0 & 0 & G(z) & 0 \\ 0 & 0 & 0 & 0 & G(z) \end{bmatrix} \quad (7)$$

where $E(z)$ and v_1 are determined using Eqs. (1) and (3), respectively. $G(z)$ is the shear modulus varying along the thickness direction which is equal to $E(z)/2(1 + v_1)$.

6 Equations of Motion

The variational principle is utilized to derive the equation of motion that can be represented mathematically as [3],

$$\frac{\mathrm{d}}{\mathrm{d}t}\left(\frac{\partial T}{\partial \dot{\Lambda}_i}\right) + \frac{\partial}{\partial \Lambda_i}\left(\Pi - \overline{W}\right) = 0 \quad (8)$$

where T, Π, and $\overline{W}$ are the kinetic energy, potential energy, and work done by the external load which are determined as follows,

Kinetic energy [3]

$$T = \frac{1}{2}\int_V \{\dot{u}\}^T \rho(z)\{\dot{u}\}\mathrm{d}V = \frac{1}{2}\int_A \{\dot{\Lambda}\}^T [m]\{\dot{\Lambda}\}\mathrm{d}A \quad (9)$$

Strain energy [3]

$$\Pi = \frac{1}{2}\int_V \{\in\}^T \{\sigma\}\mathrm{d}V = \frac{1}{2}\int_A \{\Lambda\}^T [k]\{\Lambda\}\mathrm{d}A \quad (10)$$

For Eqs. (9) and (10), the matrices $[m]$ and $[k]$ are defined as,

$$[m] = \int_{t_1}^{t_4} \rho(z)[\Re]^T\,[\Re]\,\mathrm{d}z;\ [k] = \int_{t_1}^{t_4} [\chi]^T\,[H]^T\,[Q][H][\chi]\mathrm{d}z$$

7 *Work Done by the External Load*

In the present study, underwater blast loading is adopted which is uniformly distributed over the plate's surface and varying with respect to time as [8],

$$p(t) = \frac{2p_1}{1-\psi}\left(\mathrm{e}^{-t/\upsilon} - \psi \mathrm{e}^{-\psi t/\upsilon}\right) \tag{11}$$

where p_1 is the peak pressure distributed over the surface of the plate and can be determined using the mass of TNT ($\forall$) and stand-off distance (d) as,

$$p_1 = \Omega_1\left(\frac{\forall^{1/3}}{d}\right)^{\mathrm{A}_1} \tag{12}$$

where the blast-associated parameters such as ψ, υ, Ω_1, and A_1 are referred from Ref. [8].

Hence, the work as a result of the underwater blast loading can be computed as,

$$\overline{W} = \frac{1}{2}\int\limits_{A} f_1 w \mathrm{d}A \tag{13}$$

in which $f_1 = \{0\;0\;p_1\;0\;0\;0\;0\}^T$.

8 Finite Element Analysis

In this study, Finite Element Analysis (FEA) was utilized to determine the numerical solution, i.e., the transient non-dimensional center displacement (*w*/*h*) of sandwich plates with graphene-reinforced FG porous core when exposed to underwater blast loading. The generalized coordinates and displacements based on FEA can be represented as [3],

$$\{q\} = \sum_{i=1}^{9} N_i\{\Lambda\}_i;\; x = \sum_{i=1}^{9} N_i x_i;\; y = \sum_{i=1}^{9} N_i y_i \tag{14}$$

where N_i denote the shape functions of nine-noded element.

Finally, Eq. (8) can be deduced by using Eqs. (9), (10), (13), and (14) as,

$$[M]\{\ddot{q}(t)\} + [K]\{q(t)\} = \{f(t)\} \tag{15}$$

where $[M]$, $[K]$, and $\{f\}$ are the global mass, stiffness, and force vectors, respectively. The solution of Eq. (15) has been obtained by utilizing Newmark's time integration method [3].

9 Results and Discussion

In the present study, the transient analysis has been conducted to determine the nondimensional center displacement (w/h) of GnP-reinforced FG-porous sandwich plate under the influence of underwater blast loading. The following parameters have been adopted (unless otherwise specified): width-to-thickness ratio (b/h) of 10, length-to-width ratio (a/b) of 1, porosity index (k_0) of 0.5, GnP's weight fraction (w_{GnP}) of 1%, TNT (explosive) weight (w_{TNT}) of 1 gm, stand-off distance (R) of 10 m, symmetric distribution and simply-supported (SSSS) boundary conditions applied to all edges of the plate. The finite element method in conjunction with Newmark's integration approach has been employed to solve the proposed dynamic model for which a MATLAB code has been developed. The material specifications have been taken from Table 1.

In Fig. 2, the convergence study has been carried out for GnP-reinforced FG-porous sandwich plates with the thickness (b/h) and aspect (a/b) ratios of 10, and 1, respectively. The simply supported sandwich plate has been investigated to obtain transient nondimensional center displacement (w/h) at various mesh sizes while considering porosity index (k_0) = 0.5 and weight fraction of GnP (w_{GnP}) = 1%. As depicted in Fig. 2, it has been observed that the nondimensional center displacement (w/h) attains convergence with a mesh size of 5×5, which has been utilized for the entire analysis.

Figure 3 presents the validation of transient center displacement (w) of the plate when exposed to blast loading. The length (a) and thickness (h) of the square ($a/b = 1$) plate have been considered as 508 mm, and 3.4 mm, respectively. The displacement obtained for the non-porous plate without any reinforcement has been compared with the existing results [10], and a good concurrence between the present proposed model and the existing model has been achieved (Fig. 3).

Figure 4 illustrates the influence of the porosity index (k_0) on the center displacement (w/h) of the GnP-reinforced sandwich plate. It has been noted from Fig. 4 that the displacement increases with an increase in the value of k_0. This happens due to the lowering of the structural stiffness, which allows more displacement when the sandwich plate is exposed to underwater blast loading.

Table 1 Material specification of the graphene nanoplatelets (GnP) and aluminum

Particular	Young's modulus (GPa)	Density (kg/m^3)	Poisson's ratio
GnP	1010	1062.5	0.186
Aluminum	70	2702	0.3

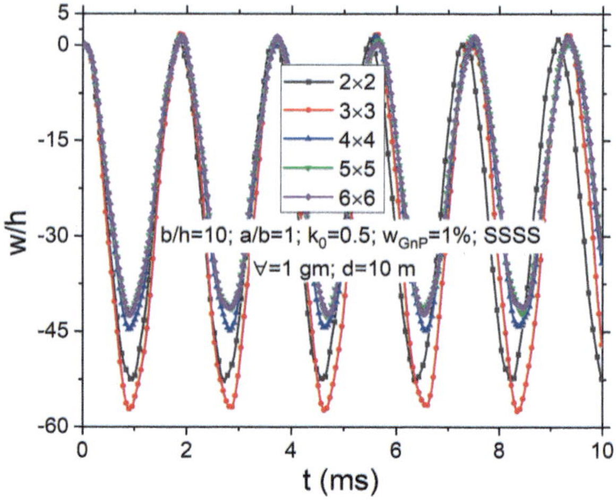

Fig. 2 Convergence study for transient *w/h* of GnP-reinforced FG-porous sandwich plate

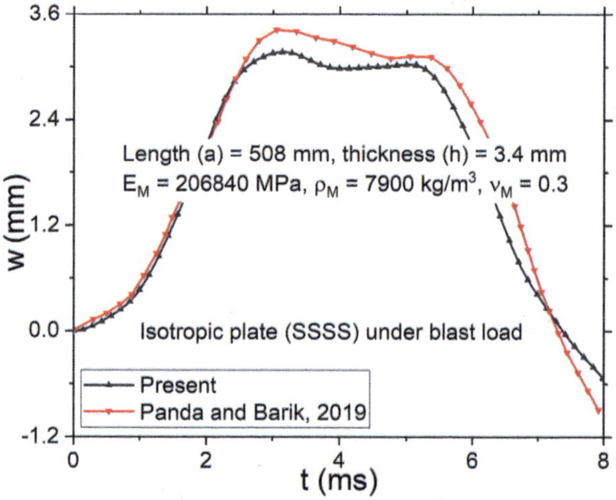

Fig. 3 Validation study of center displacement (*w*) of simply supported plate subjected to blast loading

In Fig. 5, the influence of GnP weight fraction on non-dimensional center displacement (*w/h*) of the FG-porous ($k_0 = 0.5$) sandwich plate has been demonstrated. It has been seen in Fig. 5 that the center displacement experiences a significant decrease as the weight fraction of GnP (w_{GnP}) increases. The reason for this is that the introduction of even a small quantity (i.e., 1%) of GnP into the metal matrix results in a substantial improvement in the structural stiffness of the porous plate as illustrated in Fig. 5.

Figure 6 depicts a comparison of different distributions, i.e., symmetric and asymmetric distributions as defined by Eq. (1). The symmetric distribution means the porosity distribution and GnP dispersion pattern are symmetric and conversely, it is true for the asymmetric distribution. The graphene-reinforced ($w_{GnP} = 1\%$) FG-porous ($k_0 = 0.5$) sandwich plate has been investigated for *b/h*, and *a/b* ratios as 10,

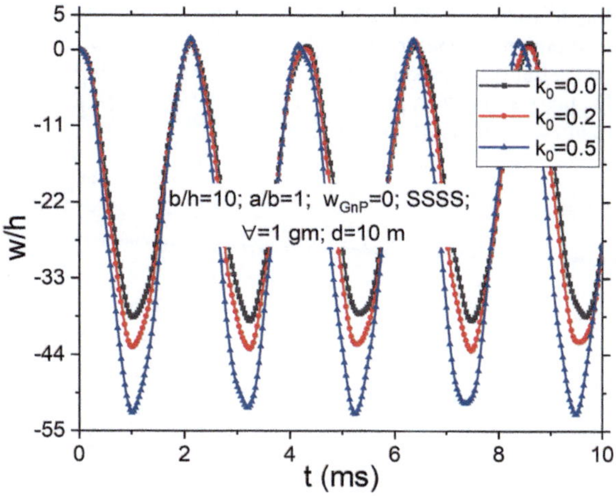

Fig. 4 Influence of porosity index (k_0) on the transient response of GnP-reinforced FG-porous sandwich plate

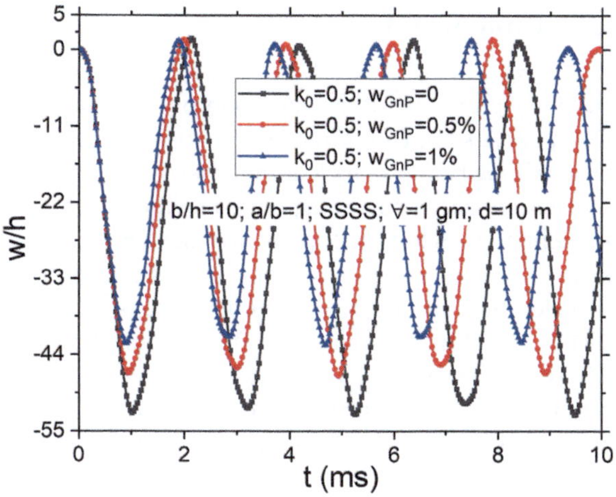

Fig. 5 Influence of GnP's weight fraction (w_{GnP}) on the transient response of GnP-reinforced FG-porous sandwich plate

and 1, respectively. It has been recognized from Fig. 6 that the asymmetric distribution has higher displacement as compared to the symmetric one because asymmetric distribution contributes comparatively lesser stiffness to the plate structures.

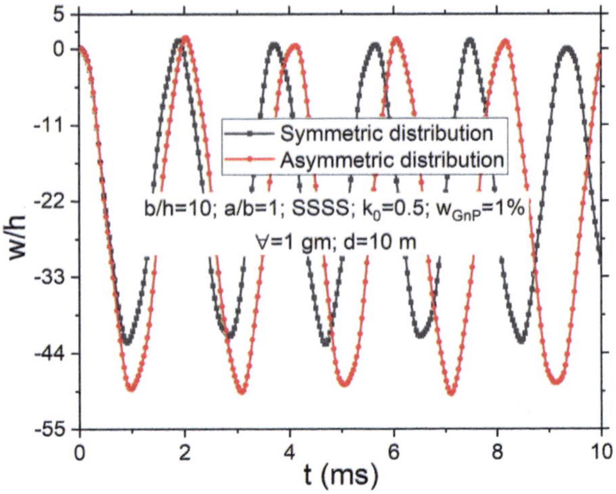

Fig. 6 Influence of porosity distribution and GnP pattern on the transient response of FG porous sandwich plates reinforced with GnP

10 Conclusion

In this study, the transient analysis of sandwich plates containing graphene nanoplatelets (GnP)-reinforced FG porous core has been conducted under the influence of underwater blast loading. The dynamic governing equation has been solved using Newmark's time integration technique, which takes inputs from finite element modeling. A higher-order plate displacement model has been adopted in the present study, which converges comparatively at a lesser mesh size of 5×5. The study has revealed that an increase in porosity leads to a significant rise in transient center displacement due to reduced stiffness. However, the addition of a small quantity of 1% weight fraction of GnP can significantly enhance the stiffness and effectively mitigate the issue. Furthermore, sandwich plates with symmetric porosity and a symmetric GnP pattern have been found to exhibit higher structural stiffness.

References

1. Birman V, Kardomateas GA (2018) Review of current trends in research and applications of sandwich structures. Compos Part B Eng: 221–240
2. Yang J, Chen D, Kitipornchai S (2018) Buckling and free vibration analyses of functionally graded graphene reinforced porous nanocomposite plates based on Chebyshev-Ritz method. Compos Struct: 281–294
3. Shakir M, Talha M (2022) Influence of material uncertainty on higher-order FG-GPLs reinforced porous spherical panels under blast loading. Thin-Walled Struct: 109319
4. Mohseni A, Shakouri M (2020) Natural frequency, damping and forced responses of sandwich plates with viscoelastic core and graphene nanoplatelets reinforced face sheets. JVC/J Vib Control: 1165–1177

5. Gao Z, Li H, Zhao J, Guan J, Wang Q (2021) Analyses of dynamic characteristics of functionally graded porous (FGP) sandwich plates with viscoelastic materials-filled square-celled core. Eng Struct: 113242
6. Quan TQ, Ha DTT, Duc ND (2022) Analytical solutions for nonlinear vibration of porous functionally graded sandwich plate subjected to blast loading. Thin-Walled Struct: 108606
7. Karakoti A, Pandey S, Kar VR (2022) Nonlinear transient analysis of porous P-FGM and S-FGM sandwich plates and shell panels under blast loading and thermal environment. Thin-Walled Struct: 108985
8. Van Long N, Ich T, Huy D, Minh T (2022) Nonlinear dynamic responses of sandwich-FGM doubly curved shallow shells subjected to underwater explosions using first-order shear deformation theory. Ocean Eng: 111886
9. Kant T, Pandya BN (1988) A simple finite element formulation of a higher-order theory for unsymmetrically laminated composite plates. Compos Struct: 215–246
10. Panda S, Barik M (2019) Transient vibration analysis of arbitrary thin plates subjected to air-blast load. J Vib Eng Technol 2019(72):189–204

Programming Both in and Outside of a Plane Auxeticity in Metamaterials Through Exploiting Hybrid Unit Cells

D. Kundu and T. Mukhopadhyay

Abstract The Poisson's effect, which is a compressive force applied longitudinally, has a limited tendency to cause engineering materials used for structural purposes to only grow in the lateral dimensions. We demonstrate varied auxeticity within the context of multi-material unit cells using the bending-dominated behaviour of elementary beams with variable curve. The hybrid unit cells (formed based on their clusters) and 3D dual ring (3DDR) unit cells can exhibit partly auxetic, solely auxetic, strictly non-auxetic, and null-auxetic activity. Less density is produced by the open cell shape of designed auxetic materials. It would be effective and scientifically informative to define multi-directional auxeticity using an analytical method. In order to create adequate confidence in the findings made using the known analytical method, extensive finite element validation has been done. Due to their enhanced mechanical powers, auxetic materials find use in highly challenging sectors.

Keywords Multi-directional auxeticity · 3D lattices · Auxetic · Non-auxetic · Poisson's ratio · Young's modulus · 3DDR unit cell · Hybrid unit cell

1 Literature Review

Poisson's ratios play a significant role in defining the fundamental constitutive mechanical properties of materials. If we compress most naturally occurring solid materials by applying force in one direction, they will try to grow in the other two perpendicular dimensions. The Poisson's ratio for these compounds is positive. Auxetic materials [1–3] with negative Poisson's ratios, which are possible through artificial microstructuring [4, 5], are now gaining popularity due to their enhanced mechanical powers in a variety of uses. It has been convincingly demonstrated in

D. Kundu (✉) · T. Mukhopadhyay
Department of Aerospace Engineering, Indian Institute of Technology Kanpur, Kanpur, India
e-mail: diptimank21@iitk.ac.in

T. Mukhopadhyay
e-mail: tanmoy@iitk.ac.in

R. K. Ray et al. (eds.), *Advances in Theoretical and Applied Mechanics*, Lecture Notes in Mechanical Engineering, https://doi.org/10.1007/978-981-97-0418-7_33

recent years [6, 7] that a negative Poisson's ratio can lead to better structural designs with control over stress distribution and deformation [8], greater energy absorption [9], enhanced shear stiffness and fracture toughness [10], variable permeability, wave propagation modulation, shape modulation, and improved actuator and sensor development. Compared to other kinds of materials, auxetic materials frequently have an open cell structure and a reduced density. Based on their improved mechanical characteristics, auxetic materials have been discovered for use in a variety of highly challenging areas, such as defence, aircraft, mechanical, and biomedical engineering.

2 Programming Auxeticity

Researchers have been continuously working on micro-scale artificial structuring to get auxetic materials with negative in-plane Poisson's ratios conspicuously at the 2D level. Hence, a strong motivation exists in this domain to structure micro-lattices for the manifestation of auxeticity the both in and outside of a plane directions at the 3D level.

To change Poisson's ratios in multiple directions, we have created ubiquitous 3D dual ring (3DDR) type periodic micro-lattices [11, 12]. In 3DDR type unit cells and their enlarged versions, multi-modal auxeticity can be accomplished in both in and outside of plane orientations. A single 3DDR unit cell can be tessellated to produce partly auxetic lattices, or those with negative and positive Poisson's ratios in two mutually vertical planes. Fully non-auxetic is possible, as well as fully auxetic (negative Poisson's ratio in both in and outside of a plane directions), fully null-auxetic (no Poisson's ratio in either in-plane or out-of-plane directions), and fully null-auxetic. In a periodic structure, one unit cell (i.e., recurring units) can be analysed with the right border conditions to determine the overall behaviour of the complete grid. Figure 1a shows a single 3DDR unit cell with two planar bands in two orthogonal planes linked by four coupled beams. The geometrical connections that connect the sides of the parent unit cell in the three global directions with a total length in the X-direction of $2a$ are represented in the picture by the non-dimensional factors m and α.

Energy-based methods are used to acquire the beam-level deformations. They control the unit cells' both in and outside of a plane displacement behaviours. A compressive force is applied along the Z-direction that allows the cell to spread along the X-direction concerning the ring in the XZ plane in order to derive the Poisson's ratios of the lattice created by tessellating the 3DDR unit cell. The cell now contracts in Y-direction with regard to the ring in the XY plane because the displacement in X-direction is the same as that of the ring in the XY plane. So, for the 3DDR lattice, we see a partly auxetic tendency. With a negative Poisson's ratio in the Y-direction and a positive Poisson's ratio in the X-direction, the 3DDR lattice displays partly auxetic activity as a result. Totally non-auxetic lattices with favourable Poisson's ratios in both in and outside of a plane orientations are what we aim to achieve. A composite unit cell has been suggested for this purpose by grouping $N_1 + N_2$ 3DDR units in a

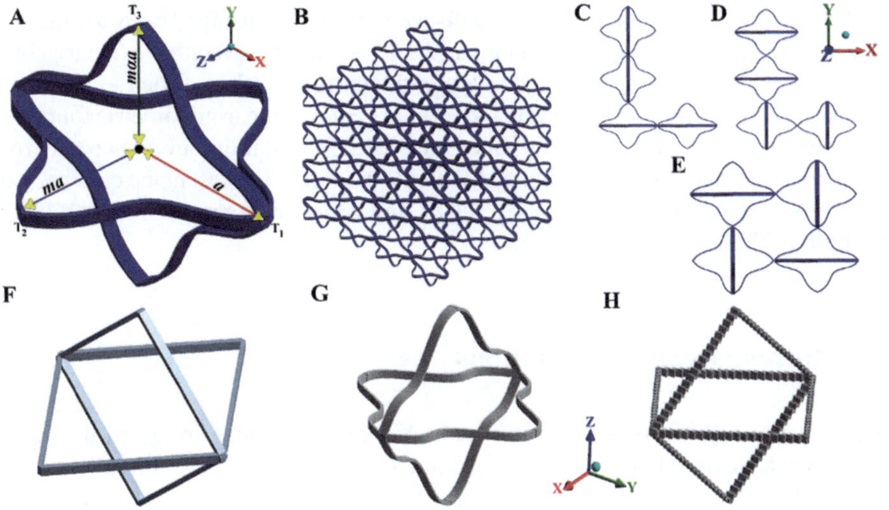

Fig. 1 *3D dual ring (3DDR) unit cell.* **a** View in three dimensions of the parent unit's 3DDR cell displaying partly auxetic activity. **b** Entire lattice developed using 3DDR unit cells. **c** A hybrid unit cell is one that has four progenitor unit cells arranged in a cluster to produce lattices with positive Poisson's ratios in both orientations. **d** Four progenitor unit cells were clustered together to create a hybrid unit cell, which then developed lattices with a negative Poisson's ratio in both orientations. **e** Four parent unit cells are clustered together to create a hybrid unit cell, which results in lattices with zero Poisson's ratio in both orientations. **f** Special case: 3DDR unit cell formed using straight-line curves. **g** 3DDR unit cell formed using trigonometric geometry induced straight-line curves with higher amplitude and less undulations. **h** 3DDR unit cell formed using trigonometric geometry-induced straight-line curves with lower amplitude and more number of undulations

unique configuration, as shown in Fig. 1b. Here, we have maintained the local X-axis at the unit cell level along the global X-axis at the lattice level for N_1 number of cells and have done the same for N_2 number of cells by maintaining the local X-axis along the global Y-axis. The hybrid unit cell is the L-shaped collection that we have lastly suggested. Be aware that changing N_1 and N_2 in addition to all other design factors of a 3DDR unit allows us to influence the Poisson's ratio values, opening up more design options for Poisson's ratio modulation. In addition, we have constructed pure auxetic lattices with negative Poisson's ratios in the both in and outside of a plane orientations. A composite unit cell has been suggested to achieve this goal by grouping $N_1 + N_2$ 3DDR units in a unique configuration, as shown in Fig. 1c. Here, the local Y-axis has been maintained along the global Y-axis for N_2 cells while the local X-axis has been put along the global X-axis for N_1 cells. We also suggest a hybrid unit cell that can exhibit null-auxeticity (i.e., zero Poisson's ratios in the both in and outside of a plane orientations) and is created by grouping four 3DDR units shown in Fig. 1d. Analyzing the individual 3DDR unit cells' displacements across the XY plane allows one to accomplish this condition.

Previously [5], similar 3DDR unit cells were developed using various geometries like linear, parabolic, elliptical, and logarithmic. It was observed that a linear geometry gave a higher Young's modulus, E_z. In the current work, we tried to introduce amplitude-controlled undulations in the form of waves using trigonometric functions as shown in Fig. 1g, h. Later in the paper, analytical formulations are shown to derive the displacements, Poisson's ratio, and Young's modulus. To present the efficiency of the trigonometric geometry-induced 3DDR unit cells, graphs are shown in relation to those of the linear 3DDR unit cell.

3 Mathematical Formulations

The equations of the geometric profiles in *XZ* and *XY* planes are given by Eqs. 1 and 2. We have used two parameters k_{amp} and k_{wave} to control the amplitude size and the number of waves over the initial straight-line curve. For further calculations, we have considered 10 mm as the value for *a*.

$$f(x, z) = -mx + ma - k_{\text{amp}} \sin\left(\frac{(2k_{\text{wave}} - 1)\left(\sqrt{1 + m^2}\right)\pi x}{a\left(1 + m^2\right)}\right) \tag{1}$$

$$f(x, y) = -m\alpha x + m\alpha a - k_{\text{amp}} \sin\left(\frac{(2k_{\text{wave}} - 1)\left(\sqrt{1 + m^2\alpha^2}\right)\pi x}{a\left(1 + m^2\alpha^2\right)}\right) \tag{2}$$

The deflection of the 3DDR unit cell in the global directions (*X*, *Y*, *Z*) is given by Eqs. 3–5. These equations are used in combination to get the overall deflection of a hybrid unit cell, which are formed by clustering the 3DDR unit cells. The numerical values of the parameter set Φ_n. (*n*: *1*–11) can be evaluated using the integral equations derived from stain energy principles, worked out in the previous work [5]. In the current study, *Z* is considered as the flexural rigidity in relation to the cross-section and material property of the rings which form the 3DRL unit cell. We can have two different Z values considering a multi-material unit cell, which is explored in [5].

$$\Delta^X_{\text{Overall}} = -\frac{F}{Z}\left(\frac{\Phi_7\Phi_{10}}{(\Phi_5 + \Phi_{10})}\right) \tag{3}$$

$$\Delta^Y_{\text{Overall}} = \frac{F}{Z}\left(\frac{\Phi_7\Phi_{11}}{(\Phi_5 + \Phi_{10})}\right) \tag{4}$$

$$\Delta^Z_{\text{Overall}} = \frac{F}{Z}\left(\Phi_3 - \frac{\Phi_7\Phi_6}{(\Phi_5 + \Phi_{10})}\right) \tag{5}$$

The negative and positive Poisson's ratios given by the partial auxetic 3DDR unit cell are given by the following equations.

$$\upsilon_{yz}^{-ve} = -\left(\frac{\Phi_7\Phi_{11}}{(\Phi_3(\Phi_5 + \Phi_{10}) - \Phi_7\Phi_6)}\right)\left(\frac{1}{\alpha}\right) \tag{6}$$

$$\upsilon_{xz}^{+ve} = \left(\frac{\Phi_7\Phi_{10}}{(\Phi_3(\Phi_5 + \Phi_{10}) - \Phi_7\Phi_6)}\right)(m) \tag{7}$$

The expression for the Young's modulus for the partial auxetic 3DDR unit cell is given by Eq. 8.

$$E_z = \left(\frac{Z(\Phi_5 + \Phi_{10})}{2(a\alpha)((\Phi_3\Phi_5 - \Phi_7\Phi_6) + \Phi_3\Phi_{10})}\right) \tag{8}$$

The clustered unit cell that gives non-auxetic behaviour gives the values of the Poisson's ratios and the Young's modulus using the following set of equations.

$$\upsilon_{xz}^{+\text{ve}} = \left(\frac{\Phi_7\Phi_{10}}{(\Phi_3(\Phi_5 + \Phi_{10}) - \Phi_7\Phi_6)}\right)(m) \tag{9}$$

$$\upsilon_{yz}^{+\text{ve}} = \left(\frac{\Phi_7(-\Phi_{11} + \Phi_{10}N_2)}{(\Phi_3(\Phi_5 + \Phi_{10}) - \Phi_7\Phi_6)}\right)\left(\frac{m}{(N_2 + m\alpha)}\right) \tag{10}$$

$$E_z = \left(\frac{Zm(\Phi_5 + \Phi_{10})}{2(N_1a(m\alpha + N_2))((\Phi_3\Phi_5 - \Phi_7\Phi_6) + \Phi_3\Phi_{10})}\right) \tag{11}$$

The clustered unit cell that gives auxetic behaviour gives the values of the Poisson's ratios and the Young's modulus using the following set of equations.

$$\upsilon_{xz}^{-ve} = -\left(\frac{\Phi_7\Phi_{11}}{(\Phi_3(\Phi_5 + \Phi_{10}) - \Phi_7\Phi_6)}\right)\left(\frac{1}{\alpha}\right) \tag{12}$$

$$\upsilon_{yz}^{-\text{ve}} = -\left(\frac{\Phi_7(-\Phi_{10} + \Phi_{11}N_2)}{(\Phi_3(\Phi_5 + \Phi_{10}) - \Phi_7\Phi_6)}\right)\left(\frac{m}{(1 + m\alpha N_2)}\right) \tag{13}$$

$$E_z = \left(\frac{Z(\Phi_5 + \Phi_{10})}{2(N_1a\alpha(N_2m\alpha + 1))((\Phi_3\Phi_5 - \Phi_7\Phi_6) + \Phi_3\Phi_{10})}\right) \tag{14}$$

4 Results and Discussion

The effective Young's modulus, $\overline{E_z}$ in Figs. 2 and 3 is calculated using the ratio of Young's modulus of the trigonometric geometry-induced 3DDR unit cell and that of the straight line geometry-induced 3DDR unit cell. Initially in Fig. 2, we try with higher amplitude functions varying k_{amp} greater than 1 keeping lower number of undulations limiting k_{wave} to 3. The values of $\overline{E_z}$ are decreasing from 1 indicating that we have not achieved any improvement in the stiffness of the 3DDR unit cell.

Secondly, in Fig. 3, we lower the amplitude factor to create minute undulations and increase the number of waves. Here, we have achieved higher stiffness along the Z direction as the value is greater than 1. However, as we increase the number of waves, the stiffness gradually decreases. Also, it can be shown that with higher α values, there is a substantial decrease in the values of effective Young's modulus. For a certain magnitude of k_{amp}, it can be observed how varying the parameter, m changes the nature of the plots (Fig. 4).

Again, in Fig. 5, when we increase the undulations, controlling k_{wave}, keeping lower amplitude, it is observed that the Poisson's ratio values are constant for this case. Also, it can be depicted that with higher α values, there is a substantial decrease in the magnitudes of effective negative Poisson' ratios, whereas the effective positive Poisson' ratios are constant with α. It is interesting to examine how changing the parameter, m, alters the nature of the graphs for a given magnitude of amplitude factor, k_{amp}.

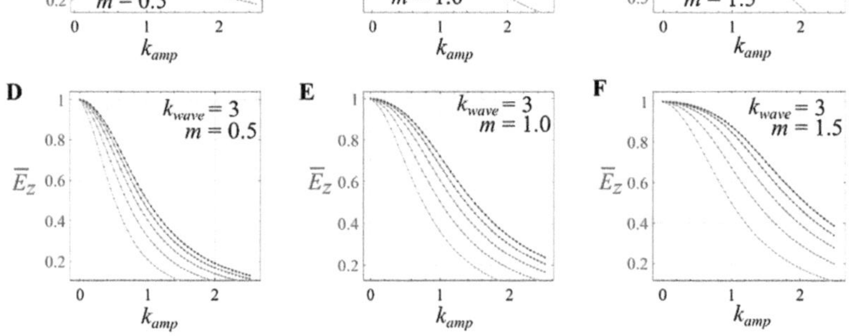

Fig. 2 *Variation of effective Young's modulus with amplitude factor.* **a–f** Decrement of effective Young's modulus with increment of amplitude factor against different geometrical parameters, m and α using trigonometric 3DRL unit cell. We can observe that $\overline{E_z}$ is decreasing with increasing amplitude of the wave while taking lower number of undulations

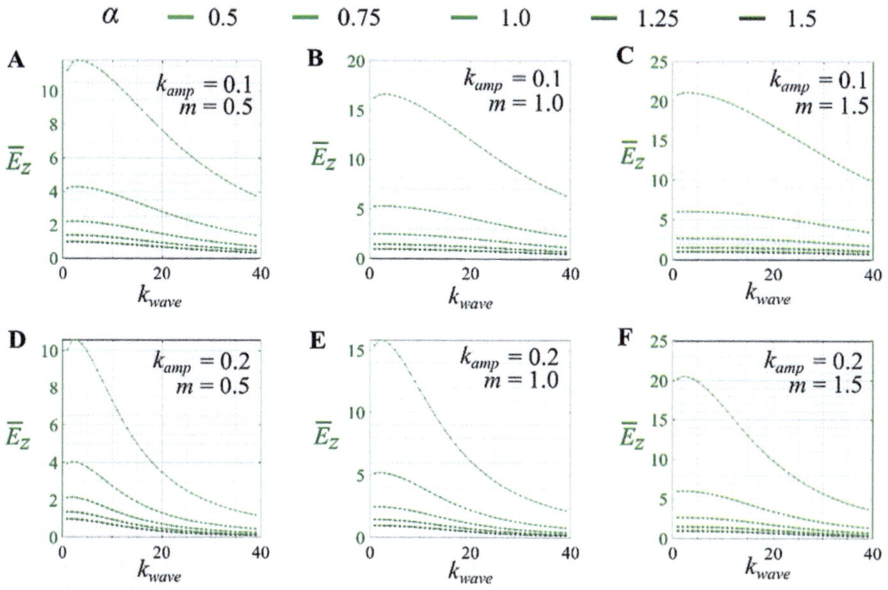

Fig. 3 *Variation of effective Young's modulus with wave factor.* **a–f** Decrement of effective Young's modulus with increment of wave factor against different geometrical parameters, m and α using trigonometric 3DRL unit cell. We can observe that $\overline{E_z}$ is decreasing with increasing number of waves while taking lower amplitude value

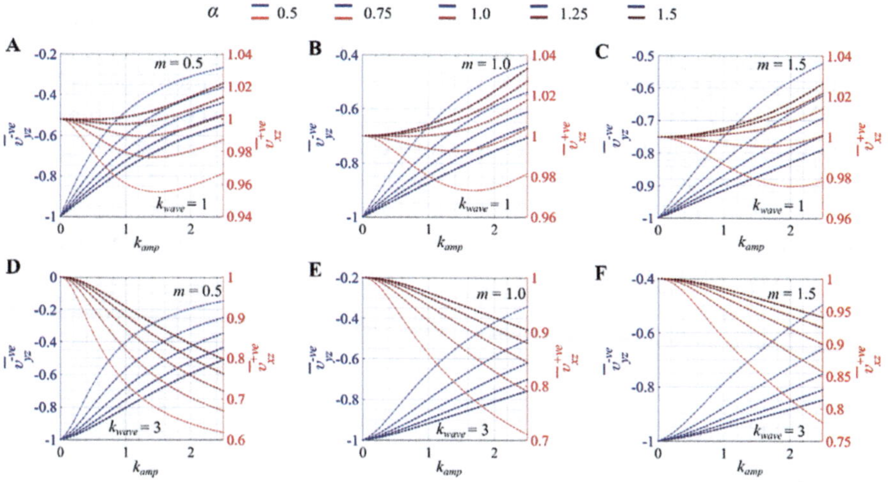

Fig. 4 *Variation of effective Poisson's ratio with amplitude factor.* **a–f** Decrement of effective Poisson's ratio with increment of amplitude factor against different geometrical parameters, m and α using trigonometric 3DRL unit cell. Both the positive and negative Poisson's ratios are varying with increasing amplitude factor

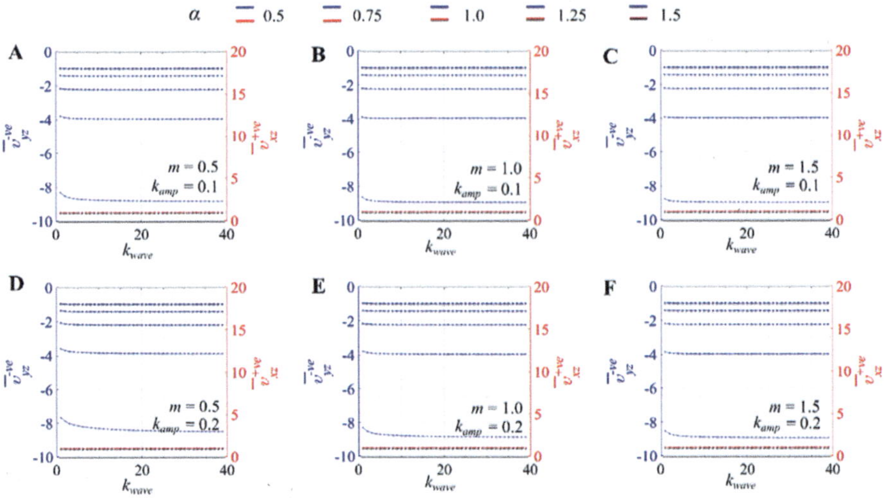

Fig. 5 *Variation of effective Poisson's ratio with amplitude factor.* **a–f** Change of effective Poisson's ratio with increment of amplitude factor against different geometrical parameters, m and α using trigonometric 3DRL unit cell. Both the positive and negative Poisson's ratios are constant with increasing wave factor

Results for lattices constituting clustered unit cells constituting the trigonometric 3DDR unit cells, showing auxetic and non-auxetic behaviour (refer to Fig. 1c, d) are shown in Fig. 6. The variation of Young's modulus and Poisson's ratios is shown with varying amplitude factor and wave factor along with controlling different geometric parameters, α. Figure 7a gives an idea of the mechanism of the clustered unit cell with four 3DRL partially auxetic units (using compatibility conditions over the connected nodes) that will constitute a lattice to provide zero Poisson's ratio. Subfigure C depicts how we can control the geometrical parameters, m and α simultaneously, such that we model the clustered unit as per the requirements. It is observed that introducing low amplitude undulations in the form of trigonometric functions gives rise to fluctuations within 2.5%.

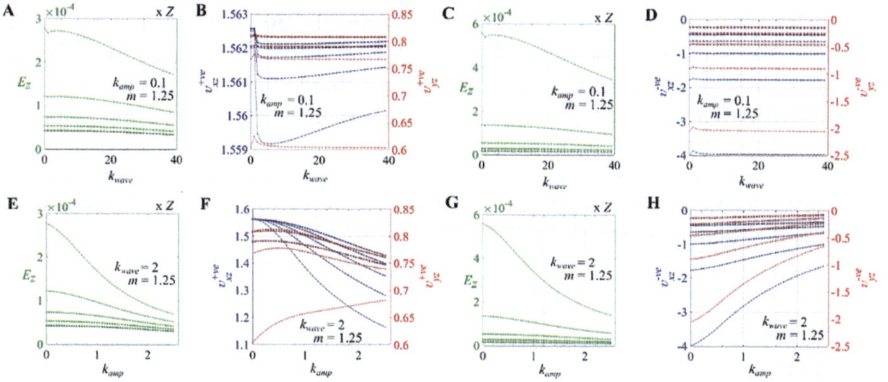

Fig. 6 *Variation of Young's modulus and Poisson's ratio with amplitude factor and wave factor.* **a–d** Change of Young's modulus and Poisson's ratio with increment of wave factor against different geometrical parameters, m and α using clustered unit cell showing non-auxetic behaviour for (**A**, **B**) and clustered unit cell showing auxetic behaviour for (**C**, **D**). **e–h** Change of Young's modulus and Poisson's ratio with increment of amplitude factor against different geometrical parameters, m and α using clustered unit cell showing non-auxetic behaviour for (**A**, **B**) and clustered unit cell showing auxetic behaviour for (**C**, **D**)

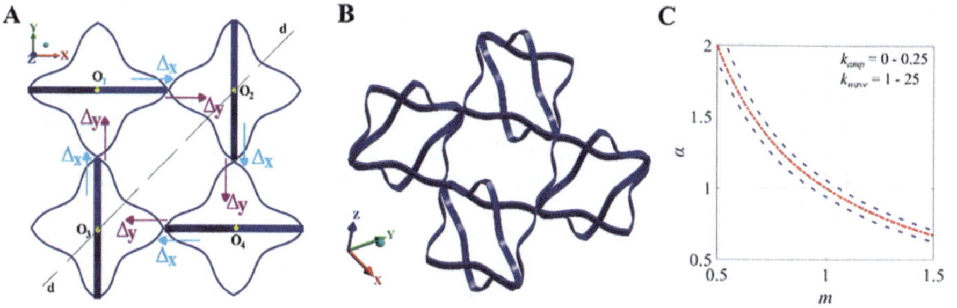

Fig. 7 *Clustered unit cell that can show zero Poisson's ratio when it constitutes a lattice.* **a** Mechanism of the clustered unit cell showing zero Poisson's ratio. **b** Isometric view of the clustered unit. **c** Red curve showing the variation of α with m for a clustered unit constituting straight line, 3DDR partial auxetic unit cell along with the blue dashed lines showing the fluctuations in the curve as we introduce undulations and increase k_{wave} up to 25 with lower k_{amp}

5 Conclusion

The suggested microstructures constituting the 3DDR unit cells and the hybrid unit cells can withstand substantial stresses, and the current investigation helps to characterise how they behave. This paper develops a general, computationally effective mathematical framework for calculating the Poisson's ratio and Young's modulus of the suggested 3D microstructure. Amplitude and wave-controlled trigonometric geometry-induced unit cells were thoroughly studied in the paper for the achievement of better stiffness along the loading direction. Using them, we develop the partially auxetic lattices and cluster them to form the auxetic and non-auxetic lattices. Finally, we also give a thought about the lattices showing zero Poisson's ratios. Hence, tailorable geometry parameters have led us to achieve mechanical metamaterials, which can be manufactured and applied to structures based on the requirements.

Acknowledgements TM would like to express their gratitude for the IIT Kanpur Initiation funding they got for the duration of this study project. DK thanks the Indian Ministry of Education for providing funding assistance in the form of a doctoral scholarship.

References

1. Evans KE, Nkansah M, Hutchinson I, Rogers S (1991) Molecular network design. Nature 353(6340):124–124
2. Scarpa F, Ciffo L, Yates J (2003) Dynamic properties of high structural integrity auxetic open cell foam. Smart Mater Struct 13(1):49
3. Mukhopadhyay T, Adhikari S (2016) Effective in-plane elastic properties of auxetic honeycombs with spatial irregularity. Mech Mater 95:204–222
4. Wang H, Zhao D, Jin Y, Wang M, Mukhopadhyay T, You Z (2020) Modulation of multi-directional auxeticity in hybrid origami metamaterials. Appl Mater Today 20:100715
5. Prajwal P, Ghuku S, Mukhopadhyay T (2022) Large-deformation mechanics of anti-curvature lattice materials for mode-dependent enhancement of non-linear shear modulus. Mech Mater 2022:104337
6. Ghuku S, Mukhopadhyay T (2022) On enhancing mode-dependent failure strength under large deformation: the concept of anti-curvature in honeycomb lattices. Compos Struct 2022:116318
7. Singh A, Mukhopadhyay T, Adhikari S, Bhattacharya B (2022) Active multi-physical modulation of Poisson's ratios in composite piezoelectric lattices: on-demand sign reversal. Compos Struct 280:114857
8. Alderson KL, Fitzgerald A, Evans KE (2000) The strain dependent indentation resilience of auxetic microporous polyethylene. J Mater Sci 35(16):4039–4047
9. Lakes R, Elms K (1993) Indentability of conventional and negative poisson's ratio foams. J Compos Mater 27(12):1193–1202
10. Choi J, Lakes R (1992) Non-linear properties of polymer cellular materials with a negative poisson's ratio. J Mater Sci 27(17):4678–4684
11. Mukhopadhyay T, Kundu D (2022) Mixed-mode multidirectional poisson's ratio modulation in auxetic 3D lattice metamaterials. Adv Eng Mater 24(5):2101183
12. Wang L, Zhu S, Wang B, Tan X, Zou Y, Chen S, Li S (2021) Latitude-and-longitude inspired three-dimensional auxetic metamaterials. Extreme Mechanics Letters 42:101142

Influence of Induction Heating on Thermal Characteristics and Residual Stress for Plasma Arc Welding of Structural Steel Plate

Sunil Kumar Biswal and Sukhomay Pal

Abstract Structural steels are widely adopted in offshore structures, high-rise buildings, bridges, storage tanks, and construction machines because of their low cost and toughness. However, structural steels are encountered various problems, such as cracking, residual stress and distortion during the arc welding process, leading to early failure of the weld joint. Currently, a finite element-based analysis is conducted by adopting a high-frequency induction heating (HFIH) source as a preheat source to relieve the residual stresses. HFIH is a non-contact, environmentally friendly heating process suitable for the scanning heating of the weld line. The present study finds that the addition of HFIH significantly affects the final residual stress distribution and improves the weld penetration.

Keywords Welding residual stress · Electromagnetic-thermal analysis · Thermo-mechanical analysis · Induction preheating · Arc welding

1 Introduction

With the advent of environmentally friendly and sustainable materials, structural steels remain significant due to their relatively low cost and versatile engineering properties [1]. Joining steel plates with efficiency and high productivity in welding operations is a challenge to the competitiveness of manufacturing industries. Arc welding is a conventional low-cost fabrication process widely adopted to fabricate and assemble structures used in the automobile, shipbuilding, and locomotive industries. However, the non-uniform simultaneous heating and cooling in the different regions of the weld induce residual stress and distortion in the welded component. This can augment the premature structural failure of the welded component under externally applied load. Therefore, to ensure the structure's safety, residual stress and distortion must be controlled prior to the installation. High-frequency induction

S. K. Biswal (✉) · S. Pal
Mechanical Engineering, IIT Guwahati, Guwahati, Assam, India
e-mail: biswa176103105@iitg.ac.in

R. K. Ray et al. (eds.), *Advances in Theoretical and Applied Mechanics*, Lecture Notes in Mechanical Engineering, https://doi.org/10.1007/978-981-97-0418-7_34

heating (HFIH) has been extensively adopted in order to reduce the residual stress in welding [2]. The HFIH is a non-contact sustainable heating source, which generates the heat in the workpiece by a high-frequency electromagnetic field. In order to predict the residual stress and distortion at the design stage, FEM simulation is a cost-effective approach [3]. Therefore, quantitative estimation and mitigation of residual stresses during the weld is a crucial concern prior to real-world fabrication [4]. Researchers adopted several approaches to minimise the weld-induced residual stress for structural steels through numerical and experimental methods. Suman et al. [5] performed a numerical and experimental study to analyse the effect of oxyacetylene gas flame preheating for the submerged arc welding of P92 steel plate. Zhang et al. [6] studied the combined effect of the laser weld source with metal inert gas welding on the distortion and residual stress distribution of the 4 mm thick copper plates. Similarly, Zhao et al. [7] experimentally report a comparative study between laser-arc double-sided welding with arc welding for 5A06 alloy. Jiang et al. [8] adopt a water-cooled heat sink pipe to cool the bottom of the weld bead during the welding to reduce the residual weld stress. Sarmast et al. [9] studied the evolution of residual stress during the repair welding of S355J2+N steel. The author found that initial preheating of the weld reduces the residual stress of the weld. Although, the published works have emphasised on various techniques to reduce the residual stress due to welding. However, the effect of induction preheating on residual stress and thermal characteristics in plasma arc welding (PAW) is limited. Therefore, the current work proposed a three-dimensional finite element (FE) model to analyse the effect of the HFIH induction source on plasma arc welding of 6 mm S355 steel plate. Subsequently, temperature history and residual stress distribution for both PAW and IPAW processes are accessed. It is observed that induction preheating improves weld penetration and reduces residual stress.

2 Finite Element Simulation

In IPAW process, a high-frequency induction non-contact heating mechanism is used to preheat the weld line prior to joining of the workpiece using a constricted plasma arc heat source. As a result, the schematic view of the IPAW process for joining of a flat plate is shown in Fig. 1. The plasma arc source and induction coil are positioned by a standoff gap of 40 mm. In addition, a coupling gap of 2 mm (as shown in Fig. 1a) is maintained between the inductor coil and workpiece for electromagnetic interaction between the inductor coil and workpiece. In current work, a two-turn co-direction current inductor coil is considered for preheating the workpiece. The dimension of the inductor coils is shown in Fig. 1b. The temperature-independent material property of the copper inductor coil is referred from the literature [10]. For the accurate FE simulation, the temperature-dependent thermal properties of the workpiece are considered from the published literature [11].

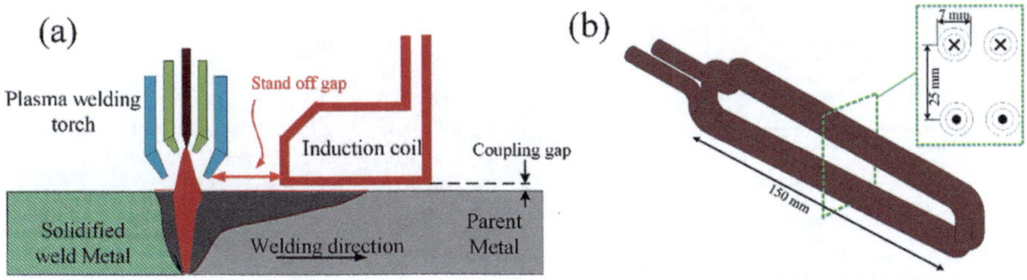

Fig. 1 **a** Schematic of induction-assisted plasma arc welding, **b** schematic of the inductor coil

2.1 Finite Element Model

The commercial software COMSOL 5.5 is used to execute the coupled electromagnetic-thermal–mechanical analysis. The temperature history generated for the induction heating is investigated by coupling the electromagnetic-thermal analysis in the FE model. The temperature history of the plasma arc welding is solved using thermal analysis. Further, the residual stress and distortion are evaluated using coupled thermal–mechanical analysis. The detailed workflow of the FE model is shown in Fig. 2a. The three-dimensional FE model consists of a full-size inductor coil, workpiece and air domain comprised of 115,250 elements as shown in Fig. 2b. The workpiece dimension of 150 mm × 150 mm × 6 mm is used for the FE simulation. A finer triangular mesh of minimum size 0.5 mm is adopted near the weld zone whereas a relatively coarse mesh of minimum 2 mm size is adopted for the other surface of the workpiece. Further, a swept mesh of a minimum size 1 mm is provided along the thickness direction of the workpiece in order to get an accurate solution. The inductor coil is meshed with tetrahedral domain mesh with a minimum element size of 1 mm. Finally, the air domain is finely meshed using a tetrahedral mesh with a minimum element size of 1 mm near the coupled gap surface between the inductor coil and workpiece. The other air domain is coarsely meshed with a minimum size of 3 mm to get an efficient solution with a lower computation cost. The temperature-dependent material properties are taken into consideration, and the equations are applied to individual discretised elements. These equations are then assembled to generate a global set of equations, which are then solved to obtain the solution at any individual discretisation. The finite element model is used to calculate the magnetic flux density, eddy current, temperature field, and residual stress. The FE model analysis is divided into three steps as follows.

2.1.1 Electromagnetic Field Modelling

A transient electromagnetic analysis is carried out, by solving eddy current generation, which is developed by solving Maxwell's Equation (Eqs. 1–4) in Cartesian coordinates.

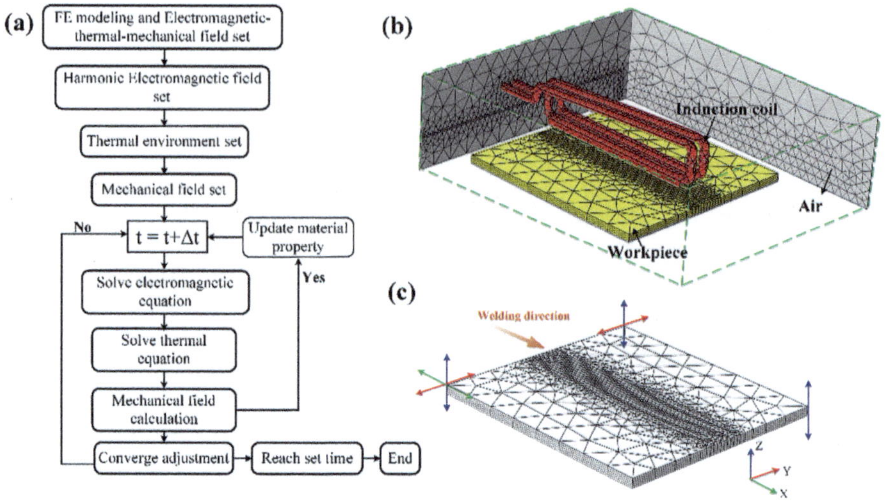

Fig. 2 **a** Flowchart of the FE simulation, **b** FE model meshing of the coupled electromagnetic-thermal-mechanical analysis, **c** boundary constraints for residual stress calculation

$$\nabla \times E = -\frac{\partial B}{\partial t} \tag{1}$$

$$\nabla \times H = J + \frac{\partial D}{\partial t} \tag{2}$$

$$\nabla . B = 0 \tag{3}$$

$$\nabla . D = \rho \tag{4}$$

where H is the magnetic flux intensity, J is the current density, D is the electric flux density, E is the electric field intensity, B is the magnetic flux density, ρ is the electric charge density.

2.1.2 Thermal Field Modelling

The transient nonlinear heat flow analysis is performed to determine the temperature distribution in the weld and heat-affected zones (HAZ) using the Fourier thermal conduction differential equation.

$$C\gamma \frac{\partial T}{\partial t} - k\left(\nabla^2 T\right) = Q_{\text{eddy}} + Q_{\text{Arc}} \tag{5}$$

The heat source density generated by eddy current per unit time (Q_{eddy}) is defined as:

$$Q_{\mathrm{eddy}} = \frac{1}{\sigma}|J_e|^2 = \sigma\left[\partial\left(Ae^{j\omega t}\right)/\partial t\right]^2 = \sigma\left(j\omega Ae^{j\omega t}\right) \tag{6}$$

The heat distribution of melt-in mode plasma arc welding by taking account of a double ellipsoidal heat source ($Q_{\mathrm{Arc}} = Q_f + Q_r$). The schematic of the double ellipsoidal heat source is shown in Fig. 3.

$$Q_f = \frac{Q6\sqrt{3}f_f}{a_f bc\sqrt{\pi}}\exp\left(\frac{-3(x - v.t)^2}{a_f^2} - \frac{-3y^2}{b^2} - \frac{-3z^2}{c^2}\right) \tag{7}$$

$$Q_r = \frac{Q6\sqrt{3}f_r}{a_r bc\sqrt{\pi}}\exp\left(\frac{-3(x - v.t)^2}{a_r^2} - \frac{-3y^2}{b^2} - \frac{-3z^2}{c^2}\right) \tag{8}$$

where $f_f(0.66)$ and $f_r(1.34)$ define the weighing function of the heat source and it satisfies the relationship of $f_f + f_r = 2$. The parameters a_f, a_r, b, and c are the double ellipsoidal beam parameter that signifies the heat source's shape in x, y, z spatial domains. The velocity (v) and time (t) define the continuity of the plasma arc along the x direction. The required dimension of the heat source is measured by ripple marking from the transverse cut section of the workpiece. The movement of the heat source is by updating the position of the heat source on each time step depending on the scanning velocity(v) as depicted in Eqs. (7) and (8).

For plasma arc welding processes, the total heat input is given by:

$$Q = \eta VI \tag{9}$$

where η is the coefficient of welding efficiency; V is the voltage of the welding arc and I is the welding current. The parameters of the heat source model for IPAW and PAW are depicted in Table 1. The required convective and radiative heat transfer coefficients are considered in the FE model [11].

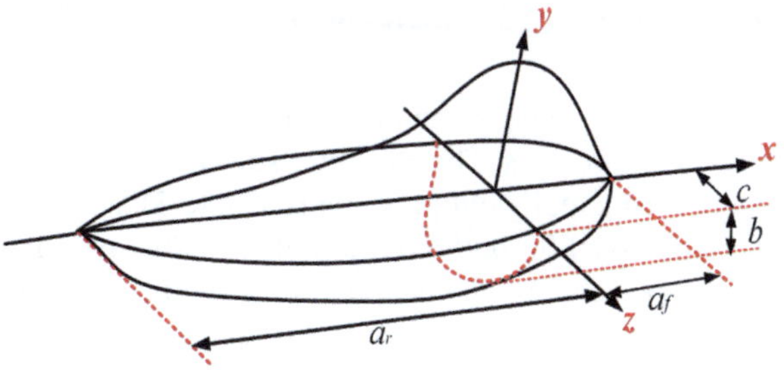

Fig. 3 Double ellipsoidal heat source model

Table 1 Parameters of the plasma heat source model

	a_f (mm)	a_r (mm)	b (mm)	c (mm)	V (volt)	I (A)	V (mm/s)	η
IPAW	4	8	6	5.5	26	140	2.5	0.7
PAW	3	6	3.5	3	26	140	2.5	0.7

2.1.3 Mechanical Field Modelling

The structural mechanic module is coupled in the model to predict the thermal distortions and residual stresses. The deformation of the welded structure results from the plastic strain at elevated temperature and the thermal expansion and contraction. The total strain ($\nabla\varepsilon^{\text{total}}$) in the workpiece is considered by summing up three individual strain neglecting the volumetric change and phase transformation strain, as shown in the equation.

$$\nabla\varepsilon^{\text{total}} = \nabla\varepsilon^{\text{elastic}} + \nabla\varepsilon^{\text{plastic}} + \nabla\varepsilon^{\text{thermal}} \tag{10}$$

where $\nabla\varepsilon$ represents the incremental strain components due to elastic, plastic, thermal changes, respectively. The elastic strain increment $\nabla\varepsilon^{\text{elastic}}$ is predicted using Hook's law with temperature-dependent Young's modulus and Poisson's ratio. After yield stress is surpassed, plastic straining occurs and a strain hardening occurs. For the plastic strain increment $\nabla\varepsilon^{\text{plastic}}$, the Von-Mises yield criterion, temperature-dependent mechanical properties and linear elastic, perfectly plastic behaviour is assumed for the numerical analysis. The clamping boundary constraints for the mechanical analysis are shown in Fig. 2c. The thermal strain increment is evaluated by considering the following:

$$\nabla\varepsilon^{\text{thermal}} = \alpha.\nabla T \tag{11}$$

where α is the thermal expansion coefficient, and ∇T is the change in temperature.

3 Results and Discussion

In this study, a square butt joint configuration is considered for the PAW and IPAW fabrication process. The characteristic geometrical sizes of weld bead are taken from the experimental weld pool shape (Table 1), and they are used as the input into FE thermal model for both PAW and IPAW processes. In case of IPAW, the inductor coil geometry (Fig. 1b) is inputted to the FE model. The induction preheating is assessed with a coil current of 550 A at a frequency of 330 kHz provided as input to the FE model. The initiation of the plasma heat source and movement of the induction coil is followed by a static induction heating of 40 s.

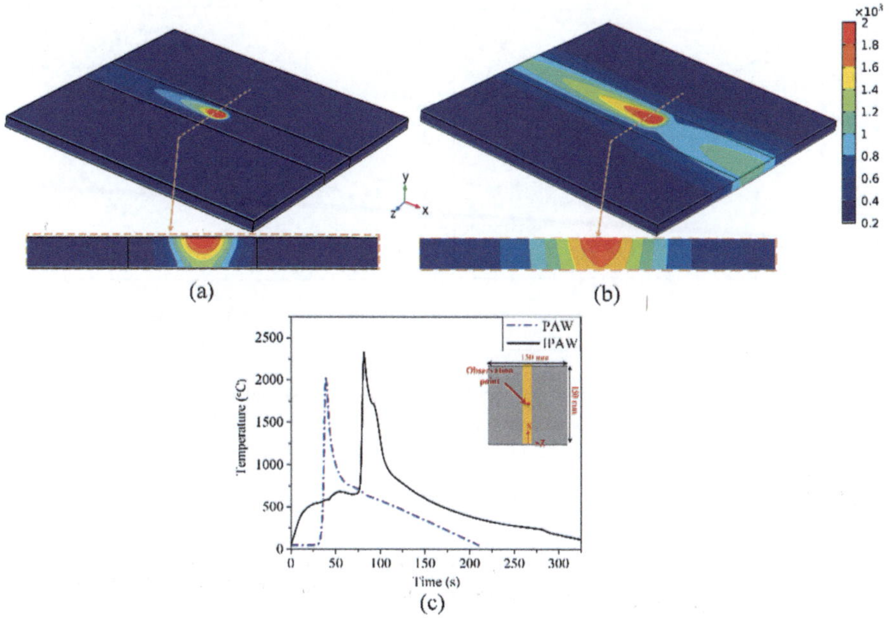

Fig. 4 Temperature distribution **a** PAW, **b** IPAW, **c** comparison of temperature–time plot between PAW and I-PAW process

3.1 *Temperature Analysis*

Figure 4a shows the temperature contour plot at 40 s. It is observed that the maximum temperature of 2060 °C is obtained near the weld pool. The melting pool is observed until the thickness of 3.7 mm, considering a metallurgical melting point temperature of 1500 °C. In the case of the IPAW process, the maximum temperature of 2330 °C (Fig. 4b) is 16% higher than the PAW process. A full weld penetration and a wider weld pool geometry are observed for IPAW compared to the PAW process. The temperature history of the IPW and IPAW is shown in Fig. 4c. It is observed that the workpiece is preheated to a maximum surface temperature of 686 °C, which helps to achieve a higher depth of penetration for IPAW. Further, it is observed that the addition of induction preheating for IPAW promotes a slower cooling rate in comparison to the PAW process, which might help to get lower residual stress disparity in the workpiece.

3.2 *Residual Stress Analysis*

FE based numerical simulations are conducted for both the PAW and IPAW processes to compare the residual stress distribution. Figure 5a shows the transverse residual

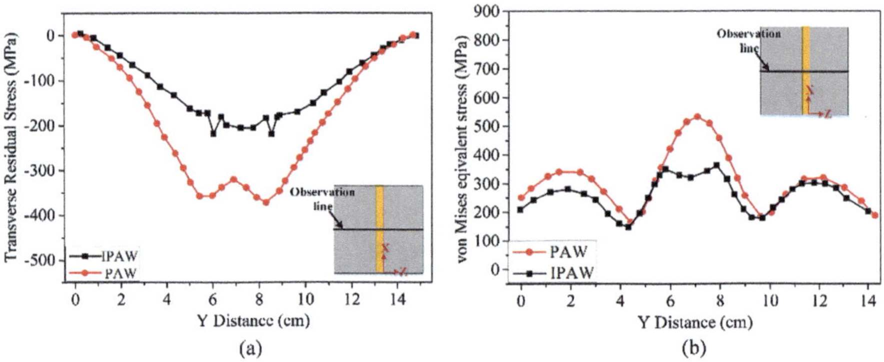

Fig. 5 Residual stress distribution on top surface. **a** Transverse residual stress, **b** von Mises residual stress

stress distribution on the top surface of the workpiece. For the IPAW process, a higher stress of 219 MPa in the weld zone is observed close to the heat-affected zone. The nature of the stress is compressive for both the weld and heat-affected zone. In the case of the PAW, a higher residual stress of 371 MPa is observed near the weld, which is close to the yield strength. It may occur due to the temperature of the fusion zone is higher than the melting temperature, and the phase transformation effect and hardening effect are ignored in the FE model [11]. Further, it is observed that the transverse stress gradually declines from the weld to edge of the weld for both IPAW and PAW processes.

Figure 5b shows a comparative plot of von Mises equivalent stress distribution on the surface of the workpiece. It is observed that the workpiece higher equivalent von Mises stress of 352 and 530 MPa is kept near the weld zone for IPAW and PAW process at room temperature. The retention of higher equivalent von Mises stress near the weld zone of PAW might be causing a premature failure near the weld under mechanical loading. Whereas for IPAW, the peak von Mises equivalent stress below the yield stress may provide a relatively better mechanical performance than the PAW process.

4 Conclusions

In this current work, a coupled electromagnetic-thermal–mechanical 3-D FE model is proposed to predict the temperature and residual stress distribution. The proposed FE model is used for a comparative study to analyse the effect of induction heating on plasma arc welding. The study explores the temperature characteristics and residual stress distribution for IPAW and PAW processes. It is observed that the addition of induction preheating is able to achieve a peak temperature of 1600 °C at the bottom surface of the weld line for the IPAW, which may confirm a complete weld penetration. Further, the addition of induction heating for IPAW effectively reduces the von Mises residual stress, which signifies a lesser weld distortion compared to PAW.

References

1. Roepke C, Liu S, Kelly S, Martukanitz R (2010) Hybrid laser arc welding process evaluation on DH36 and EH36 steel. Welding J 89(7):140–149
2. Pinto H, Corpas M, Guio JA, Pyzalla AR, Jahn A, Standfuß J (2009) Microstructure and residual stress formation in induction-assisted laser welding of the steel S690QL. Steel Res Int 80(1):39–49
3. Zhang C, Sun J, Li Y, Deng D (2022) Numerical simulation of residual stress of butt-welded joint involved in complex column-beam welded structure. J Manuf Process 83:458–470
4. De A, DebRoy T (2011) A perspective on residual stresses in welding. Sci Technol Weld Joining 16(3):204–208
5. Suman S, Biswas P (2020) Comparative study on SAW welding induced distortion and residual stresses of CSEF steel considering solid state phase transformation and preheating. J Manuf Process 51:19–30
6. Zhang L-J et al (2016) A comparative study on the microstructure and properties of copper joint between MIG welding and laser-MIG hybrid welding. Mater Des 110:35–50
7. Zhao YB et al (2011) A comparative study of laser-arc double-sided welding and double-sided arc welding of 6 mm 5A06 aluminium alloy. Mater Des 32(4):2165–2171
8. Jiang W, Zhang Y, Woo W (2012) Using heat sink technology to decrease residual stress in 316L stainless steel welding joint: finite element simulation. Int J Press Vessels Pip 92:56–62
9. Sarmast A, Schubnell J, Farajian M (2022) Finite element simulation of multi-layer repair welding and experimental investigation of the residual stress fields in steel welded components. Welding World 66(6):1275–1290
10. Biswal SK, Pal S (2022) Numerical investigation of the dimension factor of hairpin coil for sustainable induction heating. In: Recent advances in manufacturing modelling and optimization. Springer, Singapore, 11–19
11. Sun J et al (2019) Solid-state phase transformation and strain hardening on the residual stresses in S355 steel weldments. J Mater Process Technol 265:173–184